THE

JOURNAL

OF THE

ROYAL GEOGRAPHICAL SOCIETY

OF

LONDON

VOLUME 9

PART II

Elibron Classics
www.elibron.com

Elibron Classics series.

ISBN 0-543-82024-6 (paperback)
ISBN 0-543-82023-8 (hardcover)

This Elibron Classics Replica Edition is an unabridged facsimile of the edition published in 1839 by John Murray, London.

THE

JOURNAL

OF THE

ROYAL GEOGRAPHICAL SOCIETY

OF

LONDON.

VOLUME THE NINTH.

1839.

PART II.

LONDON:

JOHN MURRAY, ALBEMARLE-STREET.

Price 5

London: Printed by WILLIAM CLOWES and SONS, Stamford Street.

CONTENTS OF PART II., VOLUME IX.

Royal Geographical Society.

1839.

AT THE

ANNUAL GENERAL MEETING, May 27, 1839,

The following Report from the Council was read :—

The Council has again to congratulate the Society upon the steady increase in the number of its Members, affording a gratifying proof, it is willing to believe, of the interest felt by the public in the advancement of geographical science.

Since our last anniversary, sixty-three new Members have been elected; twelve vacancies by death and resignation have occurred; and the Society now consists of 651 Members, besides sixty Foreign Honorary and Corresponding Members.

During the past year we have to lament the loss of a distinguished Foreign Honorary Member, the Baron de Hamelin, an Admiral in the French navy, well known as having been one of the French expedition to the coasts of Australia, about the beginning of this century; and, more lately, as having presided over the Dépôt de la Marine de France, and to whose liberality we are indebted for some of the most costly and valuable works which adorn our library.

Finances.—The state of the finances of the Society, the details of which are annexed, continues satisfactory. The expenditure for the past year has been considerable, in consequence of 225*l.* having been paid towards the South African expedition (now terminated), and 150*l.* towards the expedition to Kurdistán; yet it is gratifying to add, that these heavy demands have been paid out of the current income of the Society, leaving untouched the capital of 4800*l.* invested in the funds—that stock representing the amount of the compositions of 275 compounding members now living.

Publications.—The Geographical Journal for 1838 consists of three Parts—the first published in February, the second in May, and the third in November; of Vol. IX., the first Part has also been published, and the second Part is now laid upon the table.

It is gratifying to add that not less than 70 papers on subjects connected with geography have been contributed to the Society since our last anniversary.

The first edition of Part II. Vol. III. of the Journal having been exhausted, while the demand for it still continues both by the public at large, and by new Members desirous of obtaining complete sets of the Transactions, the Council has ordered it to be reprinted; and directed that in future the first edition of the Journal be augmented from 1250 to 1500 copies, in order to meet the increased demand on the part of the public.

Her Majesty's Donation.—Her Majesty having been graciously pleased to continue to the Society (as announced at the late anniversary) the annual donation of fifty guineas, as originally granted by her Majesty's royal predecessor for the promotion of geographical science and discovery, the Council have resolved that this sum should be converted into two gold medals of exactly equal value, to be designated the Founder's Medal and the Patron's Medal, respectively.

Acting upon this resolution, the Council has awarded the Founder's Medal for the year 1838 to Mr. George Simpson, of the Hudson's Bay Company's service, who, in conjunction with, and under the immediate orders of, Mr. P. W. Dease, traced the hitherto unexplored coast to the west between Return Reef and Point Barrow, in 1837; and during the past year has discovered 90 miles of coast eastward from Point Turnagain of Franklin, on the northern shore of America.

In making this award, the Council cannot refrain from expressing their admiration of the spirited and liberal conduct of Governor Pelly, and the Committee of the Hudson's Bay Company—so honourable to British commercial enterprise—under whose orders this service has been executed.

The Council has in like manner awarded the Patron's Medal to Dr. Edward Rüppell, of Frankfort, in testimony of the services he has rendered to geography by his travels and researches in Nubia, Kordofán, Arabia, and Abyssinia.

Original Expeditions.—Of the expeditions more immediately under the control of the Society, that of Mr. Schomburgk, into British Guayana, has now nearly completed its fourth year, and the detailed report of his ascent of the rivers Essequibo, Berbice, and Corentyn, has appeared in Vols. VI. and VII. of this Journal; during the two last seasons, which he has spent in the interior of the country, he has explored the source of the Essequibo, crossed the equator, and penetrated into 0° 12′ S. lat.; returning thence to Pirará, he crossed the Brazilian frontier to Fort San Joaquim descended about 30 miles of the Rio Branco, and explored the Carumá mountains on its eastern bank. Returning again to Pirará, Mr. Schomburgk has travelled in a N.W. direction to the sources of the river Caruni, and to the mountains of Rorima, and at the date of his last letter, 20th Nov. 1838, was about to proceed to the westward to the head-waters of the Orinoco, with the intention of descending that river to Esmeralda, thence to cross over to the Padaviri, a tributary of the Amazons, and thus work his way back again to Demerara. Both Mr. Schomburgk and his party had suffered much from fever and climate, but his zeal in the cause of discovery appears unabated.

The objects of the expedition to Kurdistán, undertaken in conjunction with the Society for the promotion of Christian Knowledge, were announced last year, and the instructions given to the travellers are appended to this Report. Quitting England in June last, Mr. Ainsworth made a series of observations for magnetic dip and intensity, in his way across the continent of Europe to Constantinople; there, joining Mr. Rasám, the party left that city on 20th September, and, taking the road by Izmíd, Erekli, Bóï-ábád, and ascending the course of the Halys, reached Angora in the beginning of December, 1838. A detailed report of this journey is printed in the volume now laid on the table.

Foreign and Colonial Correspondence.—Five Foreign Corresponding Members have been added to the list since our last anniversary—namely, Colonel Skribaneck, Director of the Dépôt Topographique at Vienna; Colonel Oberreit, Chief of that at Dresden; Professor C. C. Rafn, Secretary to the Society of Northern Antiquaries at Copenhagen; Dr. Von Schlieben, head of the Statistical Society of Saxony; and Don Eduardo Carrasco, Hydrographer in Peru—and the Council have great pleasure in noticing the gradual increase of the Foreign and Colonial Correspondence of the Society.

Library.—A list of accessions made to the library, consisting of upwards of 400 volumes and 500 maps and charts, is printed with this Report. For many of these works the Society is indebted to the liberality of Foreign Academies; but more especially to the Dépôts de la Guerre et de la Marine in France; to the Academies of St. Petersburg, Berlin, Copenhagen, Lisbon, and Paris; and to the Dépôts Topographiques of Austria and Wurtemburg, for several valuable donations.

The insufficiency of the apartments at present occupied by the Society is a great and increasing evil. The endeavours which the House Committee have made during the last year to remedy this evil have been unremitting, and the Council regret to add, hitherto unsuccessful.

BALANCE-SHEET FOR 1838.

Dr.

	£.	s.	d.
Balance in hand, January 1st, 1838	30	18	11
Entrance of 65 Members at 3*l.*	195	0	0
Composition of 27 Members at 17*l.*	459	0	0
Subscription of 257 Members at 2*l.*	514	2	0
Arrears paid up	50	0	0
Dividends on Stock (4800*l.* reduced 3½)	168	0	0
Royal Premium	52	10	0
Journals sold in 1838	128	17	6
Christian Knowledge Society for Kurdistán Expedition	500	0	0
	£2,098	8	5

Cr.

	£.	s.	d.
House rent	110	0	0
Salaries	300	0	0
Printing Journal, Vol. VII.	285	0	0
Reprinting Journal, Vol. III. Part I.	24	0	0
Illustrations for Vol. VII.	170	14	6
——— for Reprint of Vol. III.	16	7	6
Kurdistán Expedition	298	0	0
South African ditto	223	0	0
Guayana Expedition	5	5	0
Royal Premium	52	10	0
Books and Maps	64	5	8
Collector	21	0	0
Alteration in Die of Medal	10	0	0
Stationery	11	15	4
Firing and Lights	10	6	3
Porterage, Carriage of Parcels, &c.	15	18	0
Attendance and Wages of Housekeeper	17	17	0
Postage of Letters	18	1	0
Expenses of Evening Meetings	16	13	6
Insurance, Advertisements	4	10	0
Furniture and Repairs	1	6	0
Subscriptions returned	6	0	0
	1,682	9	9
Balance in hand, Dec. 31st, 1838	415	18	8
	£2,098	8	5

(Signed) JOHN BIDDULPH, Treasurer.

The above Accounts have been compared by us with the corresponding vouchers, and found to be correct.—10th May, 1839.

(Signed) { JAMES MEEK.
FRANCIS BECKFORD.

ESTIMATE FOR 1839.

Dr.

		£.	s.	d.
Balance in hand (1st January, 1839)		415	18	8
Probable Amount of Subscriptions, viz.				
Entrance of 60 Members	£180			
Composition of 25 ditto	425			
Subscription of 250 ditto	500			
		1,105	0	0
Arrears likely to be recovered		100	0	0
Dividends on Stock		168	0	0
Sale of Journal		140	0	0
Royal Premium		52	10	0
		£1,981	8	8

Cr.

	£.	s.	d.
House rent	110	0	0
Salaries	300	0	0
Printing Journal Vol. VIII.	320	0	0
Illustrations for Journal	150	0	0
Reprinting Journal, Vol. III., Part II.	42	0	0
Illustrations for Reprint	20	0	0
Kurdistán Expedition	500	0	0
Guayana Expedition	100	0	0
White Nile Expedition	50	0	0
Books and Maps	100	0	0
Office Expenses, including Firing, Lights, Refreshments at Meetings, Attendance, Postage, &c.	100	0	0
Royal Premium	52	10	0
Sinking Die of Patron's Medal	52	10	0
Collector	21	0	0
	£1,918	0	0

(Signed) JOHN WASHINGTON,
Secretary.

At the Anniversary Meeting held on the 27th May, 1839, the President presented the gold medals, awarded respectively to Dr. RÜPPELL, of Frankfort, and Mr. THOMAS SIMPSON, of the Hudson's Bay Company, in the following words :—

"Chevalier BUNSEN,—It is peculiarly gratifying to me that, as President of the Royal Geographical Society, it has fallen to my lot to place a gold medal, the gift of our most gracious Sovereign, and awarded by the Council of this Society to my much valued friend Dr. Edward Rüppell, of Frankfort, in the hands of another highly valued friend, the Chevalier Bunsen, for the purpose of its being transmitted to its deserving owner. This medal is awarded to Dr. Rüppell in testimony of the high sense which the Council feel of the services rendered by him for the advancement of Physical Geography: and though Dr. Rüppell, together with his habits of minute accuracy, his innate love of truth, and the truly philanthropic spirit which he carried with him to Arabia Petreæa, and to the burning sands of Abyssinia, and Kordofan; although Dr. Rüppell, besides his assiduity and exactness in using his mathematical instruments for astronomically fixing the sites of every important place he visited, was also a distinguished and indefatigable naturalist, we love to regard him chiefly as a geographer; and as such to proclaim, as we do, our high sense of his merits as a large benefactor to the science we profess to cultivate. Dr. Rüppell made three journies into Africa: his first was of short duration, for he soon left Egypt to return to Europe, where alone he could hope to attain sufficient proficiency as a practical astronomer, to travel with advantage through unknown regions. He placed himself under the tuition of the lamented Baron von Zach; he returned to Egypt, and visited Arabia, the gulf of 'Akabah (which he was the first to explore geographically), Abyssinia, and Kordofan. All the produce of his travels in the department of zoology, which absorbed a large half of the pecuniary means at his disposal, Dr. Rüppell nobly and generously made over on his return home to his native city. The same has been the result, and the same the sacrifice, of his second journey into the same countries: and the city of Frankfort is mainly indebted to the disinterested conduct of one of the most illustrious of her citizens, for the high rank she holds, even in Germany, in the possession of a splendid museum of natural history, and a learned director at the head of it. The most valuable portion of Dr. Rüppell's geographical discoveries and *data* are to be found in the pages of his own luminous and learned narratives (the latter part of which is now in the course of publication), or in those of the *Correspondance Astronomique*, edited at Genoa by his friend, Baron von Zach; and they will long continue to be the chief guides of those who may undertake hereafter to follow Dr. Rüppell's steps in the honourable route which he has laid open.

"Chevalier Bunsen,—The Royal Geographical Society of London are highly gratified with the opportunity thus afforded them of presenting, through the channel of one, to whom all branches of literature are so highly indebted, the royal medal, awarded by the Council to Dr. Rüppell, your distinguished countryman."

To which Mr. Bunsen replied:—

"Sir,—In receiving the medal your Society has awarded to Doctor Eduard Rüppell, I beg to express the thanks of that learned countryman of mine, to whom I shall be happy to forward this honourable token of your high esteem. In expressing these thanks, and at the same time the warm interest I personally take in the distinctions decreed by you to my countryman, I know I speak, also, the feeling of the illustrious city to which he and his collections belong, and those of our common country, Germany, which is proud of the successful efforts of one of her most meritorious sons, and will equally be so of the honour England has at this moment bestowed upon him, through this Society. And, indeed, this encouraging proof your nterest and of your admiration is well adapted to recall to the mind recollections of a particularly pleasing nature to both countries. For when I look back to the distinguished travellers and discoverers whom my native country has produced in this and in the last century, I see before me an imposing series of illustrious names, connected with this country by the ties of hospitality and friendship, and powerfully assisted in their noble endeavours by the generous zeal of your learned and patriotic societies. Allow me only to mention Forster, the natural philosopher and distinguished writer, who accompanied Captain Cook on his great voyage of discovery: and of later times, Hornemann and Burckhardt, who received in this country the means of executing their bold plans, to penetrate into unknown parts of Arabia and of Africa. It has not been forgotten, Sir, in my country, and will never be forgotten, that England followed with anxious interest the course of those intrepid travellers, and that it mourned over the cruel fate that put an end to their enterprize, as if they had been her own native children. Let me only add, that he, who, for his admirable simplicity and the clearness of his observations, may be called the modern Herodotus, and who, with Pococke, Alexander von Humboldt, and a few others, ranks among the most accomplished travellers of all ages—that Niebuhr, the elder, I say, found among the English in India and in this country, the most effectual assistance, and the most encouraging interest, as he himself has recorded, and as his great son has so thankfully acknowledged in the biography of his father.

"As to your Society, Sir, in particular, it has by its very statutes acknowledged, and by its actions invariably sanctioned, the great and elevating truth, that science and virtue have their home and their sanctuary in every country, where these pillars of humanity are duly appreciated. I name both together, because they ought to be inseparable, and I am particularly happy to find that, in this very instance, you have equally honoured both: for as you yourself have so feelingly remarked in your discourse, the distinction awarded to Dr. Rüppell has been equally given to the intellectual merit and efforts of the traveller, and to the noble disinterestedness and the generous patriotism of the good citizen, who, after having sacrificed his fortune for the advantage of science, offered the gift of his rich and valuable collections to his native town. It is unnecessary to enlarge further on this subject; but it will be gratifying to you to know, that the generosity of Dr. Rüppell has not been bestowed upon a barren ground. That city, which may well be held up as a model for other rich and commercial towns, on account of the noble institutions she has raised for science and literature,—mostly monuments of the patriotism of her children,—is uniting those collections in a magnificent museum, worthy of its contents, of its founders, and of the free town that produced Goethe.

"As to myself, I can only say in answer to the extremely partial mention you have made of my name, that, deeply feeling how little I have done to merit such partiality, I trust I shall never be found deficient in gratitude for the kindness which has inspired it."

The President then addressing the Deputy Governor of the Hudson's Bay Company, said:—

"MR. HARRISON,—In the absence of Mr. Thomas Simpson, to whom the Council of the Royal Geographical Society of London have awarded their medal in testimony of the deep gratification which they feel in the success which has attended the enterprise and exertions of Mr. Simpson on the north coast of North America, and in the absence of Mr. Pelly, Governor of the Hudson's Bay Company, I am happy that you have been selected by that gentleman to accept the friendly office of receiving this medal in the name of Mr. Simpson. The services which Mr. Simpson, and his friend and senior officer, Mr. Dease, have rendered to Geography, in advancing, almost to its completion, the solution of the great problem of the configuration of the northern line of the North American continent, are, in the opinion of the Council of this Society, eminently meritorious; and Mr. Simpson and Mr. Dease, whatever may be the result of their further labours, have already earned for themselves a high place amongst those who have added to the fame and glory of British enterprise. In the summer of 1837, Messrs. Dease and Simpson, under directions from the Hudson's Bay Company, starting from the Great Slave Lake, followed the steps of Franklin as far as the point called Franklin's Farthest, whence they traced the remainder of the coast to the westward to Point Barrow; by which they completed our knowledge of this coast the whole way west of the Coppermine River, as far as Behring's Straits. During the following summer, the same adventurous officers, encouraged by the generous spirit of their employers in their attempts to brave the obstacles which a hard and rugged nature was interposing in the way of our knowledge of that part of the American continent, and anxious that this palm of victory should also, with so many others, encircle the brow of Britannia, again started from their winter quarters, as early as the season would permit, and, descending the Coppermine River, they again followed Sir John Franklin's route to Cape Turnagain—his eastern extreme—in longitude 109° nearly; from which point Mr. Simpson proceeded to the eastward about ninety miles. From the spot he had then attained, and which has since received the appropriate name of "Simpson's Farthest," he could descry a further line of coast about thirty miles in extent; and he had the gratification of thence discovering a fine open sea to the north and east: his supply of provisions here obliged him to retrace his steps. The result of these two expeditions is, that the northern shores of America,—all the acquisition of British hardihood, perseverance, and judgment,—can now be accurately laid down on our maps, from Behring's Straits to the 106th degree of longitude, forming a continuous line of coast of upwards of sixty degrees; and a fair prospect is opened, that another season may go far to complete our knowledge of the whole.

"Sir,—I beg to place this medal in your hands, to be given to Mr. Simpson by Mr. Pelly, the Governor of your prosperous and high-spirited association; and I beg that Mr. Simpson may be assured that this Society warmly participate in the honourable reward which Messrs. Dease and Simpson may expect in the gratitude of their country, and that we shall be happy to give them a welcome reception within these walls on their return to England."

To which the Deputy Governor replied:—

"Sir,—Highly gratifying as it must be to me personally, to receive, on the part of Mr. Simpson, this mark of the approbation of the Royal Geographical Society of London, I cannot but regret that, owing to an unavoidable engagement, Mr. Pelly, the Governor of the Hudson's Bay Company, is prevented from being present; as I am sure he would have been peculiarly gratified, on this occasion, having taken so large a share, in conjunction with Mr. George Simpson, in planning the operations and directing the arrangements that have led to this successful result on the arctic shores of America. Mr. P. W. Dease and Mr. Thomas Simpson, the two officers of the Hudson's Bay Company who have been employed on this expedition, were selected by the Governor and Committee, as combining the vigour, zeal, and ardour of youth with the experience, judgment, and discretion of riper years, and the result has justified the choice. The time of conferring on these gentlemen the distinction of a Royal Premium seems particularly happy, as it is at a moment when the parties themselves might apprehend, from not having done all they hoped to do last summer, that their work would be considered incomplete. But the award of this evening will prove to them that the medals of this Society are bestowed not only for works already performed, but as an encouragement for future exertion in the cause of discovery. It is a great satisfaction that the Hudson's Bay Company, as a commercial Company, have been able to extend their discoveries not only *within*, but *beyond* Her Majesty's dominions; and that you, Sir, on the part of the Geographical Society, should have expressed your approbation that they have not limited or restricted their endeavours, in time or expense, whenever they could aid the great cause of the advancement of geographical science and discovery."

Instructions of the Council of the Royal Geographical Society of London addressed to the Leaders of the Expedition for Exploring in Kurdistán. 1st June, 1838.

Gentlemen,—Having laid before the President and Council of the Royal Geographical Society the paper signed by you jointly, a few days before Mr. Rasám left England, in which you enumerate the various objects to which you understand that your attention is to be directed during the expedition into Kurdistán, I am instructed by the Council to express to you their general approbation of the details contained in that document. But, as a considerable portion of it is occupied with a reference to the special objects contemplated by the Society for Promoting Christian Knowledge, I am now, by the desire of the Council, to call your particular attention to those points of geographical research, of which it is more peculiarly the object of this Society to promote the investigation.

The Council are aware that Mr. Rasám quitted England about the middle of February, and would proceed immediately to Malta, whence he will embark for Constantinople, so as to meet Mr. Ainsworth there about the last week in July.

Mr. Ainsworth will leave England in the month of June to reach Constantinople by a land journey, taking such a route, and visiting such parts of the continent on his way, as he may think expedient with a view to furnishing himself with such additional information and materials for his further proceedings as he may be desirous to obtain; and should he traverse the Turkish provinces, the passages and heights of the Balkán (of which we know very little) will be particularly deserving of his attention.

The mission will be furnished with letters of introduction and recommendation to her Majesty's ambassador at the Porte; and the Council have no doubt that his excellency will readily provide them with the proper ferman, to insure their good reception in the Turkish provinces of Asia Minor, as well as with letters to her Majesty's agents at the several towns in the interior, where they may be resident.

It is fully understood between the two Societies who provide the funds for this expedition, and on whose joint account it is undertaken, that although Mr. Ainsworth will more especially conduct the inquiries for the Royal Geographical Society, and Mr. Rasám those for the Society for Promoting Christian Knowledge, and although both Societies are convinced from the well-known character of these gentlemen, and from the experience they have already had of each other as travelling companions, that the utmost cordiality and community of opinion will subsist between them, yet, to preserve a general unity of action and to provide for possible events, the primary controul and direction of the expedition is to rest with Mr. Ainsworth, in whose hands will be placed the power and discretion of drawing upon England for the necessary funds, in such proportions and on such occasions as they may be required.

In proceeding across the continent of Asia Minor to the banks of the Tigris, the first point which the expedition will make for is the city of Ḳaïṣaríyah, or Cæsarea ad Argæum. For this purpose Mr. Ainsworth will be guided by the information which he may obtain at Constantinople as to the facilities that may offer themselves on landing on the southern coast of the Black Sea; that is, whether he may be induced to prefer the beaten track by Angora, or whether by being landed at or near Sinope, he could not trace the whole, or nearly the whole, course of the Halys (Ḳizil Irmák) previous to reaching Ḳaïṣaríyah. The more accurate determination of the upper course of this river and also of its source (if this can be accomplished without inconvenience), is considered as coming within the objects of the expedition.

By whatever line Mr. Ainsworth may reach Ḳaïṣaríyah he will endeavour to ascertain the elevation of as many points as he can along the line of his route; and particularly of the high table lands and their principal summits, such as Páshá Ṭágh, and Karájah Ṭagh, between Angora and Aḳ-serái; to trace whether volcanic, or a continuation of the saliferous deposits extend between Aḳ-serái and Yúẓḳát; to establish the sites of any ruins said to exist in that district; and, if within reach, to fix the northern and southern extremes of the lake of Ḳóch-ḥiṣár.

The table lands, which form the chief characteristic feature in the physical conformation of this peninsula, attracted the attention of the Greek geographer Strabo, to whom we are still indebted for some of our most valuable information as to Asia Minor.

It has been suggested that the travellers, when at Ḳaïṣaríyah, should ascend the Argæus. This has lately been effected to within a few feet of the summit, by one of our latest travellers in that country. Mr. Ainsworth will of course enter into communication with that gentleman in order to ascertain how far it may be requisite to repeat the ascent, and the occasion will not be omitted to examine particularly the effects produced in this neighbourhood by the earthquake of 1835.

When in the neighbourhood of Ḳaiṣaríyah, the attention of the expedition will be particularly called to the sources and course of the Melas (Koremos, or Ḳará Sú), which in most of our maps has been made to join the Euphrates a little below Malatíyah; but both in the sketch map of Mr. Brant, who visited Garún and Ḳaiṣaríyah in August, 1835,* and more recently from the obser-

* Journal, vol. vi. p. 213.

vations of Mr. W. I. Hamilton, who was there in August, 1836, it is shown that no river flowing eastward to the Euphrates rises to the N.W. of Mount Argæus; and the latter traveller has proved that the Ḳará Sú, probably the Melas of Strabo, which rises about 5 miles to the westward of Ḳaïṣaríyah, flows undoubtedly towards the Halys.* From Mr. Brant also we learn that the Tokmah Sú has one of its sources not far to the N.W. of Garún, but as the exact spot is not ascertained, and as there has long remained so much doubt with respect to the sources and course both of the Tokmah Sú flowing eastward to the Euphrates, and of the Kará Sú flowing westward to the Kizil Irmak or Halys, it is desirable that Mr. Ainsworth should bestow much attention on the topography of the neighbourhood of Ḳaiṣaríyah, in order to set the question completely at rest; and should he follow the valley or valleys of the Tokmah Sú to Malatíyah, he should notice how far this river is navigable above its junction with the Euphrates.

It would also be desirable to connect with this, if practicable, the identifying of the towns of Castabala, Comana Cappadociæ, and others, as marked in the Peutingerian tables, upon the Roman road from Cæsarea Mazaca to Melitene.

From Malaṭíyah Mr. Ainsworth will have to decide whether the expedition shall proceed direct to Móṣul, by the track through Diyár-Bekr, and down the valley of the Tigris, by Ḥiṣn Keáïfá and the Jebel Túr; or whether they will take the more westerly line, and after visiting and examining first the Núshan pass, where the Euphrates is hemmed in by, and passes through, the chain of the Taurus mountains; and secondly, the cataracts above Samosata, proceed by O'rfah, Harrán, and the Ḳarájah Ṭágh, or Mons Masius of the ancients, to Márdín, Niṣíbín and Sinjár, exploring on their way the Rás el 'Aïn, and 'Aïn el Jebel on the Khábúr to the W., and thence, if possible, reach Móṣul by Ḥadhr on the Hawálí river. This last place has, we believe, only been visited by one European. It is supposed that several ancient sites will be found on this line with ruins and inscriptions; and those of O'rfah and Ḥárran should be well examined for reasons connected with sacred history. The fixing of the positions (where not already determined in the Euphrates expedition) by astronomical observation, and the measuring of the elevation above the sea of all places, where such astronomical observations shall be made, will be of essential importance.

Arrived at Móṣul, the principal attention of the expedition will be turned to the country of the Nestorian, Chaldæan, and Jacobite Christians on the eastern banks of the Tigris, and generally, as far as time and circumstances will admit, to the whole

* Vide the annexed Memorandum, by Mr. W. I. Hamilton.

range of the Carduchian and Gordiæan mountains, now comprehended under the name of Kurdistán.

For more detailed instructions respecting the inquiries to be pursued, especially by Mr. Rasám, but jointly and severally also by the two travellers, into the state of the Nestorian and Jacobite Christians dispersed over this country, we refer you to those which you will receive from the Society for the Promotion of Christian Knowledge; but the Council of the Royal Geographical Society are by no means insensible to the importance of this object of your research, as eventually conducive to the introduction of civil and religious instruction amongst a Christian people so long established in this most interesting part of the Asiatic continent; and they presume that you will make it your business to inform yourselves of the general state and condition of these communities, of their religious establishments, of their population, wealth, agriculture, commerce and occupations; the state of education prevailing amongst them, the liturgies they use, the peculiar tenets they profess, and the various languages and dialects which they write or speak, and what historical records they may possess of their original emigration.

Special attention is recommended to the accurate orthography of the names of places, rivers, mountains, &c., stating them in the Arabic character throughout your whole journey; noting likewise, whenever it may occur, the difference of names in the different languages spoken in the same place by the natives; and copying accurately any inscriptions, particularly Greek, which may be found.

You will be empowered to employ a portion of the funds, placed at your disposal, for the purpose of purchasing or copying any MSS. of historical or religious interest which you may be of opinion it may be desirable to have here. This Society, however, understands that the funds available for that special object will be separately provided by the Society for Promoting Christian Knowledge.

It would be superfluous here to enumerate in minute detail the particular sites,—whether towns, rivers, mountains, or lakes, which it is hoped you may be able to visit and explore during your journeyings in Kurdistán. This must necessarily depend upon circumstances beyond the controul and cognizance of the Society; and you must throughout be guided by the information you receive as to the lawless or peaceful state of the different tribes of independent Kurds. Suffice it to say that the points which occur to the Society, as the chief desiderata to improve our knowledge of the geographical features of Kurdistán and its vicinity, are the E. and W. banks of the Tigris between Mósul and Diyár-Bekr; the obtaining distinctly the names of all its tributaries and the precise points of junction with their recipient; the range and heights of the Jáwár, and most remarkable mountains; the course and sources of the great and little Záb rivers, and their various feeders; of the Khábúr and its junction with the Tigris (as there is reason to believe that all our maps are wrong on this point); the lakes which they may rise from, or which are traversed by them, and the more precise determination of the place, apparently a little N. of Jezírah, where the Tigris is hemmed in on the E. and W. by the high lands which fix one of the most important points in the route of the "Ten Thousand." Mr. Ainsworth will, of course, pay particular attention, wherever he may come upon it, to the illustration of this campaign, so important to the geography and history of this part of the continent, which on his homeward route he may be enabled to trace to the sea at Trebizond. Major Rennell's Geography of Western Asia, and his "Retreat of the Ten Thousand," will be most useful in directing his attention to the passages in ancient writers, which will receive elucidation from his researches, and in pointing out many matters for inquiry; and all the topographical notices of Strabo relating to this district should be minutely compared with present appearances on the spot.

The expedition will also explore and ascertain the site of Sért, where ruins and inscriptions may probably be found, proving the existence there of an ancient city (quære Tigranocerta), those of 'Amádíyah (the seat of the noblest Kurd tribes), with its lead-mines—Júlámerk; Ráwánduz; the approaches to, and the field of Arbela. We shall be glad to have, also, all the details you can collect of the Nestorian and Chaldean tribes of the Tiyárí and the Hákarí district; of the iron and lead mines in the neighbourhood of Júlámerk; and it is recommended generally to examine the mineral resources of this district with a view to its supplying coal for the navigation of the Tigris by steam; or if not coal, would it afford other fuel?—to ascertain the highest point of the range of hills between the Mús mountain and the Aïágha Ṭágh or Zagros;—to trace the direction and limits of the Zákú range, and to determine the position, and measure the elevation of the Peak of Ráwanduz, distant about 70 miles, E. by N., of Mósul, and probably in sight from that city, and also from Arbela. It need hardly be mentioned, that attention should be paid throughout your journey to objects of natural history, and especially to geology and botany.

If, by means of joining a caravan, or by other arrangements, you should be able to cross the mountains which form the boundary between the dominions of Turkey and those of Persia, and which are said to bear oaks producing the finest gall-nuts of the East, it would be very desirable that you should visit, and, if possible, make the circuit of the great lake of Urmíyah, in the province of Azerbáïján; and determine, astronomically, the latitude of its north and south points. Various tribes of Nestorian and other Christian communities are believed to dwell upon its western shore.

The political and moral state of the tribes of Mohammedan Kurds throughout this district, their languages, superstitions, and other peculiarities, will also, of course, not escape your observation; and it would be desirable to verify the reports concerning the Yezídís, or Fire Worshippers, or *Shaïṭan perest*, or Devil Worshippers.

The next object of your attention, and one to which the Society attaches great interest, is the lake Ván, Arsissa Palus of Ptolemy, and the Lacus Mantianus of Strabo, occupying a large and elevated plain, said to exceed 5000 feet above the level of the sea, between some of the principal feeders of the Tigris and those of the Murád Sú, or main confluent of the Euphrates.

The elevation of the surfaces of these two lakes above the sea; their depth, at any point where it can be ascertained, and the specific gravity of their waters; the astronomical determination of the N., S., E., and W. limits of Lake Ván, and the elevation above it of Sipán Ṭágh, will here be among the chief objects of your inquiry.

Having completed the circuit of this lake, and noted its various peculiarities, and having explored the Bitlís and Se'rt rivers, you will cross the Arghí Ṭágh, or eastern extremity of Mons Niphates, endeavouring to ascertain whether there are any easily practicable passes through this range, and generally, what was the most probable route from Mount Ararat to the plains of Mesopotamia. If circumstances are favourable to the attempt, you will endeavour to ascend one or both of the heights of Ararat, if possible from the S.W. side, as Parrot has already effected this from the eastern side. Descending thence through the Báyazíd country, you will follow the Murád Ṣú, availing yourselves of such opportunities as the country may offer, to visit occasionally the heights forming the southern boundary of the valley of Ṭaráberan; some of which are supposed to be of a volcanic nature; also the sources of the Erzen, and its subterraneous course between the lakes of Arethusa and Zoroanda. One of the passes through the heights will lead you to Arghanáh Ma'den and the principal source of the Tigris; the forests in whose neighbourhood are said to have furnished Alexander and Trajan with the wood for building their floats. From this point you will have no difficulty in regaining Malaṭíyah and the points on the Euphrates from which you may have started.

Although the Society cannot contemplate your having sufficient time at your disposal for extending your journey into Armenia, or even to Erzrúm, it would, of course, be very desirable that you should do so, if it be found practicable. At all events you will avail yourselves of the opportunity of communicating with Her Majesty's Consul at Erzrúm.

Another object of research has also been suggested, namely, Khúzistán, or Susiana, a considerable part of which has lately been visited by Major Rawlinson, but not under those advantages, in a scientific point of view, which will be at your disposal. The same observations, however, which I have made in reference to your proposed visit to Armenia, will apply also to this district.

It is very possible, likewise, that the time and means at your command may be inadequate to the accomplishment of all that has been pointed out to you; of this you will be the best judge when you are in the country: the selection of the least known and most important points is therefore left to your discretion, as well as the order in which the several objects of research are to be attempted, remembering always, that the chief objects are, the making acquaintance with the Nestorian Christians, and the examination of the country which they inhabit.

You and Mr. Rasam are both such experienced travellers, and you both already possess so much general knowledge of the country you are about to visit, and Mr. Ainsworth, in particular, is so well aware of the objects which will mainly interest the Royal Geographical Society, that it is quite unnecessary to enumerate more particularly the points deserving your attention. Nor would the Council wish to say more upon the subject than to express their confident hope, that the result of your inquiries, in addition to the rich harvest of general and physical knowledge, which you will acquire, will furnish also much new matter, specially illustrative of portions, at least, of the memorable expeditions of Xenophon with the "Ten Thousand;" of those of Alexander, Lucullus, Trajan, Julian, and Heraclius. With this view it will be very desirable that, previous to setting out, Mr. Ainsworth should be provided,—if not with the best ancient and modern works illustrative of the countries which he will visit, at least with transcriptions of the most important passages of the ancient writers which refer to these localities: a frequent comparison of these and also of modern authorities, with actual observations made on the spot, will generally point to matters for investigation which might otherwise be overlooked.

Extracts of names of places from Ptolemy, and from the several ancient itineraries, with the distances, will also be of great use, as it is presumed that many of these places will be identified by the coincidences of the present names with those in the Greek geographer, and other remote authorities. Accurate attention to this part of his instructions, assisted by a well-directed study of the details of ancient history, and of the revolutions to which this part of the world has so frequently been subject, will enable Mr. Ainsworth to illustrate in many instances what is now obscure, and will ensure him the grateful thanks of all men of learning.

Travellers have also mentioned that many ancient coins are to be collected in Kurdistán. It does not enter into the views of the Royal Geographical Society to make any acquisition of this description; and they would not willingly see that any considerable portion of the funds at your disposal were diverted from their special object to the purchase of these or similar objects of antiquity. At the same time they are aware that monuments of that kind are sometimes of importance in illustrating points of comparative geography; but for this purpose a special note must be kept of the localities where they are found: you will therefore use a sound discretion when opportunities may offer to purchase coins, and you will bear in mind that every outlay on this head must be held in subservience to the main purpose of your expedition.

In regard to astronomical and other instruments, the Society understand that Mr. Ainsworth is already in possession of, or will be furnished with, the following, viz.:—

1 Kater's azimuth and altitude circle, 3½ inches in diameter, and stand.

1 sextant of 7 inches radius.

1 pocket sextant.

3 chronometers (Molyneux), Arnold and Dent.

1 three-feet telescope, mounted (Dollond).

2 artificial horizons; 1 of mercury, 1 with a spirit-level.

2 mountain barometers (Newman, Troughton and Simms), both compared with the standard at the Royal Society.

6 thermometers; 2 of which fitted for measuring heights by boiling water.

1 Kater's compass, with azimuth sights.

3 pocket-compasses.

1 clinometer.

The requisite apparatus, with 3 pairs of needles, for measuring magnetic dip and intensity.

Nautical Almanacks for 1838, 1839, and 1840.

Astronomical Tables and the most useful books of reference; and the various maps connected with the country.

Road books, note books, squared paper, &c.

It is strongly recommended that all observations be entered immediately in a book kept for that purpose, with the state of the weather—barometer and thermometer—and a note as to whether a good or indifferent observation;—that a meteorological journal be constantly kept, registering the height of the barometer at least twice a-day (9 A.M. and 3 P.M. will be the best hours), and a series of hourly observations, for 36 hours at the equinoxes and the solstices; observations for variation are also of great importance to enable us to trace the curves of magnetic variation throughout this part of the world.

The 3 feet telescope will, it is hoped, enable Mr. Ainsworth to obtain his longitude at Móṣul by occultations, when it must be remembered that a *central meridional* occultation is the most favourable observation of this sort; if by moon-culminating stars, it must not be forgotten that the telescope should be adjusted to the meridian, or the observation will be liable to error; but until leisure be afforded for such observations, it will be better to assume the longitude of Móṣul (his-head quarters) at 43° E. of Greenwich, as determined by Mr. Rich, and to measure meridian distances from it by his watches. Thus all positions will be *relatively* right, and must move together with Móṣul, should its position hereafter be found to be erroneous.

You are aware that the funds disposable for this expedition are 1000*l.*;—500*l.* of which are provided by the Society for the Promotion of Christian Knowledge, and 500*l.* by the Royal Geographical Society. This sum of 1000*l.* will be at Mr. Ainsworth's disposal, for the objects of the expedition as above described; but you will be pleased to bear in mind, that it is understood that this sum is to supply the whole of your expenses for the space of 2 years from the period of your quitting England for Constantinople; and Mr. Ainsworth will regulate his drafts accordingly.

By order of the Council,

John Washington,
Secretary.

1st June, 1838.

Note by W. I. Hamilton, Esq.

By landing at Ereḳlí or Heraclea Pontica, Mr. Ainsworth might proceed to Angora through an interesting and almost totally unknown country; in which case I should recommend his proceeding along the seashore to Amáṣerah or near the mouth of the Parthenius or Bártín, and thence to turn inland to Za'farán Bólí, which I have reason to believe is a large and highly flourishing place.

Between Angora or Engúrí and Ḳaïṣsár the principal object of investigation will naturally be the course of the Halys and its tributaries; but it would also be highly interesting to ascertain the nature of the rocks which form the hills of Páshá Ṭágh, and perhaps also Rayah Ṭágh, which rise above the great central plateau of Asia Minor between Engúrí and Aḳ-seráï, and which are the summer residence of the Kurd tribes which inhabit the plain of Ḳóniyeh during the winter, and to ascertain whether they are volcanic, or a continuation of the saliferous deposits which extend between Aḳ-seráï and Yúz-ḳát.

Between Ḳír-shehr and Nemb-shehr it would be well to inquire for a place called U'cháyúk, upon the mountain of Bóz-tágh, where, I was informed by a Turkish Imán, were some curious remains and many columns.

The red sandstone formation through which the Kizil Irmak flows between Angora and Ḳaïṣariyah, as well as the deposits of rock salt, will of course claim some share of Mr. Ainsworth's attention; and I would particularly point out the rock salt deposits of Hájí Bektásh, which are also situated between Kír-shehr and Nemb-shehr.

If Mr. Ainsworth determines to ascend Mount Argæus it can only be done by procuring some Armenian guides from Everek-köi, at the S.E. foot of the mountain, and about 10 or 12 hours distant from Kaïsa-riyah: I do not believe the ascent to be practicable from any other side. If he then wishes to reach the extreme pinnacle of the mountain, he must be provided with a scaling ladder 30 or 40 feet high, or with a grappling of some kind, which he may heave up so as to fix itself in the rock above, and then swarm up the loose knotted rope attached to it. Neither of these apparatus will he be able to procure upon the spot, and perhaps hardly at Kaïsariyah.

With regard to the effects produced in the neighbourhood of Cæsarea by the earthquake of 1835, there is not much to be seen in the town itself, with the exception of the walls of some houses which are cracked; but about 6 or 7 miles from Cæsarea, on the road to the Greek convent of Yanár Tásh or Taxiarcha, there is a cliff overhanging a small village called Belí-Yasí, a large portion of which was thrown down during the earthquake upon the village, burying several houses and killing many of the inhabitants.

The Kará-sú, which rises in the marshy plain 5 or 6 miles to the westward of Cæsarea, and which is probably the Melas of Strabo, flows undoubtedly into the Halys. The river, which falls into the Euphrates at Malatíyah, if it has its origin anywhere near the eastern foot of Mount Argæus, must rise at a considerable distance from the mountain, for, in proceeding round Mount Argæus on the east side, I perceived no trace of any stream or water except such as flowed N.W. or S.W. Mr. Ainsworth would probably find the sources of the river, which flows east-ward, to the south of the road leading from Cæsarea to Malátíyah.

The priests at the Greek convent, 3 hours E. by S. from Cæsarea, where the bishop resides, spoke of some considerable remains at a place called Eúrán-shehr, 3 days' journey from Cæsarea to the eastward (can this be Garun?); and ten hours beyond that, of two lofty rocks 40 or 50 feet high covered over with Greek inscriptions.

If Mr. Ainsworth took the overland road from Constantinople he might, beyond Eskí-shehr, visit the interesting mines or pits of Meer-schaum or Ecume de Mer, between Eskí-shehr (Dorylæum) and Sevrí-hisár; and from thence striking south he might explore the *two* sources of the Sangarius, one of which must be near the place marked Pœma-nene in Colonel Leake's Map, and W.S.W. from Sevrí hisar, and which joins the other branch flowing from the south between Akekám and Bálá hisár. The tracing of these rivers would throw great light upon the march of Manlius in his Gallo-Grecian expedition.*

If, on the other hand, Mr. Ainsworth lands at Sinope, he cannot do better than trace the Halys as nearly as possible the whole way from Báfirah to Cæsarea.

1*st June*, 1838. W. I. H.

---

* I understood from the Turkománs of Hamah Hájí, 6 hours S.E. from Alekám, that there was a route across the Harmáneh to Akserai.

ACCESSIONS TO THE LIBRARY,

27th MAY, 1839.

EUROPE.

Titles of Books.	*Donors.*
AUSTRIA.—Travels in Austria, Russia, and Turkey, by the Rev. C. B. Elliott, M.A. 2 vols. 8vo. London, 1838	REV. C. B. ELLIOTT.
BADEN.—Baden Geographisch und malerisch beschrieben, von A. J. V. Heunisch und A. Schreiben. 8vo. Stuttgart, 1838	
BELGIUM.—Essai sur la Statistique Générale de la Belgique, par Xavier Heuschling. 8vo. Bruxelles, 1838 .	M. P. VANDERMAELEN.
——— LIMBOURG, Dictionnaire Géographique de la Province de. 8vo. Bruxelles, 1837. . . .	IDEM.
——— LUXEMBOURG, Dictionnaire Géographique du. 8vo. Bruxelles, 1838	IDEM.
——— DICTIONNAIRE des Hommes de Lettres, des Savans, et des Artistes de la Belgique. 8vo. Bruxelles, 1837	IDEM.
BRITISH ISLES.—The Silurian System, founded on Geological Researches in the Counties of Salop, Hereford, Radnor, &c., by R. I. Murchison, Esq. 2 vols. 4to. London, 1839	R. I. MURCHISON, ESQ.
——— DEVONSHIRE, Classification of the older Stratified Rocks of, by Professor the Rev. A. Sedgwick and R. I. Murchison, Esq. Pamphlet, 1839 . .	IDEM.
——— ENGLAND and WALES, A new and comprehensive Gazetteer of, by James Bell. 4 vols. 8vo. Glasgow, 1837	G. G. CUNNINGHAM, ESQ.
——— TABLES of the Revenue, Population, and Commerce of the United Kingdom. Folio. London, 1838	G. R. PORTER, ESQ.
——— TIDE Tables for the English and Irish Channels, and River Thames, for 1839. Hyd. Off., London	CAPTAIN BEAUFORT, R.N.
——— REPORT of the Commissioners for the Commutation of Tithes. 1838. Pamphlet, 4to . .	CAPTAIN W. BULLER, R.N.
FAROE ISLANDS.—The History of their Inhabitants, by C. C. Rafn. 8vo. Copenhagen, 1833 . . .	PROF. C. C. RAFN.
FRANCE.—Annuaire des Marées des Côtes de France, pour l'An 1839. 12mo. Paris. Dép. de la Mar. .	MR. A. M. R. CHAZALLON.
——— Mémorial du Dépôt Général de la Guerre. Tomes III., IV., V., VI. 4to. Paris. . . .	LE DEPÔT DE LA GUERRE.
GERMANY.—Das Germanische Europa, von Dr. E. Mendelssohn. 8vo. Berlin, 1836	.
——— The Spirit of her History, Literature, Social Condition, and Natural Economy, by Dr. Bisset Hawkins. 8vo. London, 1838.	DR. BISSET HAWKINS.

Titles of Books.	*Donors.*
GERMANY.—FRANKFORT.—Kurze Physisch-Geographische beschreibung der umgegend von Frankfort, von Dr. G. L. Kriegk. Pamphlet, 8vo. Frankfort, 1838 .	DR. KRIEGK.
GREECE.—The Spirit of the East, by D. Urquhart, Esq. 2 vols. 8vo. London, 1839	MR. COLBURN.
ICELAND.—Voyage en Islande et au Gröenland en 1835-6, publié sous la direction de M. Gaimard. 7 livraisons des planches. Paris, 1838	LE DEPÔT DE LA MARINE DE FRANCE.
——— RELATION d'un Voyage dans la Mer du Nord aux Côtes d'Islande, du Gröenland, de Ferro, &c., en 1767 et 1768, par M. de Kerguelen Tremarec. 4to. Paris, 1771	
ITALY.—Classical Tour through, by Sir Richard Hoare, Bt., 4to. London, 1819	
——— Dell' attuale Condizione della Scienza Statistica in Italia, in 1834, 5, and 6	COUNT GRÅBERG AF HEMSÖ.
——— TUSCANY.—Dizionario Geografico, Fisico, Storico della Toscana, da E. Repetti. Firenze, 1838 (*in continuation*)	
——— PAPAL STATES.—Statistica degli Stati Pontefici ed Estensi dal Col. Conte Serristori. Firenze, 1838	
SICILY.—Annali Civili del Regno delle Due Sicilie. 4to. Napoli, 1838 (*in continuation*)	COLONEL VISCONTI.
——— NAPLES.—Relazione delle Operazione Geodetiche nelle Provincie di Napoli, da F. Fergola. Pamphlet, 4to	IDEM.
——— ——— Del Sistema Metrico della Citta di Napoli, da F. Visconti. 8vo. Naples, 1838 . .	IDEM.
PRUSSIA.—Statistisch, Topographisch, Historisch, uebersicht des Preussichen Staats, von Dr. F. Förster. 8vo. Berlin	
RUSSIA.—Observations Météorologiques et Magnétiques faites dans l'étendue de l'Empire de Russie, redigées par M. T. A. Kupffer. No. 2. 4to. St. Petersburg, 1837	THE COUNT DE CANCRINE.
——— Excursions in the Interior of Russia, by Robt. Bremner, Esq. 2 vols. 8vo. London, 1839 . . .	MR. COLBURN.
——— BLACK SEA.—Mémoire ur un Nouveau Péryple du Pont Euxin, par le Comte J. Potocki. 4to. Vienna, 1796	
SAXONY.—Mittheilungen des Statistichens vereins das Königreich Sachsen (*in continuation*). 1838. Dresden	DR. VON SCHLIEBEN.
——— METEOROLOGICAL Report of Saxony from 1827 to 1838 ——— MEDICAL Statistics in Saxony for 1838. .	THE STATISTICAL SOCIETY OF SAXONY.
SWITZERLAND.—Savoy and Piedmont. Handbook. Murray, 1838	

AFRICA.

ABYSSINIA.—Reise in Abyssinien, von Dr. Eduard Rüppell. Frankfort, 1838	DR. E. RÜPPELL.
——— REISE in Abyssinien im 1836, A. von Katte. 8vo. Stuttgart, 1838	
ALGIERS.—Voyages dans les Régences de Tunis et Alger, par Peyssonnel et Desfontaines. Publiés par M. Dureau de la Malle. 2 vols. 8vo. Paris, 1838. . .	
DESCRIPTION Nautique des Côtes de l'Algérie, par M. A. Bérard, Capitaine de Corvette . .	LE DEPÔT DE LA MARINE DE FRANCE.

Titles of Books.	*Donors.*
Africa, North.—Excursions dans l'Afrique Septentrionale. Paris, 1838. 8vo.	Major Sir Grenville Temple, Bart.
——— A Journey to Marocco in 1826, by Capt. G. Beauclerk. 8vo. London, 1828 . . .	
——— West.—Travels in 1818-21, from the River Gambia to the Niger, by Major W. Gray and Surgeon Dochard. 8vo. London, 1835	
——— Fragmens d'un Voyage en Afrique en 1785-6-7, par S. M. Y. Golberry. 2 vols. 8vo. Paris, 1802	
——— Travels into the Inland Parts of Africa, containing a description of the several nations for 500 miles up the Gambia, by F. Moore. 8vo. London, 1738	
——— Travels in the Interior of Africa to the Sources of the Senegal and Gambia, by G. Mollien. 4to.	
——— Central.—A Geographical and Commercial View of Northern and Central Africa, by Jas. Macqueen. 8vo. Edinburgh, 1821	Ja. Macqueen, Esq.
——— South.—Expedition of Discovery into the Interior of Africa, through the countries of the Great Namâquas, Boschmans, and Hill Dâmaras, by Capt. Sir James Alexander. 2 vols. 8vo. London, 1838 . .	Sir James E. Alexander.
——— Narrative of an Expedition into, in 1836-7, from the Cape of Good Hope to the Tropic of Capricorn, by Capt. W. C. Harris. 8vo. Bombay, 1838 .	Captain W. C. Harris.
——— Report of the Meteorological Committee of the South African Literary and Philosophical Institution, July, 1836	Sir J. W. Herschell, Bart.
——— East.—Memoir respecting British Interests, and of the Native Tribes at and near Natal, by S. Bannister, Esq. Pamphlet, 1839	Saxe Bannister, Esq.
Egypt.—Voyage à Méroé, et au fleuve blanc en 1819—22, par M. F. Caillaud. 4 tom. 8vo. avec Atlas. Paris, 1827	A. T. Holroyd, Esq. and A. Th. D'Abbadie, Esq.
Egypt.—Belzoni's Travels in Nubia and Egypt. 2 vols. 8vo. London, 1825	
——— Lithograph of Hieroglyphics on the Coffin of Mycerinus, in the third pyramid of Gizeh. August, 1837	Colonel Howard Vyse.
Egypt and Mohammed Ali Pasha in 1837, by A. T. Holroyd, Esq. Pamphlet, 8vo	A. T. Holroyd, Esq.
Madagascar.—History of, by the Rev. Wm. Ellis. 2 vols. 8vo. London, 1838	
Red Sea.—On the Locality and Passage of the Red Sea by the Israelites, and on the Situation of Mount Sinai. Pamphlet, by C. Beke	Dr. C. Beke.

ASIA.

Asia.—Erdkunde von Asien, von Carl Ritter. Vol. V. 8vo. Berlin, 1838	Professor Ritter.
——— Asiatic Society of London, Journal of. No. IX.	Royal Asiatic Society.
——— Proceedings of the Committee of Agriculture and Commerce	Royal Asiatic Society.
——— Nouveau Journal Asiatique, 1831-9	Société Asiatique de Paris.

Titles of Books.	*Donors.*
ASIA.—ASIATIC Society of Bengal, Journal of, to October 1838	
—— MAGASIN Asiatique, par J. Klaproth. 2 vols. 8vo. Paris, 1828	
ARABIA.—Voyages dans l'Arabie, et Description de l'Arabie, par Carsten Niebuhr. 3 vols. 4to. Amsterdam, 1778	
—— PINKERTON'S Voyages and Travels in various parts of Asia. 4 vols. 4to. London, 1802 . .	
ASSYRIA.—Researches in Assyria, Babylonia, and Chaldea, by William Ainsworth, Esq. 8vo. London, 1838	W. AINSWORTH, ESQ.
—— TRAVELS in Assyria, Media, and Persia, by J. S. Buckingham. 2 vols. 8vo. London, 1830 .	
—— L'EUPHRATE et le Tigre, par M. D'Anville. 4to. Paris, 1779	
AVA.—Journal of an Embassy to the Court of, in 1827, by John Crawford, Esq. 4to. London, 1829 .	
CAUCASUS.—Travels in Caucasus and Georgia in 1807-8, by Julius Von Klaproth. Translated from the German. 4to. London, 1814	
—— VOYAGE autour du Caucase, par F. Dubois de Montpéreux. Tom. I. et II. 8vo. Paris, 1839.	
—— TRAVELS in the Western Caucasus, by E. Spencer. 2 vols. 8vo. London, 1838	MR. COLBURN.
CHINA.—Courses of the Hurricanes and Typhoons of the China Sea. 8vo. New York, 1838 . . .	W. C. REDFIELD, ESQ.
INDIA.—The History, Antiquities, Topography, and Statistics of Eastern India, by R. M. Martin. 3 vols. 8vo. London, 1838	
—— JOURNEY to the North of India, by Arthur Conolly. 2nd edition. 2 vols. 8vo. London, 1838 .	CAPTAIN A. CONOLLY.
—— THE Expediency and Facility of establishing a Meteorological and Monetary System throughout India. By Major T. B. Jervis, E.I.C. Engineers. 8vo. Bombay, 1836	MAJOR T. B. JERVIS, E.I.C. ENGINEERS.
—— CUTTOCK.—A Geographical, Historical, and Statistical Account of Orissa Proper or Cuttock, by A. Stirling. Calcutta, 1830	THE COURT OF DIRECTORS OF THE EAST INDIA COMPANY.
—— NAGPORE.—Report on the Territories of the Rajah of, by Richard Jenkins. 4to. Calcutta, 1827 . .	
—— RAJWARA.—Personal Narrative of a Tour through the Western States of Rajwara, &c., in 1835, by Lieut. Boileau, Bengal Engineers. 4to. Calcutta, 1837	MRS. BOSANQUET.
—— DIE STUPAS oder die Architectonischen Denkmale an der Indo Baktrischen Königsstrasse und die Colosse von Bamiyan, von Carl Ritter. 8vo. Berlin, 1838	PROF. C. RITTER.
—— VOCABULARIES of Seven Languages spoken in the Countries West of the Indus, by Lieut. Leech of the Bombay Engineers. Bombay	CAPTAIN MELVILLE GRINDLAY.
JAPAN.—Voyage au Japan en 1823-30, par P. F. Siebold. Tom. I. 8vo. Paris, 1838	
—— JOURNAL of an Expedition from Sincapore to Japan, by Dr. Parker. 12mo. London, 1838 . . .	
PALESTINE.—Reisen durch Syrien und Palästina nach Cypern und durch Kleinasien, in 1766, von Carsten Niebuhr. 4to. Hamburg, 1837 . . .	JOHN MURRAY, ESQ.
—— PALASTINA, von Carl Von Raumer. 8vo. Leipzig, 1838	PROF. CARL VON RAUMER.

Titles of Books.	*Donors.*
PALESTINE Reise in das Morgenland, in 1836-7, von Dr. G. H. Von Schubert. 8vo. Erlangen, 1838 . .	
——— A brief Account of the Countries adjoining the Lake of Tiberias, the Jordan, and the Dead Sea, by F. Seetzen. 1810	W. R. HAMILTON, ESQ.
PHILIPPINES.—Estado de la Poblacion de las Filipinas, correspondiente a el Año 1818. Manila, 1820 .	DON JOSE M. CARDANO.
TARTARY.—Relation des Mongols ou Tartares, par le Frère Jean du plan de Carpin, précédé d'une Notice, par M. D'Avezac. 4to. Paris, 1838	M. D'AVEZAC.
NOVAIA ZEMBLA.—Expédition à, et en Laponie, par M. Baer. St. Petersbourg. July, 1837	M. KUPFFER.

AMERICA.

AMERICA.—Antiquitates Americanæ, sive Scriptores Septentrionales rerum ante Columbianarum in America. Folio. 1837. Copenhagen. Published by the Royal Society of Northern Antiquaries. . . .	THE ROYAL SOCIETY OF NORTHERN ANTIQUARIES.
——— EXAMEN Critique de l'Histoire de la Géographie du Nouveau Continent aux 15me et 16me Siècles, par le Baron A. von Humboldt. 5 vols. 8vo. Paris, 1838	
———, NORTH.—Fauna Boreali Americana; or the Zoology of the Northern Parts of British America, by Dr. Richardson. 4to. London, 1836	DR. RICHARDSON.
——— NARRATIVE of an Expedition in H.M. ship Terror, in 1836-7, by Captain Sir George Back, R.N. 8vo. London, 1838	CAPTAIN SIR G. BACK, R.N.
NEWFOUNDLAND.—Voyage of H.M. ship Rosamond to, by Lieut. E. Chappell, R.N. 8vo. London, 1818	
NEW BRUNSWICK.—Remarks on the disputed N.W. Boundary of, by Major Yule, R.E. London, 1838 . .	MAJOR P. YULE, R.E.
——————— Report on the State and Condition of, by E. N. Kendall. 8vo. 1835 . . .	LIEUT. E.N. KENDALL, R.N.
NEW YORK.—Report on the Geological Survey of .	J. E. WORCESTER, ESQ.
HUDSON RIVER.—Exploring Visits to the Sources of the, by W. C. Redfield. New York, 1838	W. C. REDFIELD, ESQ.
MASSACHUSETTS.—Second Annual Report on the Geology of the Public Lands, &c.	
PENNSYLVANIA.—Second Annual Report on the Geological Exploration of. 8vo.	J. E. WORCESTER, ESQ.
UNITED STATES.—Report of the Committee of Agriculture relative to the introduction of tropical plants into	
——————— MEN and Things in America, by A. Thomason. 8vo. London, 1838	A. BELL, ESQ.
CUBA.—Compendio de la Geografia de la Isla de, par Felipe Poey. Habaña, 1836.	
——— Geografia de la Isla de Pinos, par Don A. H. Danier. Habaña, 1836	W. MACLEAY, ESQ.
GUATEMALA.—Narrative of an Official Visit to, by G. A. Thompson. 8vo. London, 1829	
AMERICA, SOUTH.—Voyage dans l'Amérique Méridionale, par Don F. de Azara. 4 vols. 8vo. Paris, 1809 .	
BOLIVIA.—Das Land Otoquis, von Dr. D. L. Kriegk. Pamphlet. 8vo. Frankfort, 1838	DR. KRIEGK.

Titles of Books.	*Donors.*
——— NOTICE de Deux Tentatives d'Ascension du Chimborazo, par A. von Humboldt. Paris, 1838 . COLOMBIA.—Ueber die hochebene von Bogota, von A. von Humboldt. Pamphlet	BARON A. VON HUMBOLDT.
CHILE.—The Geographical, Natural, and Civil History of, translated from the Italian of Don J. Molina. 8vo. London, 1809	
PARAGUAY.—The Reign of Dr. J. G. de Francia in, by Messrs. Rengger and Longchamp. 8vo. London, 1827 .	
——— Letters on, by J. P. and W. P. Robertson. 2 vols. 8vo. London, 1838	
LA PLATA.—The Provinces of La Plata, by Sir Woodbine Parish. 8vo. London, 1839	SIR WOODBINE PARISH.
PATAGONIA.—Surveying Voyages of H.M. ships Adventure and Beagle, 1826-36 (Patagonia, Chile, and Peru), by Captains P. P. King and R. Fitz Roy, R.N., with Extracts from the Journal of C. Darwin, Esq. 4 vols. 8vo. London, 1839	CAPTAIN FITZ ROY, R.N.
MAGNETIC Observations made during the Voyages of H.M. Ships Adventure and Beagle, 1826-36. Discussed by Major Sabine. Pamphlet, 1838 . . .	MAJOR SABINE, R.A.

POLYNESIA.

AUSTRALIA.—Three Expeditions into the Interior of Eastern Australia, by Major Sir T. L. Mitchell. 2 vols. 8vo. London, 1839	
——— Journey of Discovery to Port Phillip in 1824-5, by W. H. Hovell and H. Hume, Esq. Pamphlet. Sydney, 1837	SAXE BANNISTER, ESQ.
——— Six Months in Australia, with some Account of Port-Philip and Portland Bay, by T. H. James. 8vo. London, 1838	T. H. JAMES, ESQ.
SOCIETY AND SANDWICH ISLANDS.—Polynesian Researches during a residence of eight years in the, by W. Ellis. 4 vols. 8vo. London, 1836	
NEW ZEALAND.—Narrative of a Voyage to, in 1814-15, by J. L. Nicholas. 2 vols. 8vo. London, 1817 .	
——— A Narrative of Travels and Adventures in New Zealand, between 1831-7, by J. S. Polack. 2 vols. 8vo. London, 1838	J. S. POLACK, ESQ.

MISCELLANEOUS.

ABORIGINES Protection Society, Proceedings of, No. 1 .	THE SOCIETY.
ACADEMIA real das Sciencias de Lisboa, discurso lido em 15 de Maio de 1838 na sessao publica da, par J. J. de Macedo. Lisbon, 1838	COUNSELLOR J. J. DA COSTA DE MACEDO.
ACADEMIE des Sciences à Paris. Comptes rendus des Séances de l'. 4to. Paris, May, 1838—May, 1839 .	THE ACADEMY OF SCIENCES AT PARIS.
——— Impériale des Sciences à St. Petersbourg, Mémoires de, 1830-1837. Bulletin Scientifique, III., IV., V.	THE IMPERIAL ACADEMY OF SCIENCES AT ST. PETERSBURG.

Titles of Books.	*Donors.*
Akademie der Wissenschaften zu Berlin; Abhandlungen der Königlichen. 1832-36. Berlin. 4to.	The Royal Academy of Sciences at Berlin.
Almanac, Nautical, for 1841. 8vo. London	Captain Beaufort, R.N.
——— American, for 1839. 12mo. Boston.	J. E. Worcester, Esq.
——— British, for 1839. 12mo. London	
——— Geographische für 1839, von H. Berghaus. 12mo. Berlin	
American Antiquarian Society, Transactions and Collections of the. 8vo. Vol. II. Massachusetts, 1836	Albert Gallatin, Esq.
——— Philosophical Society, Proceedings of the, for 1838. Vol. I.	The Phil. Soc. of Philadelphia.
——— Journal of Science and Art, for 1838, by Dr. Silliman	
Annalen, der Erd-Volker und Staatenkunde von Dr. H. Berghaus, for 1838. 8vo.	
Annales des Voyages; to April 1839. 8vo. Paris	
Antiquaries.—Annals and Memoirs of the Royal Society of Northern Antiquaries at Copenhagen. First Series. 8vo. 1838	The Royal Society of Northern Antiquarians.
Arts.—Transactions of the London Society of Arts. Vol. LII. Part I. 8vo. 1838	The Society of Arts.
Asiatic Society, Catalogue of the Chinese Library of the	The Royal Asiatic Society.
Astromomical Society of London, Memoirs of the. Vol. X., for 1838	The Royal Astronomical Society.
Annuaire du Bureau des Longitudes, pour 1838. Paris	Le Bureau des Longitudes.
Athenæum Journal to May 1839. 4to. London.	The Editor.
Bibliographical Essay on the Collection of Voyages and Travels edited and published by Livinus Hulsius and his successors, from 1598 to 1660. 4to. London and Berlin, 1839	Mr. A. Asher.
Bibliothèque Universelle de Genève, from 1830 to 1839. 8vo.	M. de la Rive.
British Association for the Avancement of Science. Vol. VII. 8vo.	The British Association.
Botanical Society, Proceedings of the London. Vol. I. Part I. 1839	The Botanical Society.
Caillié.—Notice Historique sur la Vie et les Voyages de René Caillié, par M. Jomard. 8vo. Paris, 1839	M. Jomard.
Catalogue over Islands stifts Bibliothek. Copenhagen, 1828	Royal Society of Northern Antiquaries.
——— Méthodique d'un Choix de Globes et Sphères, d'Atlas et de Cartes, redigé par C. Picquet. 8vo. Paris, 1837	M. C. Picquet.
Colonial Policy of the British Empire, by R. M. Martin. Pamphlet. Part I. 8vo. London, 1837	R. M. Martin, Esq.
D'Anville, Notice des Ouvrages de. Pamphlet. Paris, 1802.	
Dictionnaire Etymologique et Comparatif des Langues Teuto-Gothiques, par Henri Meidinger. Frankfort. 8vo. 1833	Mr. H. Meidinger.

Titles of Books.	*Donors.*
DICTIONNAIRE Turc et Français, par J. D. Kieffer et T. X. Bianchi. 2 vols. 8vo. Paris, 1835 . . .	
——— Français et Arabe, par E. Bocthor. Revu par A. Caussin de Perceval. 2 vols. 4to. Paris, 1829	
——— Français et Wolofe, et Français et Bambara, avec un Dictionnaire Wolofe et Français, par J. Dard. 8vo. Paris, 1825	
DICTIONARY, Lempriere's Classical. 8vo. London, 1836	
——— Morell's Latin and English. 4to. London, 1783	
DIE Deutschen Volkstämme; geographisch und geschichlich beleuchtet mit besonderer Beruchsichtigung der Sprache, von H. Meidinger. 8vo. Frankfort, 1833	MR. H. MEIDINGER.
Die Heuschreckenplage der Länder der alten Welt. Pamphlet. 12mo.	PROFESSOR C. RITTER.
EDINBURGH, Transactions of the Royal Society of. Vols. V. to XIII., from 1829 to 1838 . . . . ——— Proceedings of the Royal Society of. I.-XII. 1832 to 1838	THE ROYAL SOCIETY OF EDINBURGH.
——— Review for October 1838	
ENGINEERS, Transactions of the Institution of Civil, Vol. III., Part 1, 4to., 1839 . . . . ——— Proceedings of the Institution, 8vo. .	THE INSTITUTION OF CIVIL ENGINEERS.
FOREIGN Quarterly Review to April, 1839	
FRANKLIN Institute of Pennsylvania, Journal of the, Vol. XXI., 1838, Philadelphia	THE FRANKLIN INSTITUTE.
FRIENDLY Contributions for the Benefit of the Royal Schools of Industry at Kensington and Shepherdsbush. London, 1838, 12mo.	COLONEL FOX.
FUNFTE jährliche Uebersicht der Thätigkeit der Gesellschaft für Erdkunde in Berlin, bis 5. Mai, 1838, von H. Lichtenstein	PROFESSOR C. RITTER.
GEOGRAFIA, Tratado Elementar de Geografia Antiqua e Moderna, por José de Urcullu, Tom. III., 8vo. Porto, 1839.	DON JOSE DE URCULLU.
GEOGRAPHIE.—Bulletin de la Société de, à Paris, to April, 1839, 8vo.	LA SOCIETE DE GEOGRAPHIE DE PARIS.
GEOGRAPHICAL Society of London, Journal of the Royal, Vol. VIII., Part 3, and Vol. IX., Parts 1 and 2. .	
——— of Bombay, Proceedings of the, for May, August, and Nov. 1838. 8vo. Bombay .	THE GEOGRAPHICAL SOCIETY OF BOMBAY.
——— of Frankfort—Jahresbericht des Geographischen Vereins zu Frankfurt. 1837-8 .	GEOGRAPHICAL SOCIETY OF FRANKFORT.
——— Dictionary, Ancient and Modern, by J. Conder, 8vo. London, 1834	
GEOGRAPHY, Uber eine geographische Producten kunde, von Carl Ritter	PROFESSOR CARL RITTER.
——— Quantum geografia profecerit, &c. Auctore F. C. A. Hasse. Lipsiæ, 1838.	VICE CHANCELLOR F. C. A. HASSE.
——— a System of, Popular and Scientific, by Jas. Bell, 6 vols., 4to. Glasgow, 1838	
——— Physical Geography, by Professor Traill. 8vo. Edinburgh, 1838	
——— Physikalische Geographie, von Frederick Hoffmann, Vol. II. 8vo. Berlin, 1838 . . .	

Titles of Books.	*Donors.*
GEOGRAPHY, Guthrie's Geographical Grammar. 8vo. London, 1808.	SIR WOODBINE PARISH.
——— a Compendium of, from the 12th century, containing some account of a residence in Vincland in 1007, and of a voyage to Florida in 1027. Copenhagen, 1837	THE ROYAL SOCIETY OF NORTHERN ANTIQUARIES OF COPENHAGEN.
GEOLOGICAL Society, Transactions of the	THE GEOLOGICAL SOCIETY.
——— Proceedings of the, for 1838. 8vo.	
——— President's Address at Anniversary Meeting of. 8vo. 1839	
GRAMMAIRE Wolofe, par J. Dard. 8vo. Paris, 1826 . .	
——— Recherches Philosophiques, sur la Langue Oulofe, par le Baron Roger, 8vo. Paris, 1829	
GRAMMAR of the Mandingo Language, by the Rev. R. M. Macbrair. 8vo. London, 1838	REV. DR. BUNTING.
——— of the Bechuana Language, by the Rev. Jas. Archbell. Graham's Town, 1838	REV. JAS. BEECHAM.
GOSPEL of St. Matthew in the Mandingo language. 12mo. London, 1837	IDEM.
HISTORY of the Decline and Fall of the Roman Empire, by E. Gibbon. 1 vol., 8vo. London, 1836 . .	
LITERARY Gazette to May, 1839	THE EDITOR.
LONDON and Edinburgh Philosophical Journal of Science, No 77, for June, 1838	JOHN HOGG, ESQ.
——— and Edinburgh Philosophical Magazine . .	
MADRAS Literary and Scientific Society, Journal of, No. XVI., 8vo. Madras, 1838	THE LITERARY SOCIETY OF MADRAS.
MAGNETISM.—Report on the Variations of the Magnetic Intensity at different points of the earth's surface, with plates, by Major Edward Sabine, R.A. 8vo. London, 1838	MAJOR SABINE, ROYAL ARTILLERY.
MEMOIR of the Life and Writings of Wm. Marsden, Esq., written by himself. 4to. London, 1838 .	MRS. WM. MARSDEN.
MEMOIRES de la Société de Physique et de l'Histoire Naturelle de Genève. Vol. VIII. 8vo. Geneva . .	THE SOCIETY OF NATURAL HISTORY OF GENEVA.
METEOROLOGICAL Committee of the Cape of Good Hope, Report of, for 1836-7	SIR JOHN F. W. HERSCHEL, BART.
MISSIONARY Register to May, 1839. 8vo.	
NAUTICAL Magazine to May, 1839. 8vo. London . .	
OBELISQUE Egyptien de Paris, Traduction et Analyse Grammaticale des Inscriptions sculptées sur l', par F. Salvolini. 4to. Paris, 1837	
PERU.—An Exposé on the Treaty of Paucarpata, by a Peru-Bolivian. Pamphlet. Andover, 1838. . .	DON VICENTE PAZOS.
POSITIONS Géographiques des Principaux Lieux du Globe, par M. Daussy. 8vo. Paris, 1838 . . .	M. DAUSSY.
QUARANTINE Laws, their Abuses and Inconsistencies, by A. T. Holroyd. London, 1839	A. T. HOLROYD, ESQ.
QUARTERLY Review to April, 1839 (*in continuation*) .	JOHN MURRAY, ESQ.
RAILROADS, How to Measure the Earth by the Assistance of, by Oliver Byrne. Newcastle, 1838 . . .	OLIVER BYRNE, ESQ.
RAPPORT sur les Opérations de la Campagne de la Corvette l'Astrolabe, en 1838, par M. Dumont D'Urville. Paris	M. D'AVEZAC.

Titles of Books.	*Donors*
RASK'S [R. K.] Samlede tildels forhen utrykte Afhandlinger, af H. K. Rask. 3 vols. 8vo. Copenhagen, 1838	THE ROYAL SOCIETY OF NORTHERN ANTIQUARIES.
RECORDS of Ancient Science exemplified and authenticated in the primitive universal standard of Weights and Measures, by Major T. B. Jervis. 8vo. Calcutta, 1835	MAJOR T. B. JERVIS, E. I. C. ENGINEERS.
ROYAL SOCIETY of London, Philosophical Transactions of the, for 1838. 1 vol. 4to.	THE ROYAL SOCIETY OF LONDON.
——— Catalogue of Scientific Books in the Library of the. 8vo. London, 1839	
——— List of portraits in possession of, and account of instruments belonging to the Royal Society	
——— List of Fellows to the 30th Nov. 1838	
SCHREVELIUS'S Greek and English Lexicon. 8vo. London, 1831	
SEIDENZUCH in Asien, Ueber'ausbreitung der. 8vo. Berlin, 1838.	PROFESSOR C. RITTER.
SINGALESE Characters, by Prof. R. Rask. Colombo, 1821	THE ROYAL SOCIETY OF NORTHERN ANTIQUARIES.
SOUTH POLE.—A Voyage towards the, in 1822-24, by Jas. Weddell, R.N. 8vo. London, 1825	
STATISTICAL Society of London, Journal of the. 8vo. Vol. I.	THE STATISTICAL SOCIETY OF LONDON.
STATISTIQUES sur les Colonies Françaises, Partie 1 et 2. Paris, 1837. 8vo.	M. ST. HILAIRE.
STEAM Navigation in the Pacific, Statements and Documents in favour of. Pamphlet, 8vo. 1838	THE DIRECTORS OF THE PACIFIC STEAM NAVIGATION COMP.
STRABONIS Rerum Geographicarum, libri XVII. Græcè et Latinè. 2 vols. folio. Amsterdam, 1707	W. R. HAMILTON, ESQ.
STORMS.—An Attempt to develop the Law of Storms, by Lieut. Col. W. Reid. 8vo. 1838	LIEUT-COL. W. REID, R.E., C.B.
TARTARISCHEN sprachen, Versuch über die, von Dr. W. Schott. 8vo. Berlin, 1836	
TRAVELS, Book of, containing—1. Journey from Aleppo to Damascus, in 1736. 2. Condamine's Voyage to South America, in 1747. 3. Journey through Russia into Persia, in 1742	
TIDE GAUGE, description of a new, constructed by Mr. T. G. Bunt, erected on the eastern bank of the river Avon, 1837	CAPT. BEAUFORT, R.N.
TIMES Newspaper, Monthly Index and Register to, from May, 1838, to May, 1839	MR. JAMES WYLD.
TITHE Commutation Tables, Annual Supplement to Willich's, for 1839	C. M. WILLICH, ESQ.
VASCO DA GAMA.—Roteiro da viagem que, em descobrimento da India pelo Cabo da Boa Esperança, fez Dom Vasco da Gama em 1497. 8vo. Porto, 1838	DON JOSE DE URCULLU.
VOYAGE round the World. Cook's 1st, 2nd, and 3rd Voyages, 1766-72, in 8 vols. 4to., with Atlas	
——— round the World in 1803, by Admiral Von Krusenstern, Imp. Russ. Navy. 2 vols., 4to. London, 1817	C. M. WILLICH, ESQ.
——— round the World in 1690, by C. Dampier. 8vo. London, 1827	

Titles of Books.	*Donors.*
Voyage autour du Globe, de la Thétis et l'Espéran e, en 1824-6, par le Baron de Bougainville. 2 vols. 4to. Paris, 1837	Le Depôt de la Marine de France.
——— autour du Monde en 1817-20, sur les Corvettes l'Uranie et la Physicienne, par J. Arago. 2 vols. 8vo. Paris, 1822	W. R. Hamilton, Esq.
——— autour du Monde, en 1790-2, par E. Marchand. 5 vols. 8vo., et 1-4to. Paris, 1799 . .	
——— du Sieur Paul Lucas, fait en 1714, dans la Turquie, l'Asie, Sourie, Palestine, Haute et Basse Egypte, &c. 6 vols. 12mo. Amsterdam, 1720 . . .	
———Reise um die Erde, 1815-18, auf der Brigg Rurik, Kap. Otto v. Kotzebue. Von A. Von Chamisso. Leipzig, 1836	
Wernerian Natural History Society, Memoirs of the. Vol. VII. 8vo. Edinburgh, 1838	The Wernerian Society.
Zoological Society of London, Transactions of the. Vol. II. Part 2, 1838	The Zoological Society.
——— Proceedings of the. Oct. 1837, to July 1838	

MAPS, CHARTS, &c.

EUROPE.

Maps, Charts, &c.	*Donors.*
Austria.—Three sheets of the Environs of Vienna, in 28 sheets. Quart.-Mast.-Gen. Off. Vienna, 1838 .	Baron C. Hugel.
British Isles.—The Ordnance Survey of England and Wales. Sheets 49, 50, 66, 67, 68.	Master-General of the Ordnance.
Teignmouth Harbour, Devon, by Mr. Spratt. 1838	Hydrographic Office.
Pembroke Reach, by Capt. Denham, R.N. 1838 .	
Port-dyn-laen, by Lieut. Sheringham, R.N. .	
Shetland Isles, by Mr. G. Thomas, R.N., scale 2 miles to an inch. 1837	
Ireland.—Index to the Townland Survey of 15 counties. Dublin, 1838	The Lord Lieut. of Ireland.
——— Map of, in 6 sheets, scale ½ an inch to a mile, to show the Railroads. By Lieut. Larcom, R. E. Dublin, 1839	
——— Map of, in 4 sheets, scale 4 m. to an inch. Dublin, 1838, drawn by J. and C. Walker .	
——— Map of, in 2 sheets, scale 10 m. to an inch. By the Diff. of Use. Know. Soc. London, 1837 .	
Ballyshannon Harbour, by Capt. W. Mudge, R.N. 1838.	Hydrographic Office.
Lough Erne, by Lieut. Wolfe, R.N. . .	
Geological Map of the British Isles, by James Wyld. London, 1828	Mr. James Wyld.
An Index Geological Map of the British Isles, by Professor Phillips. London, 1838	Professor Phillips.
Environs of Edinburgh, scale 8-10ths of an inch to a mile. By the Diff. of Use. Know. Soc. London, 1838 .	

Maps, Charts, &c.	*Donors.*
Denmark.—Jutland, the Agger Channel, by Captain J. B. Bluhme, Danish Navy	Hydrographic Office.
France.—Carte Topographique de la France; 48 feuilles Echelle 1–80,000th. Dépôt de la Guerre, Paris, 1834-8	Le Depôt de la Guerre a Paris.
Corsica, by Captain Hell, Mar. Fr. London, 1838	Hydrographic Office.
Germany.—Topographische Karte der Umgegend von Frankfurt, am Main. Von A. Ravenstein, scale 1–75,000th. 1837 . . . . Model in relief of the town and environs of Frankfort, and of the Niederwald. By A. Ravenstein, 1837	Mr. A. Ravenstein.
Italy.—Carte générale du théâtre de la guerre, en Italie en 30 feuilles, par Bacler Dalbe. Paris, 1790 . Elba.—Carte de l'Isle d'Elba, par C. Picquet. Paris, 1824	Mr. Jas. Gardner.
Sardinia.—Cagliari by Capt. W. H. Smyth, R.N. 1838 Channel of San Pietro ,, ,, ,, Porto Conte and Alghero ,, ,, ,,	Hydrographic Office.
Naples.—Carta topographica ed idrografica dei contorni di Napoli—Scala 1-25,000th. 9 sheets. Naples	Colonel F. Visconti.
——— Plan de la Baie de Naples. Paris, 1838	Depôt de la Marine de France.
Russia.—Military Map of Russia, in 16 sheets—scale 1–680,000th. Dépôt Topog. St. Petersburg, 1837	Baron C. Hugel.
Saxony.—Topographical Map of the Kingdom, in 20 sheets. Scale 1–57,000th. Dépôt Topog. Dresden, 1836-8, 5 sheets published	
Geographische Historische Atlas von Europa, Von C. Desjardins. Vienna, 1838	Baron C. Hugel.
Karte von Ural-gebirg	
Switzerland.—Triangulation Primordiale de la Suisse .	Col. Jackson.
——— Map in 20 sheets, by MM. Woerl and Heck. Scale 1-200,000. Freiburg, 1837	Dr. Woerl.
——— Relief pittoresque du Sol Classique, en 9 feuilles	Mr. Jas. Gardner.
——— Model in relief of the Canton of Neuchatel, by J. J. Boscawen Ibbetson, Esq.	J. J. Boscawen Ibbetson, Esq.
——— Map of, scale 12 miles to an inch. By the Diff. of Use. Know. Soc. London, 1838. . .	
Wurtemburg.—Topographical Map of the Kingdom of, in 57 sheets (25 published) scale 1–50,000th. Topog. Off. Stuttgardt, 1833	Bureau Topographique a Stuttgardt.
Wurtemburg and Baden, Map of, in 12 sheets, by MM. Woerl and Heck. Scale 1-200,000. Freiburg, 1837	Dr. Woerl.

AFRICA.

Algiers.—L'Algérie, en 3 feuilles. Echelle 1-1,000,000. Paris, 1838 . . . . Province d'Alger. Echelle 1-400,000th ,, ,, de Constantine ,, ,, ,, ,, d'Oran ,, ,, ,, Plan d'Alger et des Environs. Echelle 1–2,500th	Le Depôt de la Guerre a Paris.

Maps, Charts, &c.	*Donors.*
Carte de la Côte Septentrionale d'Afrique, en 3 feuilles, par M M. Bérard, Dortet de Tessan, &c. Paris, 1836 Les attérages d'Alger, par M. A. Bérard. Paris, 1831 ALEXANDRIA, Plan du Port et des Mouillages d', Paris, 1838	LE DEPÔT DE LA MARINE DE FRANCE.
CARTHAGE, Restauration de la Ville . .	SIR GRENV. TEMPLE, BART.
MAROCCO.—Carta del Maghreb el Aksà o l'Impero di, dal Conte Gråberg di Hemsö. Firenze, 1835 .	COUNT GRÅBERG AF HEMSÖ.
TANGIER, Plan de la Baie de, par M. le Saulnier de Vauhello, 1835	LE DEPÔT DE LA MARINE DE FRANCE.
WEST AFRICA.—Map of, in 2 sheets. Scale $\frac{9}{10}$ of an inch to a degree, by the Diff. of Use. Know. Soc. London, May, 1839	
QUORRA and Chadda, Map of the rivers, by Lieut. W. Allen, R.N. Scale $\frac{1}{3}$ inch to a mile. London, 1837 GAMBIA, the river, as far as Pisania, in 3 sheets, by Lieut. R. Owen, R.N. Hyd. Off., 1826 . .	HYDROGRAPHIC OFFICE.
MADAGASCAR.—Le Canal de Mozambique, et l'Ile de. Le port du Mozambique. Le Mouillage à la Côte S.O. de Zanzibar. Paris 1838 . . . . Les Iles situés à l'est et au N.E. de Madagascar. Paris, 1838	LE DEPÔT DE LA MARINE DE FRANCE.
SOUTH AFRICA.—Les Côtes Méridionales d'Afrique et de l'entrée du canal de Mozambique. Plan du Port Natal, et de la Baie Delagoa	IDEM.
PORT NATAL, by W. T. Haddon. London, 1838 . SEYCHELLES Islands, by Capt. W. F. Owen, R.N.	HYDROGRAPHIC OFFICE.

ASIA.

Maps, Charts, &c.	*Donors.*
ARRACAN.—Route from Aeng to the Irawadi, by Captain B. Pemberton. Calcutta, 1837	CAPTAIN BOILEAU PEMBERTON, E.I.C. ENGINEERS.
CAUCASUS.—Karte des Kaukasus, by F. G. Morin. Scale $2\frac{1}{2}$ inches to a degree. Berlin, 1838 . . . . CENTRAL ASIA.—Tartary, Khiva, Bokhara, &c. Scale $\frac{4}{10}$ of an inch to a degree. By the Diff. of Use. Know. Soc. London, 1838 . . . . Bokhara, Cabul, and Beluchistán. Scale 1 inch to a degree. By the Diff. of Use. Know. Soc. London, 1838 Map of the Countries between Tartary and Birmah, in 4 sheets. Scale $\frac{7}{10}$ of an inch to a degree. By J. Wyld. London, 1838	MR. JAMES WYLD.
CHINA.—La Mer de Chine, dressé par M. Daussy, en 1838. Scale 1 inch to a degree. Paris . . . INDIA.—La Mer des Indes, par MM. Daussy et P. E. Wissocq. Scale $\frac{3}{10}$ of an inch to a degree. Paris, 1837 . Les Entrées de la Rivière Hoogly, &c. Paris .	LE DEPÔT DE LA MARINE DE FRANCE
RAJAMANDRY.—Sheet 95 of the Indian Atlas. Scale 4 miles to an inch. By the E. I. Company. London, 1838 COCHIN.—Sheet 62 of the Indian Atlas. E. I. Company, 1838 . . . . HOOGHLY.—Entrance of the river, from False Point, Palmiras, to Saugur Roads. In 3 sheets. Corrected to 1838	THE COURT OF DIRECTORS OF THE EAST INDIA COMPANY.
INDIA in 1 sheet, showing the Military Stations, by James Wyld. Scale $1\frac{8}{10}$ of an inch to a degree. London, 1837 Index to Wyld's Map of India. 8vo. London, 1838	MR. JAMES WYLD.

Maps, Charts, &c.	*Donors.*
CUTCH—Map of, with the eastern mouth of the Indus, by Lieut. A. Burnes. Bombay, 1825-8. MS.	MAJOR JERVIS, E.I.C.ENGINEERS.
GOA—Map of the Portuguese territory of, by Lieut. Darling. M.S.	
JAVA.—Straits of Madura and Bali. Scale 6 miles to an inch; by J. Swart. Amsterdam, 1837	JACOB SWART, ESQ.
JERUSALEM—Plan of the city of, by F. Catherwood. 1835	MR. W. HUGHES.
MALDIVAS.—Trigonometrical Survey of the Maldivas, by Captain R. Moresby and Lieut. F. T. Paul, Indian Navy. Scale 4 miles to an inch. E.I.Co. London, 1838	THE COURT OF DIRECTORS OF THE E.I. COMPANY.
MANIPUR.—Map of the Manipur Territory, with part of Kubo Valley and the Burmese Frontier, by Captain Pemberton. Calcutta, 1837	CAPTAIN BOILEAU PEMBERTON, E.I.C. ENGINEERS.
PALESTINE.—Karte von Palästina, von D. Seetzen. Gotha, 1810	DR. PRICHARD.
CANAAN during the time of the Patriarchs, and as divided among the tribes of Israel. In 2 sheets, illuminated. Drawn by W. Hughes. Published by C. Knight. London, 1838	
PERSIAN GULF.—Bushire Port and Karak Island, by Lieuts. Brucks and Cogan, I.N., Hyd. Off. London, 1838	HYDROGRAPHIC OFFICE.
SIBERIA—Eastern and Western, in 2 sheets. Scale ½ inch to a degree. By the Diff. Use. Know. Soc. London, 1838	
SUMATRA.—Karte von Sumatra, von H. Berghaus. Scale 2 inches to a degree. Gotha, 1837	B. HEBELER, ESQ., PRUSS. CONS. GENL.
SUNDA—Chart of the Straits of, by J. Swart. Scale 4 miles to an inch. Amsterdam, 1838	JACOB SWART, ESQ.
Sunda, oder Borneo-Meere, von H. Berghaus. Scale 2 inches to a degree. Gotha, 1838	
AMERICA.	
AMERICA, NORTH.—Carte de l'Amérique du Nord, par L. Brué, revue par L. Picquet en 1838. Une feuille. Paris	M. PICQUET.
CANADA.—The Gulf and River of St. Lawrence, by Captain Bayfield, R.N. In 40 sheets. London, 1838	HYDROGRAPHIC OFFICE.
Eastern Townships of Lower Canada. In 2 sheets. Scale 5 miles to an inch. By John Arrowsmith. London, 1839	MR. JOHN ARROWSMITH.
NEWFOUNDLAND.—Les Attérages de l'Isle de Terre Neuve, entre Cap Raze et les îsles de Pierre et Miqueton. Paris, 1838	LE DEPÔT DE LA MARINE DE FRANCE.
UNITED STATES—Bradshaw's illustrated Atlas of the. Boston, U.S., 1838	
MASSACHUSETTS.—Map of Vinland or Massachusetts and Rhode Island. Copenhagen. 1837	THE ROYAL SOCIETY OF NORTHERN ANTIQUARIES.
MEXICO—Gulf of, Cayronas, and Alacran Shoal, by Captain Barnett, R.N. London, 1838	THE HYDROGRAPHIC OFFICE.
APALACHICOLA Bay, by Captain Roals. Scale ½ inch to a mile. London, 1838	
WEST INDIES.—Bahama, Western Elbow of Cay Sal Bank, by Capt. R. Owen. London, 1838	
CUBA.—Colorados and Sancho Pardo Shoals, from Span. MSS., and Capt. R. Owen, R.N. 5 inches to a mile. London, 1838	
TRINIDAD.—Bocas de Dragos, by Captain Columbine, R.N. Scale 2 inches to a mile. London, 1838	

Maps, Charts, &c.	*Donors.*
TRINIDAD Map of the Island of, by J. Basanta. London, 1838	J. BASANTA, ESQ.
JAMAICA—Map of, by John Arrowsmith. Scale 12 inches to a degree. London, 1839	MR. JOHN ARROWSMITH.
WINDWARD ISLANDS, in 1 sheet. Scale 5 inches to a degree. By John Arrowsmith. London, 1839 . LEEWARD ISLANDS, in 1 sheet. Scale 5 inches to a degree. By John Arrowsmith. London, 1839 .	
GUATEMALA—Map of the State of, in 8 small sheets, by M. Rivera. Guatemala, 1830 . . . XIQUILISCO—Plan of the Bay of San Salvador de ISTAPA—View of, with plan of Izabal . .	DON CARLOS MEANY.
VERA PAZ—Map of the Territory of, published by the Vera Paz Agric. Assoc. London, 1837 . .	J. FLETCHER, ESQ.
GOLFO DULCE, in 2 sheets, by Capt. Owen, R.N. London, 1838 . . . . YUCATAN.—Mugeres Harbour, by Capt. Barnett, R.N. London, 1838	HYDROGRAPHIC OFFICE.
SOUTH AMERICA—General Map of, comprising the whole of the late survey of the coast, by J. Arrowsmith. One sheet. Scale 1 inch to 3 degrees. London, May, 1839	MR. JOHN ARROWSMITH.
BRAZIL.—Para, the River of, from surveys by Capt. Forster in 1831, and Capt. Sir Ev. Home, R.N., in 1835. London, 1838 . . . . GUAYANA.—Entrance of the Corentyn River, by R. H. Schomburgk in 1837. Scale 1 inch to a mile. London, 1838	HYDROGRAPHIC OFFICE.
LA PLATA—Map of the Provinces of, by John Arrowsmith. Scale 1 inch to a degree. London, May, 1839	MR. JOHN ARROWSMITH.
PATAGONIA—Map of, by the Diff. of Use. Know. Soc. Scale ·65 of an inch to a degree. London, 1838 .	

POLYNESIA.

AUSTRALIA—General Map of, in 2 sheets, comprising all the recent discoveries. Scale $\frac{8}{10}$ of an inch to a degree; by John Arrowsmith. London, 1838 . . SOUTH EASTERN portion of. Scale 1½ inch to a degree; by John Arrowsmith. London, 1838 . . SPENCER and St. Vincent's Gulfs, &c. Scale 3 inches to a degree; by John Arrowsmith. London, 1838 ADELAIDE, the district of, divided into sections from a survey by Col. Light, by John Arrowsmith. London, 1839	MR. JOHN ARROWSMITH.
PORT PHILLIP, by Lieuts. Symonds and Henry, and Mr. Shortland, R.N. Scale ½ inch to a mile. London, 1838 . . . . TORRES STRAITS—Eastern Entrance to, by Capt. P. P. King, R.N. London, 1838 . . . . . NEW ZEALAND—Chart of, combining all recent observations. Hyd. Off., 1838	HYDROGRAPHIC OFFICE.
Map of, by the Diff. of Use. Know. Soc. Scale 1 inch to degree. London, 1838	
ARAFURA Sea, by G. W. Earl. Hyd. Off., 1838 .	HYDROGRAPHIC OFFICE.

Maps, Charts, &c.	*Donors.*
MISCELLANEOUS.	
ATLANTIC.—Carte reduite des Mers du Nord, depuis 48° jusqu'à 75° de lat., par M. P. Daussy. Scale $\frac{4}{10}$ of an inch to a degree of longitude. Paris, 1838 . .	MONS. DAUSSY.
ATLAS Universel, ancien et moderne, en 65 feuilles, par L. Brué ; revue en 1837 par Ch. Picquet. Paris	TIMOTHY CURTIS, ESQ.
ATLAS containing 25 Maps, by Nicholas Visscher. Amsterdam, 1690	
ATLAS de la Navigation autour du Globe de la *Thétis* et de *l'Espérance* en 1824-6, sous les ordres du Baron de Bougainville. Paris, 1837	LE DEPÔT DE LA MARINE DE FRANCE.
ATLAS.—Physikalischer Atlas, von Dr. H. Berghaus. Parts I., II., and III. Gotha, 1838	
PACIFIC.—L'Océan Pacifique, par MM. D'Urville et Lottin; revue en 1834. Dép. de la Mar. Paris. Scale $\frac{4}{10}$ of an inch to a degree.	MONS. DAUSSY.
ACTÆON Group, by Mr. G. Biddlecomb, R.N. Hyd. Off., 1838	CAPTAIN BEAUFORT, R.N.
POLAR SEA.—South Circumpolar Chart, comprising all recent discoveries. Hyd. Off., 1839	
Carte des Routes des Corvettes L'Astrolabe et la Zélée, levée par M. Vincent Dumoulin, sous les ordres de M. Dumont d'Urville, Capitaine de Vaisseau. Dép. de la Mar. Paris, 1838	MONS. D'AVEZAC.
PORTULANO, or Chart of part of the coasts of Africa and Spain. *fac simile* of a Map existing in the library of San Lorenzo at Florence, dated 1315	COUNT GRÅBERG AF HEMSO.
WORLD.—Map of the World, showing the languages and dialects into which the British and Foreign Bible Society have translated the Scriptures ; by James Wyld. London, 1838	MR. JAMES WYLD.

PORTRAIT of Mr. James Weddell, R.N., the Antarctic Navigator. Painted by Mr. Peake, 1838 . .	JOHN BROWNE, ESQ.
PORTRAIT of a Mandingo, painted by Mr. W. Carpenter. 1838	CAPTAIN WASHINGTON, R.N.
PORTRAITS of Eminent Circumnavigators, lithographed. Paris, 1836	M. D'AVEZAC.
PORTRAIT of the late Mr. John Davidson, lithographed. London, 1835	THOMAS DAVIDSON, ESQ.
PORTRAITS of Sir J. Franklin and Sir G. Back, lithographed. London, 1836.	CAPTAIN WASHINGTON, R.N.
VICTORIA REGIA; coloured drawing of the flower, &c.	MR. SCHOMBURGH.
VICTORIA REGIA, coloured print of, and description of the flower	DR. LINDLEY.
VIEWS in Fernando Po, by Captain W. Allen, R.N. .	CAPTAIN W. ALLEN, R.N.

LIST OF MEMBERS

OF THE

ROYAL GEOGRAPHICAL SOCIETY.

Aberdeen, the Earl of, K.T., F.R.S., L.S., Pres. S.A.
Abinger, Right Honourable Lord
Acland, Sir Thomas Dyke, Bart., F.G.S., H.S.
Adam, Lieut.-Gen. Sir Frederic, G.C.B.
Adam, Vice Admiral Sir Chas., K.C.B.
Adamson, John, Esq., F.S.A., L.S., A.S.
Adare, Viscount, M.P., F.R.S.
Ainsworth, William, Esq.
Alcock, Thomas, Esq.
Aldam, William, Esq.
Alderson, Lieut.-Colonel, R.E.
Alexander, Captain Sir James Edward
Alexander, James, Esq.
Allen, Captain William, R.N.
Alsager, J. M., Esq.
Alsager, Richard, Esq., M.P.
Alves, John, Esq.
Antrobus, Sir Edmund, Bart.
Archer, Major
Arnold, Rev. Dr., F.R.S.
Arrowsmith, Mr. John, M.R.A.S.
Arrowsmith, Mr. Samuel
Ashburton, Lord
Askew, Henry William, Esq.
Atkins, John P., Esq.
Attwood, Wolverley, Esq., M.P.
Auldjo, John, Esq., F.G.S.

B.

Back, Captain Sir George, R.N.
Backhouse, John, Esq., F.G.S.
Baily, Francis, Esq., F.R.S., &c.
Baily, Arthur, Esq.
Baillie, David, Esq., F.R.S.
Baker, Lieut.-Colonel
Ball, Major, 49th Foot
Bandinel, James, Esq.
Bankes, W. H., Esq.
Baring, the Honourable Francis, M.P.
Baring, Francis Thornhill, Esq., M.P.
Baring, John, Esq.
Barrow, Sir John, Bart., F.R.S., L.S.
Barrow, John, Esq.
Bate, Mr. R. B.
Bateman, James, Esq.
Batty, Lieutenant-Colonel, F.R.S.
Beaufort, Captain F., R. N., F. R. S., Corr. Inst. France
Beaufoy, J. H., Esq., F.R.S., L.S.
Beckett, the Right Honourable Sir J., Bart., LL.D., M.P., F.R.S.
Beckford, Francis, Esq.
Bedford, His Grace the Duke of, K.G., F.G.S., &c.
Bedford, Grosvenor Charles, Esq.
Becher, Lieutenant A. B., R.N.
Beechey, Capt. Frederick, R. N., F.R.S.
Belcher, Captain Edward, R.N., F.G.S.
Bell, James C. C., Esq.
Bennett, Frederic Debell, Esq.
Bennett, John Joseph, Esq.
Bennett, James, Esq.
Bennett, William, Esq.
Bentham, George, Esq., F.L.S.
Bentinck, Lord William
Bentley, R., Esq.
Berens, Joseph, Esq.
Best, Captain the Hon. Thomas, R.N.
Best, the Hon. and Rev. Samuel
Bethune, Captain C. Drinkwater, R.N.
Bethune, Edward Drinkwater, Esq.
Bexley, Lord, M.A., F.R.S., &c.
Biddulph, John, Esq., F.H.S.
Birch, Jonathan, Esq.

Bird, James, Esq.
Biscoe, John, Esq., R.N.
Blackburne, Colonel Sir William
Blackwood, Capt. the Hon. Price, R.N.
Blackwood, Capt. F. P., R.N.
Blake, William, Esq., M.A., F.R.S., &c.
Blanshard, Henry, Esq.
Blauw, William H., Esq.
Blewitt, Octavian, Esq.
Blunt, Joseph, Esq.
Boddington, Samuel, Esq.
Borradaile, Abraham, Esq.
Botfield, Beriah, Esq., F.R.S., G.S.
Botfield, Thos., Esq., F.R.S., G.S., H.S.
Bowles, Captain William, R.N., C.B.
Bowles, Colonel
Brandreth, Captain H. Rowland, R.E.
Brayley, E. W., Esq., F.S.A.
Brecknock, Earl of, M.A.
Brereton, Rev. Dr., F.S.A.
Breton, Lieutenant W. H., R.N.
Brisbane, Sir Thomas M., G.C.B., F.R.S. L. and E., F.L.S.
Broadley, Henry, Esq.
Brockedon, William, Esq., F.R.S.
Brodie, Sir B. Collins, Bart., F.R.S.
Brooke, Sir Arthur de Capell, Bart., M.A., F.R.S., G.S., L.S.
Brooke, James, Esq.
Browne, John, Esq.
Browne, George, Esq.
Brown, Robert, Esq., Hon. D.C.L., F.R.S., L. and E. and R.I.A., V.P.L.S., Corr. Inst. Fr., Ac. St. Petersb., Ac. Berlin
Brown, Thomas, Esq., F.R.S., L.S.
Brown, Wade, Esq.
Brumell, George, Esq.
Brunel, M. I., Esq., F.R.S., &c.
Buckland, Rev. William, D.D., F.R.S., L.S., Pres. G.S.
Buckle, J. William, Esq.
Buller, Captain Wentworth, R.N.
Bullock, Captain F., R.N.
Burlington, Earl of, M.A., F.R.S., G.S.
Burnes, Lieut.-Col. Sir Alex., F.R.S.
Burney, Rev. Chas. P., D.D., F.R.S., SA., L.S., G.S., &c.
Burton, Decimus, Esq., F.R.S., SA., G.S.
Burton, Alfred, Esq.

C.

Cabbell, B. B., Esq., F.R.S., F.S.A.
Campbell, James, Esq., Jun.
Carey, John, Esq.
Carlisle, the Very Rev. Robert Hodgson, D.D., Dean of, F.R.S., G.S.
Carnac, Captain James Rivett, R.N.
Cartwright, Samuel, Esq., F.G.S.
Carnarvon, the Earl of
Cary, Capt. the Hon. Plantagenet, R.N.
The Vice-Chancellor, the Right Hon.
Chantrey, Sir Francis, D.C.L., F.R.S.
Chapman, Captain, R.A.
Chapman, Sir M. Lowther, Bart., M.P.
Chapman, Thomas, Esq.
Chatfield, Frederick, Esq.
Chatterton, Sir William, Bart.
Chesney, Col., R.A., F.R.S.
Chichester, the Dean of
Children, J. Geo., Esq., F.R.S., G.S., &c.
Christophers, John, Esq.
Church, W. H., Esq.
Clarke, Sir Chas. M., Bart., F.R.S.
Clarke, William T., Esq.
Clarke, Captain
Clarke, William Stanley, Esq., F.R.S.
Cleiland, Major-General
Clerke, Major T. H. Shadwell, F.R.S.
Clerk, Sir George, Bart., D.C.L., F.R.S., G.S., &c.
Clinton, General Sir William, G.C.B.
Cockburn, Admiral, the Right Hon. Sir George, G.C.B., G.C.H., F.R.S.
Cockerell, J. Pepys, Esq.
Coddington, Rev. H., M.A., F.R.S., G.S.
Cogan, Captain R., Indian Navy.
Colby, Colonel, R.E., LL.D., F.R.S.L. and E., G.S., M.R.I.A.
Colchester, Captain Lord, R.N.
Cole, Viscount, M.P., F.R.S., F.G.S., &c.
Colebrooke, Colonel Sir William, R.A.
Coleridge, Henry Nelson, Esq., M.A.
Collett, William Rickford, Esq.
Collett, C. F., Esq.
Colquhoun, J., Esq.
Colquhoun, Lt.-Col. J. N., R.A., F.R.S.
Colquhoun, Gideon, Esq.
Conybeare, the Rev. W. D., M.A., F.R.S.
Cooke, W., Esq.
Cooley, W. D., Esq.
Craik, G. L., Esq.
Craufurd, W. P., Esq., F.G.S.
Craufurd, Captain W., R.N.
Crawfurd, J., Esq., F.R.S., G.S., &c.
Cross, Mr. J.
Cubitt, William, Esq., F.R.S.
Cumming, William, Esq.
Cunningham, Allan, Esq.
Cunningham, George Godfrey, Esq.
Curties, John, Esq.
Curtis, Timothy, Esq.
Curzon, the Hon. Robert
Curzon, the Hon. Robert, Jun.

D.

Dale, the Rev. T. Aquila
Damer, Colonel the Hon. George L. D., M.P.
Darwin, Charles, Esq.
Dawnay, the Hon. William Henry
Dawson, Lieutenant R. K., R.E.
De Beauvoir, R. Benyon, Esq.
Denison, Lieutenant W. T., R.E.
Denman, Captain the Hon. Joseph, R.N
Derby, the Earl of, M.A., Pres. Z.S.
De Roos, Captain the Hon. J. F. Frederick, R.N.
Dickson, George Frederick, Esq.
Dickenson, Francis H., Esq.
Dickenson, J., Esq., F.R.S.
Dickson, Major-General Sir Alexander, G.C.B., R.A., F.S.A.
Dilke, Charles Wentworth, Esq.
Disney, John, Esq., F.R.S.
Doddington, Samuel, Esq.
Dollond, George, Esq., F.R.S.
Donaldson, John W., Esq.
Donkin, Lieut.-Gen. Sir Rufane Shaw, M.P., K.C.B., G.C.H., F.R.S., &c.
Doran, Captain J. G.
Doratt, Sir John, M.D.
Douglas, Edward, Esq.
Doyle, Colonel Carlo
Drummond, Major-General, R.A., C.B.
Drummond, Edward, Esq.
Drummond, Captain Thomas, R.E.
Du Cane, Captain, R.N.
Duckett, Sir Geo., Bart., M.A., F.R.S., G.S., &c.
Dundas, the Hon. Captain R. S., R.N.
Durham, the Earl of, G.C.B., F.G.S., F.H.S.

E.

Edwards, Thomas Grove, Esq.
Edwards, Rev. J., M.A.
Egerton, Lord Francis, M.P., F.G.S.
Eliot, Lord, M.P.
Elliot, Rear-Admiral the Hon. George, F.R.S.
Elliot, Rev. C. B., F.R.S.
Elphinstone, J. F., Esq.
Elphinstone, the Hon. Mount-Stuart
Enderby, Charles, Esq.
Enderby, George, Esq.
English, Henry, Esq.
Estcourt, Thomas G. Bucknall, Esq., M.P., Hon. D.C.L.
Estcourt, Lieut.-Col. J. B. B., 43rd Regt.
Evans, Captain George, R.N.
Evans, W., Esq. (a)
Evans, William, Esq. (b)
Evans, Rev. Henry Herbert
Everett, Joseph, Esq., F.A.S.

F.

Falconer, Thomas, Esq.
Fanshawe, Colonel, R.E., C.B.
Fellows, Charles, Esq.
Findlay, Alexander, Esq.
Fisher, Rev. George, M.A., F.R.S.
Fitton, William Henry, Esq., M.D., F.R.S., G.S., L.S.
Fitz Roy, Captain, R.N., F.A.S.
Fitzwilliam, the Right Hon. Earl
Forster, Edw., Esq., F.R.S., L.S., &c.
Fowke, Lieutenant Thomas, R.N.
Fox, Colonel C. R.
Franklin, Captain Sir J., R.N., F.R.S., D.C.L., F.G.S.
Fraser, Colonel John
Fraser, Jas. Baillie, Esq., F.G.S., R.A.S.
Frere, Bartholomew, Esq.
Frere, George, Esq., Jun.
Frere, Rev. Temple
Freshfield, J. W., Esq., F.R.S., F.G.S.
Fyler, George, Esq.

G.

Galloway, Colonel, E.I.C.S.
Gardner, Mr. James
Garry, Nicholas, Esq., Dep. Gov. Hudson's Bay Company, F.H.S.
Gascoigne, Capt., Ceylon Rifle Brigade
Gawler, Lieutenant-Colonel, K.H.
Gibbs, Charles, Esq.
Gilbert, Davies, Esq., D.C.L., F.R.S., G.S., L.S., S.A., H.S., A.S., Hon. M.R.S.E., M.R.I.A.
Gipps, Major Sir George, R.E.
Gladdish, William, Esq.
Goding, James, Esq.
Goldsmid, Isaac Lyon, Esq., F.R.S., G.S., &c.
Gooden, James, Esq., F.S.A.
Goold, Nathaniel, Esq.
Gordon, Adam, Esq.
Gosling, Captain George, R.N.
Gould, Captain Frederick A.
Gowen, James Robert, Esq., F.G.S.
Graham, the Right Hon. Sir James, Bart., M.P., F.R.S., &c.
Gray, John Edw., Esq., F.R.S., F.G.S., H.S.
Graves, Lieutenant, R.N.
Greene, Thomas, Esq., M.P.
Greenough, G. B., Esq., F.R.S., L.S., V.P.G.S.

Grenville, Right Hon. Thomas, F.S.A.
Gresley, Richard N., Esq.
Grey, Earl de, F.S.A.
Grey, Lieutenant George, 83rd Regt.
Grey, Rt. Hon. Sir Charles E., M.P.
Gresswell, Rev. Richard, M.A., F.R.S.
Griffith, John, Esq.
Griffiths, George R., Esq.
Griffiths, Richard Clewin, Esq.
Grindley, Captain R. Melville, E.I.C.S.
Guillemard, John L., Esq., M.A., F.R.S., G.S., L.S.
Gurney, Hudson, Esq., F.R.S., V.P. S.A.,

H.

Halford, the Rev. Thomas
Halford, Sir Henry, Bart., M.D., G.C.H., F.R.S.
Hall, Captain Basil, R.N., F.R.S., L. & E., Hon. D.C.L., F.G.S.
Halifax, Thomas, Esq.
Hallam, Henry, Esq., M.A., F.R.S.
Hamilton, J. J. E., Esq.
Hamilton, Terrick, Esq.
Hamilton, William Richard, Esq. F.R.S., V.P.S.A., M.R.S.L.
Hamilton, Captain H. G., R.N.
Hamilton, W. I., Esq., F.G.S.
Hammond, Edmond, Esq.
Hammond, George, Esq.
Hammond, William, Esq.
Hammersley, Charles, Esq.
Hanmer, Sir John, Bart., F.R.S.
Harding, Colonel George, R.E., C.B.
Hardy, Vice-Admiral Sir T. M. Bart., G.C.B.
Harriott, Captain T. G.
Herries, Right Hon. C. J., M.P.
Harrison, Sir George, K.C.H., F.R.S., S.A.
Harrison, Benjamin, Esq.
Harrison, Wm., Esq., F.R.S., G.S., &c.
Harvey, Captain Edward
Hatcher, H., Esq.
Hathorn, George, Esq.
Hatchett, Charles, Esq., F.R.S., R.S.E., &c.
Hawdon, Jos., Esq., (Sydney, N. S. W.)
Hawtrey, Rev. Dr.
Hawkins, Dr. Bisset, F.R.S.
Hay, Robert William, Esq., F.R.S., &c.
Hay, E. W. A. Drummond, Esq.
Hay, J. Holman, Esq.
Hearne, John, Esq.
Henderson, J., Esq.
Henry, Dr. Charles
Herbert, Captain Thomas, R.N.
Herbert, Hon. E.
Herbert, Jacob, Esq.
Hill, Lord Marcus
Hill, Henry, Esq.
Hoare, Sir H. H., Bart.
Hoare, Charles, Esq., F.R.S.
Hobhouse, H. W., Esq.
Hobhouse, Right Hon. Sir John Cam, Bart., M.P., M.A., F.R.S.
Hodgkin, Thos., Esq., M.D.
Hogg, John, Esq., M.A., F.L.S.
Holland, Doctor Henry, M.D., F.R.S.
Hollier, Richard, Esq., F.S.A., G.S.
Holmes, Nathaniel, Esq.
Holmes, James, Esq.
Holroyd, Arthur Todd, Esq., M.D., F.L.S.
Hooker, Sir Wm. Jackson, Ph.D., F.R.S, S.A., L,S.
Horton, Right Hon. Sir R. W., G.C.H., F.R.S.
Hotham, Admiral Sir William, K.C.B.
Howse, J., Esq.
Hudson, John, Esq.
Hughes, Mr. William
Hume, Edmund Kent, Esq.
Huntley, Captain H. Vere, R.N.

I.

Inglis, Sir R. H., Bart., M.P., LL.D., F.R.S., &c.
Irby, Captain the Hon. C. Leonard, R.N.

J.

Jackson, Colonel, K.S.W. and S.S.
James, J. Horton, Esq.
Jenkins, Richard, Esq.
Jerdan, W., Esq.
Jervis, Major T. B., E.I.C. Eng., F.R.S.
Jones, Charles, Esq.
Jones, Thomas, Esq., F.H.S.
Jones, Major-General, Sir J. T., Bart., R.E., C.B.
Jones, William H., Esq., F.H.S.
Jones, Captain H. D., R.E.
Jones, Rev. Richard, M.A.

K.

Kay, Joseph, Esq.
Kelly, William, Esq.
Keppel, Major the Hon. George, F.S.A.
King, Captain Philip Parker, R.N., F.R.S., F.L.S.
Knapp, Hambly, Esq.
Knight, Charles, Esq., F.L.S.
Knight, H. Gally, Esq., F.L.S., F.H.S.

L.

Laird, M'Gregor, Esq.
Lambert, A. Bourke, Esq., F.R.S., S.A., L.S., Hon. M.R.I.A., F.H.S.

Lance, John, Esq.
Lander, John, Esq.
Larcom, Lieutenant, R.E.
Law, William J., Esq.
Leake, Lieut.-Col. W. M., F.R.S., &c.
Lee, John, Esq., LL.D., F.R.S., S.A.
Lee, Rev. James Prince, M.A.
Lefevre, J. G. Shaw, Esq., F.R.S.
Lemon, Sir Charles, Bart., M.P., F.R.S.
Levien, Edward, Esq.
Lewis, Captain Locke, R.E., F.R.S.
Lichfield and Coventry, Right Rev. the Lord Bishop of
Lindley, John, Esq., Ph.D., F.R.S., L.S., G.S., H.S.
Lloyd, William Horton, Esq., F.L.S.
Loch, Captain
Long, George, Esq., M.A.
Long, Henry, Esq.
Lowe, Henry, Esq.
Lowry, Mr. Joseph Wilson
Lumley, Benjamin, Esq.
Lushington, Major-General Sir J. Law, G.C.B.
Lushington, Captain Franklin, A.D.C.
Lyall, George, Esq.
Lyell, Chas., Esq., M.A., F.R.S., L.S., G.S.
Lynch, Lieut. H. Blosse, Ind. Navy
Lyon, James Wittit, Esq.
Lyon, Captain David

M.

Mac Arthur, John, Esq.
Macculloch, J. R., Esq.
Macfarlane, Major J., E.I.C.S.
Macgillivray, Simon, Esq., F.S.A.
Macintosh, Colonel, K. H.
Mackenzie, Alexander, Esq.
Mackenzie, Harry, Esq.
Macdonald, John, Esq.
Macfarlane, Charles, Esq.
Mackillop, James, Esq.
Maclean, George, Esq.
Maconochie, Captain, R.N.
Mackie, Major
M'Neil, Sir John, G.C.B.
Magrath, Edward, Esq.
Malmesbury, Right Honourable Earl of, G.C.B.
Mangles, Captain, R.N., F.R.S.
Marjoribanks, Edward, Esq.
Markham, Edward, Esq.
Marshall, J., Esq.
Martin, Rev. J. W.
Martineau, Joseph, Esq., F.H.S.
Mathison, G. F., Esq.
Maughan, Captain P., Ind. Navy
Mauley, Lord de, M.P.
Maxwell, Acheson, Esq.
Meek, James, Esq.
Melville, Lord Viscount, K.T., F.R.S.
Melvill, J. C., Esq.
Mercier, Francis, Esq.
Milton, Viscount, M.P.
Mitchell, Major Sir Thomas L., F.G.S.
Montefiore, Jacob, Esq.
Montefiore, Sir Moses, F.R.S.
Monteith, Lieut.-Colonel, E.I.C. Eng.
Montgomery, James, Esq.
Moore, G. H., Esq.
Morison, James, Esq.
Morier, James, Esq., F.R.S.
Mornay, Aristides Franklin, Esq., F.L.S.
Motteux, John, Esq.
Mudge, Colonel, R.E.
Mundy, Captain G. R., R.N.
Munro, Rev. Vere
Munster, Right Hon. the Earl of, F.R.S.
Murchison, Roderick Impey, Esq., F.R.S., G.S. and L.S.
Murdoch, Thomas, Esq., F.R.S., S.A.
Murdoch, Thomas W. Clinton, Esq.
Murray, Lieut.-General the Right Hon. Sir George, G.C.B., G.C.H., F.R.S., &c.
Murray, John, Esq., M.R.S.L.
Murray, John, Jun., Esq., F.G.S
Murray, T. Laurie, Esq.
Musgrave, T. M., Esq.

N.

Neeld, Joseph, Esq., M.P., F.S.A., G.S.
Newnham, William, Esq.
Nettleship, Thomas, Esq.
Nicolson, Sir Frederick, Bart.
Nicholson, George T., Esq., B.A.
Nicol, J. D., Esq.
Northumberland, His Grace the Duke of, K.G., F.R.S., &c.
Nott, the Rev. Dr., F.S.A., M.R.S.L.
Norwich, the Right Rev. Edw. Stanley, Lord Bishop of. Pres. L.S.
Nugent, Lord, D.C.L., F.S.A.

O.

O'Brien, Captain Henry, R.A.
Ogle, Vice-Admiral Sir C., Bart.
Ogle, Nathaniel, Esq.
Oldfield, Richard K., Esq.
Oldrey, Captain William, R.N.
Ommanney, H. M., Esq.
O'Reilly, Colonel
Ouseley, the Right Hon. Sir Gore, Bart., G.C.H., F.R.S., &c.
Outram, B. F., Esq., M.D., R.N., F.R.S.

Owen, Vice-Admiral Sir E. W. C. R., K.C.B.
Owen, Captain William Fitzwilliam, R.N.
Oxendon, G., Esq.

P.

Paget, Captain Lord Clarence, R.N.
Paley, Thomas, Esq.
Palmer, G., Esq., F.G.S., H.S.
Palmer, Samuel, Esq.
Parish, Sir Woodbine, K.C.H., F.R.S., G.S.
Parker, Thomas Lister, Esq., F.R.S., &c.
Parry, Captain Sir William Edward, R.N., D.C.L., F.R.S.
Parry, Francis C., Esq., M.D.
Parson, George, Esq., R.N.
Pasley, Lieut.-Col., R.E., F.R.S., C.B.
Pechell, Captain Sir J.S.B., Bart., R.N., K.C.H., F.R.S.
Peel, the Right Hon. Sir Robert, Bart., M.P., D.C.L., F.R.S., S.A., &c.
Pelham, Captain the Hon. Dudley, R.N.
Pelly, J. H., Esq., Governor, Hudson's Bay Company, F.H.S.
Penn, Richard, Esq., F.R.S.
Pepys, W. Hasledine, Esq., F.R.S., L.S., &c.
Petit, Louis Hayes, Esq., M.A., F.R.S., G.S., &c.
Phillimore, Joseph, LL.D.
Phillipps, Sir Thomas, Bart., M.A. F.R.S., S.A., L.S., G.S., &c.
Phillips, Captain C., R.N., F.R.S.
Pigot, Henry, Esq.
Pigou, Frederick, Esq.
Planta, the Right Hon. Joseph, M.P.
Plowes, John, Esq.
Pocock, J. J., Esq.
Pollington, Viscount
Ponsonby, Honourable Frederick
Porter, G. R., Esq.
Portlocke, Capt., R.E., F.R.S., F.G.S.
Potter, William S., Esq.
Potts, Charles, Esq.
Prudhoe, Captain the Right Hon. Lord, R.N., F.R.S., S.A., &c.
Powis, Right Hon. the Earl of, M.A.
Powles, John D., Esq.

R.

Radcliffe, John, Esq.
Ramage, C. T., Esq.
Ramsay, David, Esq.
Rankin, F. Harrison, Esq.
Rawson, W. Rawson, Esq.
Reid, Lieutenant-Colonel W., R.E., C.B.
Rennie, George, Esq., F.R.S.
Rennie, Sir John, F.R.S.
Rennie, M. B., Esq.
Renny, John B., Esq.
Renouard, Rev. George Cecil, B.D., M.R.A.S.
Renwick, Lieutenant, R.E.
Rice, the Right Hon. T. Spring, M.P.
Richardson, Dr., R.N., F.R.S., L.S., &c.
Ripon, Right Hon. the Earl of, F.R.S., &c. &c.
Robe, Major, R.E.
Robe, Captain F. H.
Robinson, Lieutenant C. G., R.N.
Robinson, E. F., Esq.
Rodd, J. Rennell, Esq.
Rogers, Lieut., 90th Lt. Infantry
Roget, P. M., Esq., M.D., Sec. R.S., F.L.S., F.G.S., M.R.I.A.
Rose, the Hon. Sir George, F.R.S., LL.D.
Ross, Charles, Esq., M.P.
Ross, Major-General Sir Patrick, K.C.B.
Rous, Captain the Hon. Henry, R.N.
Rouse, John, Esq., F.S.A.
Rowlands, Dr. D.
Rudge, Edward, Esq., F.R.S., S.A., L.S.
Rumboldt, C. E., Esq., F.S.A.
Russell, the Right Hon. Lord John, M.P.
Russell, J. W., Esq., F.R.S., S.A., L.S.
Russell, Captain Robert, R.N.
Ryder, the Hon. F. Dudley

S.

Salisbury, the Marquis of
Salmon, Rev. Henry, M.A.
Salmon, Wm. Wroughton, Esq.
Sandon, Viscount, M.P.
Sandwith, Colonel, E.I.C.S.
Scarlett, Major the Hon. J. Yorke
Schald, Edward Vernon, Esq.
Scott, Claude E., Esq.
Seager, John L., Esq.
Sedgwick, the Rev. A., M.A., F.R.S., G.S.
Senior, Nassau William, Esq., M.A.
Sheringham, Lieutenant W. L., R.N.
Sherriff, Francis, Esq.
Shirreff, Captain W. H., R.N.
Shortreed, Lieutenant, E.I.C.S.
Skegg, Edward, Esq.
Skelmersdale, Lord, F.H.S.
Slater, Captain M. A., R.N.
Slater, Joseph, Esq,
Sligo, the Marquis of, K.P., F.H.S., and L.S.

Smirnove, John, Esq., F.R.S., L.S., &c
Smith, Lieutenant-Colonel Sir C., C.B., R.E.
Smith, Edward Osborne, Esq.
Smith, George Henry, Esq.
Smith, H., Esq.
Smith, George Stavely, Esq.
Smith, George, Esq., F.L.S.
Smith, James, Esq. F.R.S. L. & E.
Smith, Lieut. Webber, 48th Regt.
Smith, Octavius, Esq.
Smith, Richard Carter, Esq.
Smith, Captain Matthew, R.N.
Smyth, Lieut. Brunswick, 80th Regt.
Smyth, Captain W., R.N.
Smyth, Captain W. H., R.N., K.S.F., F.R.S., Corr. Ins. Fr.
Sotheby, Captain, R.N.
Spencer, Right Honourable the Earl
Spencer, Capt. the Hon. F., C.B., R.N.
Spottiswoode, A., Esq.
Stanley, Lord, F.R.S., S.A.
Stanley, Captain Owen, R.N.
Stanley, Lord, M.P.
Stannus, Major-Gen. Sir Ephraim, C.B.
Starling, Mr. Thomas
Staunton, Sir George T., Bart., F.R.S.
Stavely, Thomas, Esq.
Stephen, Sir George
Stephenson, Daniel, Esq.
Stoddart, Captain C., R. Staff Corps.
Stokes, C., Esq., F.R.S., S.A., L.S., G.S.
Stooks, Thomas, Esq.
Stopford, Admiral the Hon. Sir Robert, G.C.B.
Stuart, Daniel, Esq.
Sturge, J., Esq.
Sturt, Captain Charles, F.L.S.
Surtees, Stevenson V., Esq.
Sutherland, Robert, Esq.
Swinburne, Captain C. H., R.N.
Symonds, Captain Sir William, R.N.
Symonds, Capt. W. C.

T.

Talbot, Earl, K.P., F.R.S. and S.A.
Taylor, Richard, Esq., F.L.S., G.S., &c.
Taylor, John, Esq.
Temple, Major Sir Grenville T., Bart.
Templer, J. C., Esq.
Thatcher, Colonel, E.I.C.
Thornton, the Right Hon. Sir Edward, K.C.B.
Thornton, Edward, Esq.
Tindal, Lord Chief-Justice
Tooke, A. W., Esq., M.A.
Trevelyan, W. C., Esq., M.A., F.G.S., L.S.
Trevor, the Honourable G. R.
Tuckett, Frederick, Esq.
Tuffnell, Henry, Esq., F.R.S., G.S.
Turnbull, Rev. Thomas Smith, F.R.S., G.S.

U.

Urquhart, David, Esq.

V.

Vallé, A. B., Esq.
Vaughan, the Right Hon. Sir Charles, G.C.H.
Verney, Major Sir Harry C., Bart., M.P.
Vetch, Captain, R.E., F.R.S. and G.S.
Vidal, Captain, R.N.
Vigors, N. A., Esq., F.R.S., G.S., M.R.I.A., &c.
Vivian, John, Esq., M.P.
Vulliamy, B. L., Esq.
Vyvyan, Sir R. H., Bart., M.P., F.R.S., G.S.

W.

Walker, Mr. John
Walker, Mr. Michael
Walker, Lieutenant J. G., R.A.
Ward, H. G., Esq., M.P.
Washington, Captain, R.N.
Washington, Adam, Esq.
Watson, Sir Frederick B., K.C.H., F.R.S.
Weale, Mr. J.
Webb, Thomas, Esq.
Wedderburn, John, Esq.
Wellington, His Grace the Duke of, K.G., G.C.B., G.C.H.
Wells, the Very Reverend the Dean of, F.R.S., F.L.S.
Wells, Lieut.-Colonel, R.E.
West, William, Esq.
Westall, William, Esq., A.R.A.
Weston, Samuel C., Esq.
Weyland, John, Esq., F.R.S.
Whewell, Rev. W., F.R.S., S.A., G.S.
Whinyates, Lieutenant-Colonel, R.A.
White, Vice-Admiral J. C.
White, Frederick, Esq.
Wilbraham, Capt. Rich., 7th Fusileers
Wilkinson, John G., Esq.
Wilkinson, William, Esq.
Williams, Rev. David, D.C.L., F.S.A.
Willich, Charles M., Esq.
Wills, W. H., Esq.
Wilson, L. P., Esq.
Wilson, Thomas, Esq.
Wilson, Dr. Isaac, R.N.
Winterbottom, J. Edward, Esq.

Wittich, William, Esq.
Wolfe, Lieutenant, R.N.
Wortley, Hon. J. Stuart, F.R.S., G.S.
Wray, John, Esq.
Wright, the Rev. George Newnham, M.A.
Wulff, General, R.A.
Wyatville, Sir Jeffrey, R.A., F.R.S., S.A. and G.S.
Wyld, Mr. James

Y.

Yarborough, the Earl of
Yates, Rev. James, M.A., F.L.S. and G.S.
Yates, John Ashton, Esq.
Yates, Joseph Brookes, Esq.
Young, George F., Esq., M.P.
Young, James, Esq.
Young, Charles Baring, Esq.
Yorke, Lieutenant-Colonel, P.S.

Names of Individuals to whom the Royal Premium has been awarded.

1831.—Mr. Richard Lander.

1832.—Mr. John Biscoe.

1833.—Captain Sir John Ross, R.N.

1834.—Lieut.-Col. Sir Alexander Burnes, F.R.S.

1835.—Captain Sir George Back, R.N.

1836.—Captain Robert Fitz Roy, R.N.

1837.—Colonel Chesney, R.A., F.R.S.

1838.—Mr. Thomas Simpson.

Dr. Edward Rüppell.

ADDRESS

TO THE

ROYAL GEOGRAPHICAL SOCIETY OF LONDON;

Delivered at the Anniversary Meeting on the 27th May, 1839,

BY

WILLIAM R. HAMILTON, Esq., F.R.S.,

President.

Gentlemen,

In meeting you on this, the eighth anniversary of the foundation of the Royal Geographical Society, and in resigning into your hands the honourable distinction of being your President; and, in the entire confidence that you have fixed your choice on one not, I would fain flatter myself, more anxious and zealous for your prosperity and for the advancement of the science which you cultivate, but who is far better qualified, than I have proved myself, to fulfil the duties of the situation, I shall proceed to lay before you a few observations on the present state of the Society, and on the progress which geographical knowledge has made during the last year, from individual exertions, from your own labours, and from those of our rivals in the race of honour we are engaged in on the continent of Europe, and in other parts of the world; and I shall allow myself such observations as may occur to me, as not undeserving of your attention, on the increasing importance of geographical knowledge to the social and political condition of man.

The finances of a Society like the present, existing altogether on private subscriptions, will always form an object of paramount interest. I am happy to say, on reference to our balance-sheet, that we have every reason to be satisfied on that score. Our receipts in the preceding year, exclusive of the £500 received for a special purpose from the Society for Promoting Christian Knowledge, amounted to the sum of 1500*l.*; and the number of members on our list exceeds 650.

Our library, both of books and maps, has received large additions since the last anniversary. It is still, however, very far from what it ought to be; and I hope that the liberality with which members are allowed the use of it will continue to stimulate a corresponding liberality on the part of those, who are in a state to augment the collection by voluntary contributions.

We have received, among many other donations to the library, "The Silurian System," founded on geological researches in some of the western counties of England, by Mr. Murchison; some additional sheets of the Ordnance Survey of England, and the Townland Survey of Ireland; various charts from the Hydrographic Office at the Admiralty; and princely donations from the Depôt de la Guerre, and the Depôt de la Marine, of France, and from the Dépôts Topographiques of Wurtemberg and Austria, for the latter of which we are also much indebted to our zealous honorary member, Baron C. Hügel; several valuable documents from the Court of Directors of the East India Company; and we have this day received from Capt. Fitz Roy, the "Narrative of the Surveying Voyages of Her Majesty's Ships Adventure and Beagle, between the years 1826 and 1836, describing the examination of the southern shores of South America, and the Beagle's circumnavigation of the Globe;" besides many excellent maps from Mr. John Arrowsmith, the Transactions of the Imperial and Royal Academics of St. Petersburg and Berlin, Paris and Lisbon, and numerous donations which are recorded in the volume of our Journal now issuing from the press.

But here, Gentlemen, we are subject to a very serious inconvenience, the whole extent and nature of which ought not to be withheld from you. We have long suffered from the straitened room available for our public business, and most particularly for the accommodation of members who may wish to consult the library, whether for instruction or recreation. Our books and maps are piled in heaps on the chairs and tables abovestairs in most deplorable confusion; and although every pains are taken, and effectually taken too, by your Secretary and by the Librarian, to preserve some degree of order amongst the various articles of your property, which, from their high value and usefulness, do ill deserve such treatment, still, every day and hour furnish instances of the evil occasioned by the want of a larger apartment for reading, for drawing, for comparing and construction of maps, for the exhibition of instruments, for the reception of strangers, and for what is by no means the least useful or the least agreeable of our pursuits, the mutual exchange of ideas amongst the members of the Society, for the promotion of the cause which has so happily and so honourably brought us together. I took occasion in my address to you last year to advert pointedly to this subject, in its practical application to the progress of geographical inquiry: since that period the Council have been indefatigable in attempting to provide a remedy; but hitherto, I regret to say, without success. You need not be reminded, indeed you must all feel the necessity, of our not removing farther than we now are from the centre of the active movement of the metropolis, nor farther from the great public offices, from our daily intercourse with which we derive so many advantages; the suspension of which intercourse would at once deprive us of much valuable assistance, and would occasion great loss of time. The Council have visited various sites which would be eligible in themselves, if available, and we have entered into, or rather commenced, several negotiations for the attainments of this object: but all have fallen to the ground from one or the other of two causes; either the space offered was too small, or the expenditure it would entail upon us was too large. The annual balance-sheets show to the Society that we have a considerable sum of money in the Funds, which might be available for this purpose, and this, indeed, was one of the main objects the Council had in view when they resolved to fund a large portion of the receipts during the first years of our institution: and such is our conviction of the imperious necessity of providing suitable apartments for the Society, that we have even contemplated the expediency of applying the whole of this fund, in addition to a considerable increase of annual rent, for this one object; confident as we are that a very large portion of such outlay would rapidly be replaced by a large accession of new members. It might indeed for a few years cripple our means for fitting out exploring expeditions on the scale we have hitherto acted on; but we should not despair of speedily being enabled to resume that practice, should it be thought advisable, if we could once secure to ourselves a permanent, convenient, and spacious home, if we could add to our *name* a *local habitation*.

As the great desideratum in a new establishment will always be a large and commodious room for these Meetings, it has even been suggested to the Council that, in the impossibility of suiting ourselves permanently and immediately with *all* we want, we might find it advisable to take temporary apartments in this neighbourhood for the accommodation of Members generally, and for the *daily* business; whilst the courtesy of our present landlords would still secure to us the use of *this* room for the Evening Meetings.

Whatever inconvenience might occasionally be felt by such an arrangement, it would be infinitely less, and of much less consequence to the essential interests of the Society, than that which is felt now.

The hand of death during the last year has removed from our Society two persons, to whose services in the cause of Geography I must call your attention. Vice-Admiral Sir John T. Rodd, late a Member of the Council, was son-in-law of one of our most distinguished geographers, Major Rennell, who was indebted to him for many of the facts inserted in his well-known Current Charts; Sir John Rodd was also a donor of several works to our library. The Baron de Hamelin, a Foreign Honorary Member, at the head of the Depôt de la Marine of France, deserves also to be gratefully remembered by us—not only for his zeal in the promotion of scientific Geography, but for the many valuable maps and books which have been presented to us by the establishment over which he presided.

Positive proofs of the importance of accurate geographical knowledge are not wanting, many must at once occur to every one who hears me; nor are the negative proofs of less frequency, or of less weight.

The history of nations might furnish us with a long list of disastrous occurrences, owing to the ignorance of localities, in those on whom has hung the fate of armies, from the ignominious tale of the Fauces Caudinæ to the defeat of Saratoga.

What is it which has occupied all minds, exercised thousands of writers, embarrassed the most expert diplomatists of modern times, put as it were on the cast of a die an unnatural, and almost a civil war between two kindred nations, threatened to interrupt the progress of civilization over the half of the new world, and to throw into confusion the great commercial interests of the globe? What but the ignorance of the course of one or two rivers, of the range of certain hilly districts from which they derived their streams; and the uncertainty in geographical nomenclature whether the specific name of a bay, which bathes a small extent of coast, occurring in a diplomatic document, is to be considered as identical or not with the vast ocean of which it forms a part? To this must also be added the ignorance which prevailed, at a period fifty years ago, of the extent, configuration, and boundaries of our own provinces. You cannot fail to receive the allusion as stamping at once the inestimable value which, for the repose of nations, as well as for the advancement of science, we ought to set on geographical research. In another quarter of the world now under our dominion, I mean the large district of Assam, it is only, I believe, of very late years, that our botanists and naturalists have ascertained that the tea-tree, an indigenous plant of that soil, may be cultivated, so as to divert, perhaps in a few years, the entire direction of a trade which during the last half-century has poured so many millions into the Exchequer.

We all remember the fatal consequences, which ensued to our army and navy in 1827, when they had to penetrate the unknown swamps and rivers of the Birman territory.

"When at the commencement of the Birman war" (says Capt. Pemberton, of the Bengal Engineers), "our ignorance of the whole frontier became manifest, the impolicy, the shortsightedness of not having instituted, even by force, if requisite, a proper examination of the mountain passes, flashed upon the mind of even the most careless observer; and its lamentable effects were afterwards shown in an expenditure of life and treasure without parallel in the annals of Indian warfare." Another proof, if proof were wanting, to be added to those lately brought forward (by Major Jervis, in his address to the British Association, on the present state and progress of the Trigonometrical Survey in India), of the absolute necessity of a complete survey, to ensure the good government of any country.

When our troops landed in Egypt, in 1801, they suffered for a time from want of water on the coast, though, eighteen hundred years before, Cæsar had told us that it was to be found all along the coast by digging for it to a very inconsiderable depth.

On that same coast how nearly was the gallant Nelson deprived of the glorious fruits of his anxious nights and days, and indomitable perseverance, by his ignorance of the shoals, near which was anchored the enemy's fleet!

Minor misfortunes of this kind are still of every-day occurrence. There are few parts of what is called the "known world," which are yet distinctly known as they ought to be known. Every new survey corrects the thousand and one errors of omission or commission of those which have preceded it, even in the most frequented seas. And as the commerce of the world, goaded on by the thirst of adventure and profit, "æstuans angusto limite mundi," is perpetually seeking out new marts of exchange, or struggling to unlock the bars by which the jealous fears of some, and the narrow and superstitious prejudices of others, have hitherto excluded it from many of the richest countries and finest harbours in the world, this department of geographical knowledge, namely, Hydrography, becomes every day more and more indispensable, as it is that in which minute accuracy is most required.

In performing the most agreeable of the duties which devolves upon your Council, namely, in conferring the Honorary Medals provided out of the annual Royal Donation, the Council, as you have this morning heard from your Secretary, have hitherto restrained themselves to presenting one medal to the individual, whom they conceived to be the best entitled to this distinction; and this has been accompanied with the further present of the difference between the value of the metal and the amount of the Royal Donation. On the present occasion, and for the future, if the same should be considered advisable by their successors, they have resolved to present two medals of equal value and equal honour, to the two gentlemen whom, on the principles hitherto acted upon, they may judge to have rendered the most distinguished services to the cause of Geography. And, accordingly, the medals for this year have been awarded by the Council, one, which is called the Founder's Medal, to Mr. Simpson, the officer of the Hudson's Bay Company who has yet reached the farthest point on the north coast of America—east and west of the Coppermine River, and whose exertions, together with those of Mr. Dease, have been so well appreciated by you; and the other, or the Patron's Medal, to Dr. Edward Rüppell, of Frankfort, for his travels and researches in Nubia, Kordofan, Arabia, and Abyssinia.

I shall avail myself of this opportunity to lay before you a few observations on the true principles of conferring these honorary medals for distinguished eminence in the pursuit of science generally; equally applicable, as I believe them to be, to similar acts in this and other departments of knowledge.

When the founder of an honorary reward of this description shall have laid down, either by grant or bequest, certain distinct rules for realizing his donation, such rules must, of course, be religiously adhered to. But when sufficient means are annually available to a literary or scientific society, for the purpose of giving an honourable distinction, without the precise mode of its being prescribed, I should be inclined to submit to your judgment whether it might not be advisable to deviate in some manner from the course usually pursued.

This course has been in almost all instances to have a medal struck, representing on one side of it the head of the donor, and, on the reverse, either some appropriate allegorical device, or wreath, within which may be inscribed the name of the person receiving it. Each medal is thus, in one sense single, or unique of its kind; whilst, in another sense, all the medals so given are merely repetitions one of the other, the only variation being the name of the receiver. Now it is obvious that such a medal can hardly be considered as in itself *a distinction of honour.* The act by which it is conferred is, indeed, as far as it goes, an honourable distinction; but the medal itself, being necessarily locked up and preserved in the possessor's own keeping, does in no way whatever contribute to extend his fame or good name, either amongst his contemporaries or to posterity; that is, it does not spread abroad the fact of the honour having been thus bestowed upon him. It is not to him an efficient source of *honour;* and, in truth, the greater is its material value, that is, the greater the liberality to which it owes its existence, the less is it seen or known, the more carefully must it be guarded from being lost or purloined during the life of the possessor, who may strictly be described in the words of Horace, 'Magnas inter opes inops;' and, after his death, the greater is the probability that it will, sooner or later, find its way to the furnace. Under this system, many of those who hear me may, from the interest they take in Geography, recollect that medals of honour have been awarded from this chair to Ross, to Back, to Fitz Roy, to Burnes, and to Chesney; but how few, out of a particular range of study, are aware that similar marks of distinction have within a few years been bestowed, by other Societies, on a Faraday, a Hallam, an Ivory, a Scott, a Herschel, or a Davy!

To obviate these objections, I am of opinion that, instead of giving one gold medal, we ought to expend the means at our disposal, when they are sufficient for the purpose, in having the portrait of the receiver engraved upon the die; we should present him with one impression in silver, and strike off at least an hundred others in bronze, for circulation amongst the various public museums here and abroad, and also for indiscriminate sale for the supply of private collections.

By an arrangement of this character you will at once see that, through the extended circulation of such a monument, the honour, fame, name, and success of the receiver are not only made known throughout Europe, but they are perpetuated, together with a delineation of his features, possibly to the latest posterity. Such is the bronze medal, which I now hold in my hand, of the celebrated navigator, Captain Cook, which was struck, with his portrait, by the Royal Society, as a testimony of their admiration of the services he had rendered in the career of discovery. But I fear this is an unique instance of the kind in this country.

I hope, gentlemen, you may live to see this practice, originated in this country, rapidly and generally applied by our own literary and scientific societies, and soon imitated by our sister establishments on the Continent.

I would fain mention a strong additional argument in its favour, namely, the encouragement it would give to a department of the Fine Arts which has long been neglected by our countrymen, but of the revival of which, I think I see the approaching dawn. This is, indeed, I am aware, a consideration foreign to the purposes for which we are here assembled, but you will not approve it the less, because it will do good out of the sphere of your own more immediate pursuits. Science, literature, and the arts, are the great evidences by which the development of the intellectual power of men is made most manifest. Apart from a conviction of our moral duties and our religious responsibilities, these are the three Graces, which elevate one man above another, one nation above another, one era above another. In all the most prominent periods of history they have been born, have flourished, have decayed, and have revived together; and though special circumstances may now and then intervene, within limited periods, or in particular portions of the world, to the prominence of one of them, and the disparagement of another, this is not the ordinary course of national progress; it is a phœnomenon, the existence of which we should not hail with satisfaction, and we ought rather to view it in the light of a perturbation, which, like those which occur in a higher region, are righted by the action of the same principles by which they are produced.

This allusion to the execution of modern medals in aid of the pursuits of Geography prompts me to say a few words on the already intimate connection between the science of Numismatics, as pursued by the antiquary and the scholar, and the knowledge of comparative or historical Geography. In the first place, the professor of Numismatics arranges his medals upon a strictly geographical system—he begins with the most western parts of Europe, with Lusitania; he then proceeds to Hispania and its several provinces; he then arranges those of Gaul and Britain: to these succeeds Italy, first north of the Po, then to Etruria and the south of that river, and on through Umbria, Latium, Campania, Apulia, Calabria, to Sicily and its islands. Having thus disposed of Western Europe, the Numismatist begins again with the Tauric-Chersonese, and proceeds regularly through European Sarmatia, Dacia, Pannonia, the two Mœsias, Thrace, its Chersonese, to Pæonia, Macedon, and Thessaly, with their adjacent islands. He then takes the eastern coast of the Greek peninsula, through Dalmatia, Illyrium, and Epirus, with the neighbouring islands, to Acarnania and Ætolia. This brings him again to the east by way of Locris, Phocis, and Bœotia, to Attica and her islands; he then takes up the Peloponnesus, with its several divisions of Achaia, Elis, and its islands, Messenia, Laconia, Argolis, and the central Arcadia. After which come the various islands of Crete and Eubœa, and the smaller ones of the Ægean, which have always been considered as belonging to Europe.

Of the medals or coins of the Asiatic countries and towns, we have first, beginning from the most northerly, those of the Cimmerian Bosphorus, and Colchis; continuing along the coast of Asia Minor, follow those of Paphlagonia, Bithynia, Mysia, and the Troad; down the western shores of Asia Minor, through Æolis, Ionia, Caria, and Lycia, and their several islands; then eastward through Pamphylia, Pisidia, Isauria, Lycaonia, to Cilicia, and the adjoining isles; next to these are the interior regions of Lydia, Phrygia, Galatia, Cappadocia and Armenia. The Numismatist then proceeds to Syria, its several districts of Commagene, Cyrrhestica, Chalcidene, Palmyrene, Seleucis, and Pieria; then Cœlesyria, Trachonitis, Ituræa, and the Decapolis, or Haouran; concluding with Phœnicia, Galilæa, Samaritis, Judæa, and Arabia. We have then the remoter, and more inland districts of Mesopotamia, Assyria, Persia, Parthia, Bactriana, and Characene. Coming then into Africa, he takes successively the districts washed by the shores of the Mediterranean, from the Isthmus of Suez to the Straits of Gibraltar; these are Egypt, Cyrenaica, Syrtica, Byzacene, Zeugitana, Numidia and Mauritania.

I say nothing of the Consular coins struck at Rome during the Republic, or of those which are called Imperial, because, however important they may be for points of history or chronology, they are comparatively devoid of interest in reference to Geography.

The whole of the ancient world, well known to the Greeks or Romans, is thus brought successively into notice; and I believe it will be allowed by all, that the best Numismatists are also the best acquainted with the general character, as well as with the minuter details, of comparative Geography.

We may here, also, readily give the tribute of gratitude which is due to the travellers in a large portion of the districts above enumerated, for the correct fixing of many ancient sites of towns, which had before been looked for in vain, by the medals which have been collected on the spot, or in their immediate vicinity. Next to inscriptions found in situ, and bearing the name of the towns by which they were erected, these coins furnish us with the best criteria for identifying localities; and this fact should be particularly impressed on the minds of travellers, who may collect coins, that they may be aware of the importance of noting down carefully the places where they have acquired them.

Europe.—The progress of Geography in Europe can only be accurately marked by recording the accessions to the great topographical maps of each state now in progress, nearly in every country; yet, however unwilling to pass over the works of many laborious and highly deserving men, neither will time permit me, nor would it be fitting from this chair to enter into such details. It may be sufficient to remark, that several sheets have, during the past year, been added to their various surveys. In our own country, not to dwell upon the Ordnance Map of England, which is familiar to us all, and which now approaches its completion, the four northern counties only remaining to be surveyed, I would mention, that in the course of last summer several stations of the principal triangles have been completed in the north of Scotland, where the party remained till they were driven by the snow from the mountains of Cape Wrath. And during this spring they have recommenced their labours on the north-eastern coast, and in the interior of Sutherlandshire, whilst another body of surveyors will proceed to complete the secondary points along the western shore, for the purpose of affording the basis for the hydrographical survey of that coast.

In Ireland the townland survey advances rapidly, and the maps of seventeen out of the thirty-two counties are complete. But I would wish to call your especial attention to a map of that country, just published in six sheets, on the scale of half an inch to a mile, which includes the whole of the Government survey, as far as it reaches, and is completed from the best of all the other materials which were available. This map, in which the physical features of the country are well represented, is intended to show the projected lines of railroads, and reflects high credit on Lieut. Larcom, R. E., under whose immediate superintendence it has been drawn and engraved.

The labours of the Hydrographic department, under the direction of Captain Beaufort, R.N., one of your Council, and one of those who warmly and ably assisted Sir John Barrow, and others, in the original institution of this society, keep pace with the land survey of the British Isles. Parts of the east and west coasts of England, the Irish Channel, the east coast of Scotland, the coast of Wales, and the northern coast of Ireland, have been carefully surveyed; and the remainder is in progress. In addition to this, Captain Hewett is steadily engaged on his large chart of the North Sea, which, when completed, promises to be the most beautiful specimen of nautical surveying ever executed.

Although I have already called the attention of the Society generally to the presents which it has received from public institutions, or from private individuals, I cannot avoid the opportunity of thus testifying to Mr. Murchison the gratitude which, in common with all lovers of science, and of geography in particular, this Society must feel for the zeal, patience, and scientific ardour, which he has expended in the production of his very valuable and remarkable work, "The Silurian System." Mr. Murchison's chief object in this "labour of love" has, of course, been the elucidation of the geological phœnomena of that extensive district, comprehending certain border counties of England and Wales. But, besides the evidence which is brought forward to show that the Silurian region may serve as the type of a normal group of hitherto unclassified deposits, which there rise to the surface in successive ridges, and connect the coal formations and other overlying strata with the older slaty rocks, extending from the southern limits of Cheshire to the extremity of Pembrokeshire, and along with them the full survey of the operations, by which the surface of this part of our island has passed from a submarine condition into dry land, and by which the present system of drainage has been effected, the work in question embraces a great mass of valuable physical geography, not the less interesting to us from its being limited to our own homes; and the whole offers a brilliant specimen, and an example, which we may hope will not be without its imitators, of what may be effected within a few years by the earnest and honest application of an ardent mind to the elucidation of one subject. We are all aware that a single plant, the minutest corpuscle, cannot be fully described without a complete knowledge of the whole physiological structure of the animal and vegetable creations.

Carl Ritter, in his Introduction to his *Erdkunde*, a work which has at once stamped its author as the father of descriptive geography, remarks, with satisfaction, that an attentive study of his own country, from the Oder to the banks of the Rhine and the Danube, had enabled him to appreciate the description of distant regions under analogous natural circumstances, as the intimate knowledge he had acquired of the basin of one large river, the Rhine, from its source in the Alps, through its lakes, and along the course it had formed for itself to the sea, inspired him with just and accurate views respecting similar districts in other parts of the globe; and his residence in Italy and on the shores of the Mediterranean had given him a general acquaintance with the influence and vital force of volcanic action, and the several relations which unite the sea and the land, the three kingdoms of nature, and the inhabitants of the earth. He was thus enabled, by the application of his own observations to the narratives of others, to treat philosophically a subject which embraced the physical construction of the whole surface of the globe; that is, its mountains, rivers, plains, valleys, steppes, and partitions of waters.

The new edition of the "Statistical Account of the British Empire," by Mr. M'Culloch, and that of the "Gazetteer of England and Wales," by the late Mr. James Bell, of Glasgow, which embodies all the recent municipal changes, the minute "Statistical Account of Scotland," now in course of publication, the "Memoir on the County of Londonderry," the "Report of the Commissioners on the Statistics of Ireland," and the "Tables of Revenue, Population, and Commerce," by Mr. Porter, of the Board of Trade, furnish a far more complete topographical description of our country than we have hitherto possessed.

But, whilst noticing this addition to our knowledge of the British Isles, I cannot avoid remarking the great want of a General Gazetteer, or Geographical Dictionary, in the English language. Looking at the materials now flowing in from every quarter of the globe, at the great government surveys in progress throughout Europe, and the increased activity amongst travellers and explorers in all parts of the world, I cannot doubt that such a work might be undertaken with advantage. It is in vain to wait till all these surveys are completed, or all the different countries on the face of the earth are explored; geography, from its very nature, is, and must ever be, an imperfect, because a progressive, science; and the reasons for not undertaking such a work at present would be equally strong ten or twenty years hence. If a Geographical Dictionary were commenced at once, with a determination on the part of the editor to go to the best sources for his information, to adopt, and strictly to adhere to, one standard for the orthography of Arabic and other oriental names, and to endeavour to make his work a record of the present state of geographical, not statistical knowledge (as is too generally the case), there can be little doubt that it would be successful; and I should hope that the Geographical Society would not refuse to extend to it their countenance and patronage.

The great topographical map of France advances with a rapidity and regularity, which are highly creditable to the Department of the Dépôt de la Guerre, under whose orders it is executed. Sixty sheets of this national work have already been presented to our library through the liberality of General Pelet, to whom we are also indebted for the six volumes of the "Memorial," which describes all the *data* in detail, on which that survey is based. This last year has also brought to a close four folio volumes of the *Pilote Français*, a magnificent work, comprising charts of the whole of the northern and western coasts of France, and plans of the various ports on a large scale. Such works as these reflect the highest honour on a nation; and France may justly boast of that useful body of officers, the *Ingénieurs Géographes*, who, headed by MM. Beautemps Beaupré and Daussy, have raised a lasting monument of their zeal. A copy of this work has also been presented to our library by the liberality of the Director of the Dépôt de la Marine.

It will, I am sure, be gratifying to you to learn that the Austrian government are about to follow up the growing practice of forming societies for the improvement of geography, upon a very large scale; but it will be more intimately connected with the government, and will have much more extensive administrative functions, than the nature of our institutions would admit of. To this Geographical Institute, which will be placed under the direction of General Campana, aided by Colonel Skribaneck, are to be assigned the construction of maps, from trigonometrical surveys throughout the empire, and the whole of the topographical duties of the quarter-master general's department.

This department has already made great progress in publishing, amongst others, a map, in thirty-six sheets, of the provinces of Styria and

Illyria, which will be on the scale of half a mile to an inch, or 1-144,000 of nature.

The survey of Moravia is proceeding on the same scale, and the Austrian government are also engaged in mapping the kingdom of Lombardy and Venice, as well as the duchy of Modena.

Nor must I omit to mention a map of the kingdom of Wurtemberg, in fifty-seven sheets, on the scale of 1-50,000, or 1·3 of an inch to a geographic mile, which appears to unite great accuracy in detail with clearness and beauty in execution; for the donation of the twenty-seven published sheets of this map, our thanks are due to the Director of the Dépôt Topographique at Stuttgardt.

We are also indebted to Colonel Visconti for nine sheets of the beautiful map of Naples and its environs engraved under his directions. And, besides the topographical descriptions in progress in the various States of Italy, some of which I have just alluded to, we learn that a Corso di Geografia, by Signor Marmocchi, has lately been published, of which our zealous honorary member, Count Gråberg af Hemsö, gives a very favourable account. To this latter gentleman we are indebted for the fac-simile of a rare and curious Portulano of the middle ages, bearing the date of 1315, and now in the library of San Lorenzo at Florence. M. de Gråberg has also procured for the British Museum, among other portulani, one of Gracioso Benincasa, bearing the date of 1467. And, while on this subject, I may remark, that in the Egerton collection of MSS. in the Museum is also a valuable portulano, containing charts of not less than fourteen of the more noted map-makers of the middle ages.

M. Daussy still continues his useful *Additions* to his Table of Geographical Positions; but on this subject I must particularly direct your attention to some papers by Lieutenant Raper, R.N., now in course of publication in a very useful periodical, the "Nautical Magazine," in which that officer proposes to examine the *data*, on which the longitudes of the principal maritime points depend. These papers are divided into five sections, containing, I. An abstract of the principal scientific voyages and surveys from which our *data* are obtained. II. Remarks on the different modes of determining longitude. III. On the necessity of adopting *secondary meridians*. IV. On the propriety of adhering to one uniform method of placing on record chronometric determinations, that they may be immediately available for the construction or examination of charts. V. A discussion of the principal maritime points, extending to upwards of 200 positions, which are either the best determined, or the most general points of departure for ships. Such is a brief outline of the arduous task undertaken by Lieutenant Raper; and the present advanced state of hydrography imperiously demands such an examination of these points, to prevent our tables of positions from falling into utter confusion; will it be credited, that we have not less than six different positions, on apparently good authority, for the well-known place, Rio de Janeiro, in Brazil?—We sincerely hope Mr. Raper will carefully and honestly sift the evidence on which these longitudes rest, assured that a more acceptable service could not be rendered to geography in general, and especially to that branch of it which is more particularly our care, as an essentially commercial and maritime nation.

Asia.—The expired year has not been barren in its supply of new materials for the geography of the continent of Asia. One of the greatest importance, and which has already received its place in your Transactions, is the very valuable Narrative of Major Rawlinson, containing the details of his Journey from Zohab, at the foot of Mount Zagros, along the Mountains to Khuzistan (Susiana), and from thence again through the province of Luristan to Kermanshah. Few men have embarked in the pursuit of geographical knowledge better qualified, under the circumstances, than this distinguished officer. Enjoying the confidence of the prince to whose service he was attached, perfect master of the language of the country, well versed in its history, having been three years a resident on its frontier, and commanding a body of troops commissioned to visit some of the provinces of the Persian empire, Major Rawlinson, in this very important and luminous paper, has ably brought to bear, in illustration of the country he visited, the Sacred Writings, the classic authors who have narrated the various campaigns of Cyrus, Alexander, Eumenes, the two Antiochi, Mithradates, Meherdates, and the Arab conquerors, and his own acquaintance with the modern historians of the East. He has explored the gates of the mountainous range of Zagros, between the basin of the Tigris and the plains of Media; he has verified the courses of the ancient Gyndes or Diyálah, and the Choaspes or Kerkhah; ascertained the site of Holwán, as he says, one of the eight primeval cities of the world, and on the high road from Baghdad to Kermanshah; he has traced the connexion of the Jews of the Samaritan Captivity with the Gurán and Iliyat tribes; collected much information on the fire-temples of Elymais; he has discovered many architectural and sculptural inscribed monuments of the Kayanian and Sassanian dynasties at Holwán, Deira, Gilán, Zarnak, Sus, and Bisutún; he has well defined the provinces of Luristán and their several divisions, and he has been the first to throw a clear light on the very difficult questions of the successive capitals of Susiana, Susán, Sus, Jundi-Shapúr, and Shuster; as also on the names and courses of the rivers of that extensive district. The conclusions Major Rawlinson had come to on these subjects are thus stated: "I believe that in ancient times there were two cities of the name of Susán or Susa, in the province of Susiana; the more ancient, which is the Shushan of Scripture, being situated at Susán, on the Kurán or Eulæus; the other, the Susa of the Greeks, at Sus, *near*, not *on*, the Kerkhah or Choaspes. The river of Dizful I consider to be the Koprates; the Abi-zád, and its continuation the Jerábri, the Hedyphon or Hedypnus; and the united arms of the Kurán and Dizful rivers, that is, the Koprates and the Eulæus, to form the real Pasitigris."

The able arguments by which these conclusions are supported are well deserving the attention of the reader, and present a very favourable earnest of what we may expect from this enlightened traveller, when we receive the more elaborate work, in which he is engaged, on the Comparative Geography of the great Empire of Persia.

There is, however, one point in Major Rawlinson's Memoir which seems to require revision. In page 85 he accounts for no mention of the Shushan of the Scriptures, namely, that on the Eulæus or Ulai, being made by the historian of Alexander, because it did not lie upon his march from Babylon to Persepolis or to Ecbatana; and in the following page he considers that the great and rich temple in Elymais which was attacked by Antiochus Epiphanes, son of Antiochus the Great, was this Shushan of Scripture. Now it appears that the two authorities for this latter fact, which are also quoted by Major Rawlinson, are the sixth chapter of the first book of Maccabees (1—4) and the ninth chapter of the twelfth book of the Antiquities of Josephus (s. 1); and in both these authorities it is stated that Antiochus Epiphanes *was* incited to attack this city (in both places called only by the name of Elymais in Persia) by the accounts he had received of the wealth contained in it, and the coverings of gold, breastplates, and shields, suspended in its temples by Alexander of Macedon, son of Philip. We must therefore conclude, in opposition to Major Rawlinson, either that the city assaulted by Antiochus for the sake of the plunder he expected to find in it was *not* the Shushan on the Eulæus, or that this Shushan *was* conquered and occupied by Alexander. This, however, does in no way invalidate the general position of our traveller, as it is highly probable that Alexander did visit both the cities; *i. e.* the Susa near the Choaspes, on his direct route from Babylon to Persepolis, and Shushan on the Eulæus, when he took the mountain route from Susa to Persepolis, or on his subsequent march northwards from Persepolis to Ecbatana; and that his historians, as far as they are preserved to us, have shared in the ignorance common to all subsequent writers, who have confounded the two places.

The observations of Major Rawlinson, we may also add, in passing, are not more important for the sites which he visited and described in this very interesting portion of the Persian empire, than for others, which circumstances prevented him from approaching, but the value of which, for the illustration of historical geography, he has pointed out to future explorers.

One of the most valuable additions to scientific geography which have been made during this last year, is that resulting from the survey under Lieutenants Graves and Brock, R.N., in which the latter officer has constructed a chart of the Gulf of Kos, on the south-western coast of Anatolia, accompanied by surveys of the various harbours contained in it, and sketches of the most prominent points which distinguish the approaches to them. The gulf occupies an extent of nearly 60 miles from east to west, and is of great depth, no bottom being found in the middle of it with 300 fathoms of line, and even in the vicinity of the shore seldom less than from 50 to 70 fathoms. Though so near to ports which our fleets are in the constant habit of visiting, and to the common track of the eastern trade of the Mediterranean, no chart hitherto published gives any idea of its shape or extent; and the isthmus which divides the Gulfs of Kos and Symi has hitherto been laid down, apparently, by guess. The description of this isthmus by Herodotus is very exact; it is about half a mile broad, and, with a little exertion, and by digging through a hill of inconsiderable height, the Triopian promontory might be converted into an island.

Lieutenant Brock has also surveyed the harbour of Búdrún, and ascertained the site of the celebrated mausoleum of Halicarnassus. The site and massive ruins of the ancient city of Keramus have likewise been explored.

I beg also to repeat the thanks of the Society to Mr. Charles Fellows, who has communicated to the public, at one of our meetings, an outline of the journey, which he made in Asia Minor in the spring of last year. The details of this journey derive a peculiar interest from its having been performed in a northerly and southerly direction, thus crossing upon the lines of those of Keppell, Arundel, and W. Hamilton. Between Selge and Antaliyah, upon the south coast, Mr. Fellows followed in part the track of an ancient Greek or Roman road, paved with large irregularly-shaped stones, and retaining in many places the marks worn by the wheels of chariots. This precipitous road conducted him along the banks of the Catarrhactes, almost all the waters of which were lost, before they could reach the sea, in a porous substance formed, he says, of a conglomerate of pebbles and of a mass of encrusted vegetable matter. Mr. Fellows found ruins, to which he attributes the names of Isionda, Penedelissus, and Sylleum, the last with a theatre quite perfect, and with remains of paint upon the masks and ornaments, and also on the walls of the proscenium. The palm-tree he found indigenous at Phenika, on the coast of Lycia; and the valley of the Xanthus, with the extensive ruins of the city of that name, and of Tlos, also on its banks, he describes as sufficient to reward the artist, antiquary, or naturalist, for a journey simply to visit them.

Colonel Cohen, an intelligent traveller from Boston in the United States, has been kind enough to communicate to the Society the itinerary of his route through Asia Minor in the year 1833 (a part of which, as far as the ancient Iconium, he performed in company with a body of the Egyptian army), from Adana, south of the pass in the Taurus called the Cilician Gates, to Smyrna, and thence, by way of Kutayeh and Nicomedia, to Constantinople.

We are also indebted to Lord Pollington for a new itinerary from Erzrúm, by Músh, Diyár-bekr, and Birehjík, to Aleppo, a journey which he performed during the last summer.

With these and other aids already announced to you, we may hope shortly to possess a good map of this interesting portion of the Asiatic continent. Much, however, still remains to be done even in this comparatively accessible region.

The facilities which have been afforded by steam-navigation in the Mediterranean, and the readiness and security, with which the political state of the eastern coasts of that sea enables travellers to explore the interior, have of late years much increased our knowledge of Syria, and particularly of the deserts which border that remarkable country to the east and south; and we have good reason to expect that in a very few years, or perhaps sooner, we shall be able to construct a useful map of the great sandy and mountainous district, which extends from Egypt northward to the Holy Land, and from the sea-coast eastward to the hills of Idumæa. The travels of Lord Lindsay, M. de Bertou, and those of Dr. Robinson, Theological Professor at New York, and the Rev. E. Smith, an American missionary at Beïrút, in addition to those of Laborde, Linant, and others, have already cleared up many difficulties regarding this singular district, teeming as it does with memorials of great historical interest, from the Exodus of the Israelites to the later periods of the Roman Empire; and we have at length a satisfactory solution of the problem, whether the waters of the Dead Sea could ever have flowed into the Gulf of 'Akabah. M. de Bertou has shown that there is an elevated range of land, not volcanic, which forms the termination of the valley called El Ghor; this division of the waters flowing north to the Dead Sea, and south to the Ælanitic Gulf, lies between these two seas, about two-thirds of the distance from the former.

The exact level of the surface of the Dead Sea is a point of increasing interest not yet satisfactorily cleared up. Mr. Moore, by thermometric observations, has estimated it at about 500 feet *below* the level of the Mediterranean; Professor Schubert, by barometric observations, at 600 feet; whilst Mr. Russegger, an Austrian naturalist, has, also from barometric observations, recently stated it to be at a depression of no less than 1400 feet below the Mediterranean; but we trust that this point will not long remain a stumbling-block for geographers, as I am happy to acquaint you that more than a month since your Secretary, not unmindful of the interest attached to what appears to be one of the most remarkable features in the physical geography of the globe, placed an excellent barometer, made by Newman, and compared with the Royal Society's standard, in the hands of two young Englishmen about to visit Palestine, with a special request that they would endeavour to settle the point in question.

Dr. Robinson, in his track from 'Akabah to Hebron, ascended the western mountains to the height of 1500 feet, and followed the ancient Roman road nearly north. On this route he found and visited the ruins of the ancient Roman cities Eboda and Elusa, as well as the ruins and wells of Beersheba, still called Bír Sebá, at the distance of 28 miles W.S.W. of Hebron. On another journey, from Jerusalem to Gaza, by the direct ancient route, Dr. Robinson relates that he succeeded in discovering, amongst other places mentioned by Jerome and Eusebius, the site of the long-lost Eleutheropolis at Beït Jibrín, formerly Betogabris, where are the remains of a very strong Roman fortress and traces of an extensive city.

Professor Berghaus, of Berlin, is engaged in constructing a map of Palestine chiefly from materials furnished by these travellers.

Our knowledge of the peninsula of Arabia, and of some of the adjacent territories, has also received considerable additions during the last year. We are indebted to Mr. C. J. Cruttenden, of the Indian navy, for the narrative of an interesting journey which he made (during the time that one of the Company's vessels was surveying the roadstead of Mokhá), along the arid plains of Tehameh, and across the Jebel Barra to San'á, by what is called the Tarík-esh-Sham, or northern route, in the summer of 1836. This memoir, short as it is, furnishes us with a very pleasing picture of a part of Arabia, in which the generally steril nature of the soil is contrasted with the luxuriant productions of the more mountainous districts; and there is every probability that the possession of the port of 'Aden, lately ceded to our arms, will materially tend to increase our intercourse with a people, who only require to be more known, and to be introduced into commerce with the Europeans, to be better appreciated. This journey has also furnished us with some additional specimens of the Himyaritic language, which in the earlier ages was spread over these parts of Arabia Felix. The united labours of Professor Gesenius, Mr. Fresnel, and of Dr. Lepsius, will, it is hoped, soon give a clear insight into the construction of this hitherto unknown tongue; and we may perhaps obtain, from the inscriptions which have been brought home, a clue to the previous history of the country. The lamented Dr. Hulton and Mr. Cruttenden also discovered, when at 'Aden, that the remarkable structure, which had hitherto been called a Roman road, is in fact the aqueduct of Soleïman the Magnificent, extending in a general north-west direction upwards of 8 miles into the interior. It is built of red brick and stone, about 4 ft. 6 in. wide, the enclosed watercourse measuring 19 in. by 16 in.; there are no remains of arches, the ground not requiring them; and its general appearance is that of a mound about 5 feet high, and bricked over.

The abridgment of the Memoir of Captain Haines, also of the Indian navy, which accompanied his chart of the south coast of Arabia, is a highly valuable paper, and contains most useful information on a line of coast extending through 7 degrees of longitude, from the Straits of Bab-el-Mandeb to the Palinurus shoal, and which is becoming every day an object of increasing interest to our traders. This memoir was communicated to the Society by the Court of Directors of the East India Company, and you have acknowledged it as an additional proof to the many you have already witnessed of the spirit of liberality and confidence, with which that distinguished body have uniformly met all the overtures we have made to them upon subjects connected with the pursuits of this Institution.

On the coast of the Hejáz we are indebted to Mr. Thomson D'Abbadie for the very useful addition of many names of places not inserted in the published charts of the Red Sea, which this traveller obtained in a coasting voyage from Jiddah to Ras Widan (a distance of about 250 miles) during the past year. Mr. D'Abbadie, from whom we have heard, within these walls, a very animated description of his stay at Muṣawwa', and of his journey thence to Góndar, has expressed his intention of shortly returning to that country, to pursue his researches into the interior, and to make himself better acquainted with the Amarña language.

The first fruits of the expedition into Kurdistan, under the direction of Mr. Ainsworth and Mr. Raşam, will appear in the Part of our Journal now published. The expedition, on leaving Constantinople, was obliged by circumstances to take the land route along the coast of Asia Minor, from Nicomedia; and to this we are indebted for a great mass of important information respecting the modern and historical geography of the ancient kingdoms of Bithynia and Paphlagonia, as well as of part of Galatia, and as far as Angora. The former part of this route may be said to be entirely new to modern investigation, and the Society will be much gratified by the proofs Mr. Ainsworth's notes will furnish to them, of the zeal and intelligence with which he has illustrated the courses of the rivers known under the names of Hypius, Lycus, Parthenius, and the greater and lesser Halys; as well as the mountain-ranges of the Bithynian Olympus and the Olgassys of Paphlagonia; which preceding geographers, having to draw their conclusions from ancient historians and contradictory modern authorities, had left in almost inextricable confusion. Mr. Ainsworth's notices, indeed, have laid open the resources of a country rich in mineral and vegetable productions, where iron and copper-mines only wait to be worked anew, forests to be felled, and rich plains to be cultivated, in order to furnish again the materials of a wealthy and flourishing empire. The sites of many ancient cities are ascertained, and during the journey no opportunity has been lost of fixing astronomically the most remarkable positions on the route.

Our knowledge of the geography of these parts of Asia is continually receiving valuable accessions from the proceedings of the Euphrates Expedition. Although the main purpose for which this expedition was fitted out has not yet been accomplished, the obstructions which it has encountered are rather of a political than of a geographical character, and the delay may be attributed to incidental circumstances, which must ever modify speculations of this description. In the mean time, beyond the brilliant results with which we are proud to connect the name of Chesney, and the full details of which we are daily expecting from the pen of that accomplished officer, we owe to it more accurate surveys of the two great rivers of Mesopotamia, and particularly of late that of the Tigris, nearly from its source to Samarra, and thence to Baghdad, under the directions of Lieutenant Lynch, of the Indian navy. The sites of Opis and Samarra (where, after the death of Julian in his contest with Sapor, the Roman army, under Jovian, crossed this river) have been ascertained; and the Median wall, which is still crumbling in the Desert, reaching from the ancient ruins of Sipara, on the Euphrates, to the Tigris opposite to Jebbara, has been traced by the same indefatigable officer, who, I should also add, has succeeded in carrying his steam-boat through the Saglawíyah Canal, between the Tigris and the Euphrates. Another valuable result of the Euphrates Expedition has been the publication of Mr. Ainsworth's Observations on the Alluvial Plains of Mesopotamia.

Dr. Prichard, so well known for his "Researches into the Physical History of Mankind," has enriched some of the last pages of our Journal with a subject which, though not strictly geographical, is yet of the highest importance in illustrating the great question of the early migrations of mankind, the ethnography of High Asia. Dr. Prichard has proved, from the most recent researches, that there is no reason to suppose that the great nomadic nations of the Tartar race, the Nogays, the Kirghises, Turkomans, and Jakutes, were ever, like the purely Osmanli Turks, a portion of the true Caucasian race; he has given to us much curious information respecting the origin of the Turkish nation, and he has shown what new lights may be thrown upon this subject generally, by a comparison of the language of the Turkish, Mongolian, and Hungarian races. The principle known by the denomination of the Quadruple Harmony of Vowels is a very remarkable point of coincidence, which is shared by the Hungarian language in common with the Turkish, Mongolian, and Mandshú, whilst it is totally at variance with the euphony of the classic languages of the west. And we may observe in the former languages the marked contrast which exists between the great simplicity of the formation and inflexions of their words, and the complicated construction of their sentences.

The pursuit of this study by the disciples of Klaproth, Remusat, and W. Humboldt, has already, independent of characteristics of physiognomy, established the fact of several great divisions of the inhabitants of our globe. Such are the Scythian, Tartar, or Mogul race, that which is in possession of the language called Indo-Teutonic, or Indo-European, the races which have spoken, from time immemorial, the Semitic languages, the vast tribes of the continent of Africa, to say nothing of the languages of the Chinese, the Polynesians, and the Aborigines of the New World.

In connexion with this subject I cannot omit to mention the work of Mr. Du Ponceau, the venerable president of the American Philosophical Society, on Chinese Writing; to which is annexed Father Morrone's Cochinchinese Vocabulary, the first, we believe, of that language that has been published. The Geographical Society of Paris have also just published the account of the Moguls or Tartars, by Frà del Pian di Carpini, who, in the years 1245, 6, and 7, was the pope's nuncio in Tartary. This first complete edition of Carpini's narrative is preceded by an excellent introductory dissertation by M. D'Avezac, on all the ancient travels into the country of the Moguls.

Carrying our view still farther to the north and north-eastern parts of the old continent, Professor Adolph Erman, of Berlin, has lately communicated to us a map of Kamchatka, constructed from his own observations during his well-known journey across Northern Asia, and round the world, in 1828–30; from which it appears that, in some parts of that remarkable peninsula, more than one-half of the extent usually laid down on our maps must be erased. Mr. Erman has also sent us a detailed account of the geology of the peninsula, and promises that we shall soon have the personal narrative of his travels in it; the account of which, I need hardly add, is eagerly looked for by all who have read his two first very interesting and instructive volumes.

The Delta of the Indus has for the first time been made known to the European world by the survey of Lieut. Carless, of the Indian navy, comprehending the main stream of the river known by the name of Wanyání, and the two principal mouths, the Hadjámâri and the K'hedívárí. These have been trigonometrically laid down, from the sea as far as Titíyah, a distance of 35 miles. The shifting nature of the embouchure of the Indus, and the frequent changes made in its channel by the falling of its banks, have rendered this survey of very great importance; and the prospect which is opening to us of an increasing navigation of this magnificent, and, in its lower part, intricate stream, bids fair for a rapid improvement of our knowledge of the whole river itself, as well as of the extended basin which it waters. The labours of Mr. Carless will thus form a memorable and useful corollary to the adventurous and successful expedition of Burnes.

The military operations now carrying on in this part of the East cannot fail to enlarge our knowledge of its geography; and it is confidently anticipated that the result will be a very extensive development of our commerce and general intercourse with the Sindes, Seiks, and Affghans.

These events have indeed, in the first instance, served to bring into prominent relief our ignorance of the geography, not only of Asia generally, but even of those parts which immediately adjoin our own frontiers; and they have added another to the many proofs already adduced of the necessity of making ourselves thoroughly acquainted with the topography and resources of a country, if we wish either to administer the internal government justly, or to be ready to resist external aggression without an enormous expenditure of life and treasure.

Our authorities in the East appear also to partake of the general activity; and we learn, from the Report of the zealous and indefatigable

Secretary of the Bombay Geographical Society, that Sir A. Burnes had despatched the officers attached to his mission in various directions to collect information. Lieutenant Lynch had visited Candahar; Dr. Lord has been at Khundúz; and Lieutenant Wood has succeeded in reaching the source of the Oxus, which is stated to be from a sheet of water at an elevation of about 15,600 feet above the level of the sea. Having returned in safety to Khundúz, Mr. Wood had again started to examine the fords of the Oxus, within the territories of Murad Beg. The details of this expedition to a country, which had not been trodden by any European foot since the time of Marco Polo, must prove of high geographical interest; and I may be permitted to express a wish that they will be made public as soon as the services on which the officers are employed will allow it.

Of the trigonometrical survey of India the sheets of the map containing Rajahmandrí and Cochin have been published during the past year; the drawing of the Concan also has reached England, and will doubtless soon be made public.

Portions of this survey of India present instances of accuracy which are highly gratifying, one of which I may notice, connected as it is with some of the most important maritime points on the western coast of that peninsula. Lieutenant Shortreed's survey, which connects with the great series diverging westward from the Beder base, at Chorakullee and Sawurgaon, depends upon a base measured with an excellent chain by Troughton; the triangulation extends over a large tract to the eastward, southward, and westward; and in several lines and points falls in with the trigonometrical survey of the western coast by Major Jervis, depending upon a distinct base-line of 31,003 feet (10,334·3 yards) near Cushina, measured with iron rods 20 feet in length, by that officer and Captain Robinson, of the Indian Navy; the approximation of the results proceeding from two such distinct sources is surprisingly great, so much so that I cannot deny myself the gratification of quoting the following points from the official register of these operations :—

	Shortreed. Feet.	Δ	Jervis. Feet.	Δ	Diff. Inches.
Bhoja to Mera	71,859·62	82	71,858·2	43	1¼
,, to Sookhilee	54,612·67	89	54,607·5	41	5
Sookhilee to Dhunvee	99,833·89	92	99,832·7	38	1
Dhunvee to Salira	58,253·37	92	58,241·54	6	12
Mera to Karunja	99,819·25	85	99,834·5	55	14¼
,, to Lighthouse	143,723·75	87	143,742·5	66	18
,, to Sookhilee	68,420·95	88	68,409·4	181	11¼

Taking into consideration the nature of the country in which this work has been carried on, it must be admitted that the comparison of these results is highly satisfactory, as in two instances they agree within 12 and 18 *inches* respectively, in distances of 14 and 18 *miles*: thus affording a gratifying proof of the attention of the officers who conduct the detail of the survey, and the correctness of the processes employed by them.

By a communication from Colonel Everest, the Surveyor-General, in November last, he announces the satisfactory completion of the operations, both celestial and terrestrial, of the great meridional arc north of Seronj. The care and exactness with which these operations have been performed may be understood from the fact that the Seronj base, as computed, differed from actual measurement with the compensation bars, in a chain of triangles 460 miles in extent, only *seven inches and one half.*

In conducting a work of this nature, or indeed any scientific undertaking, it is manifest that either it should be done with the utmost accuracy, and the directing officer be invested with the fullest power to ensure such accuracy, or it had better be wholly abandoned. In this department of scientific investigation there are so many nice points to be taken into consideration, so many local difficulties to contend with, which no one but the responsible officer can appreciate or enter into, that the interference of any second opinion, and especially of those who from the nature of the work cannot be presumed to be capable of forming a judgment on the subject, is calculated to slacken every effort, and dishearten, perhaps, the most zealous and enthusiastic.

A retrospect of the differences of opinion between the late Major Rennell and Colonel Lambton, as related in Herbert's Life of the latter officer, and the evil consequences resulting from them, have suggested these remarks; but we may hope that such times have passed, and that the same princely liberality, which has provided funds for effecting the great survey in India, will be henceforth extended to the mode and execution of the work; that the undertaking may throughout be executed in a manner corresponding to the advanced state of science in Europe, and on a par with the execution of the great meridional arc already measured from one extremity to the other of Hindostan; and that we may see the Atlas of India worthy to take its place by the side of the Ordnance maps of England and Ireland, or the Government Surveys of France, Saxony, Austria, and Wurtemberg, and thus become a lasting monument to the munificence and enlightened spirit of the Directors of the East India Company.

The eastern frontiers of British India have been explored and described by Captain R. B. Pemberton, of the Indian Army. These frontiers comprise the great mountain-chain between Manipur and Arracan, the territories of Manipur, Assam, and Arracan, together with Poong, Kachar, Jontiyah, and the Kossiyah Hills; the growing importance of these districts to our national and commercial interests, is the best encouragement to go on as we have begun, in thus preparing the way for the progress of European influence and cultivation, by a correct knowledge of the geographical features of a country.

Lieutenants Powell and Ethersey, I.N., have also completed the survey of the Gulf of Manar, preparatory to the establishment of a navigable channel between that gulf and the Bay of Bengal, by the Pámbám passage: this latter work proceeds satisfactorily under the superintendence of Colonel Monteith of the Madras Engineers.

Amongst the maritime surveys recently executed by the officers of the Indian Navy, I must also mention the examination of part of the Chagos Archipelago, and the Saya de Malha Bank, by Captain Moresby, an account of which, together with the original and beautiful charts, and a valuable portfolio of drawings, have been communicated to the Society. It is to this officer also that we are indebted for the survey of the northern half of the Red Sea from Suez to Jiddah, and for the completion of the southern half of that sea, left unfinished by Captain Elwon: Captain Moresby has also surveyed the Laccadive Islands in 1828, and the Maldive Islands in 1834. A part of the Chagos Archipelago, named Owen's Bank, still remains to be examined, as well as about 90 miles of the N.W. portion of the Saya de Malha, and a great extent of unexplored space among the Seychelles Islands. During the last ten years Captain Moresby has surveyed upwards of 5000 miles of coast; and, if we except Captain Daniel Ross, I.N., the actual President of the Geographical Society of Bombay, few officers of the Indian Navy have rendered more essential service towards the advancement of Geography in the East.

But I have dwelt, perhaps, at too great length on the conquests of geographical science within the continent of Asia. It is, however, this part of the world which offers to the geographical inquirer more objects of high importance, considered in an historical or ethnological point of view, without taking into our consideration the more immediate but momentary interests of politics and commerce, than all the rest of the globe put together. At the same time we may strictly characterise the progress of science to which we have alluded rather in the light of recoveries than of discoveries. A long lapse of years of darkness and barbarous inroads erased from the map of the world, known to the civilised portion of mankind, immense tracts of country, once the seat of arts, learning, and triumphant dominion: we are now slowly restoring to science single parts of that lost map, and thus repairing the injuries done by our forefathers, and reconstructing in more indelible characters, we may confidently hope, the edifice which they pulled down. We reconstruct, too, in the spirit of peace and philanthropy, what the earlier dynasties erected as the monuments of sanguinary conquest, and what fell by hands armed with ruder swords, and guided by a worse spirit.

The history of the human mind, and the development of the intellect, under the various phases in which man has borne his part, as a ruler or as a subject, are, after all, as much the real and legitimate object of these inquiries as the appropriation of the knowledge we at the time acquire, to the immediate benefit of the present generation, and of those which are to come after us : for they directly tend to improve our hearts and our minds, to place before us, in the most prominent light, the superior advantages of peace over war, of rational authority over brute force, of liberty over slavery; and, whilst they should make us thankful for the benefits we enjoy from a more advanced state of civilisation, from a purer religion, and from a clearer stock of ideas in the range of the practical and the intellectual sciences, they are a perpetual warning to us that the possession of the most extended power is not in itself an earnest of its durability; that knowledge, without an enduring struggle to increase it, is no security against ignorance; that civilisation, unless based upon justice, may be overwhelmed by barbarism; that temples built upon a rock may crumble to dust; and that empires upon which the sun never sat may be forgotten.

AFRICA.—Africa still offers the same barrier to the progress of discovery, and it is feared will continue to do so, unless some systematic and well-considered plan for penetrating into the interior be devised and steadily acted upon. The recent labours of Captain Vidal, R.N., in the Bight of Benin, and on the Ashantí coast of Western Africa, and those of Lieutenant Carless, I.N., from Rás Hassún northwards to Ras Jerdassún (Cape Guardafui), and thence westward to Ras Gulwainí and Berbera, on the eastern coast, all but complete the correct outline of this great continent; but, within this coast-line, strict geographical investigation has not made very great advances during the past year.

In Northern Africa the French government have carried a triangulation over the territory of Algiers including Bonah, Kostantinah, and Storah, whence the longitude of the Kasbah of Kostantinah appears to be 6° 37′ E. of Greenwich, differing only 7 miles of longitude from the position of that place, determined by our learned countryman Shaw in 1726.

In Marocco we find, from the rough note-book of the lamented Davidson, that, following the steps of the British mission to that country in 1830, related in the 1st Vol. of the Geographical Journal, he proceeded from the city of Marocco across the plain in a S.S.E. direction into Atlas, as far as the ruined town of Tasremút, at an elevation of 3000 feet above the sea; thence turning to the westward he continued along the valleys of Atlas by a route not laid down in any of our maps, and which we are enabled partially to trace only in that of M. de Gråberg; passing within a few miles of the site of Aghmát Warikah, he appears to have issued from the mountains beyond a place called Amishmish, perhaps Imizmizi of our maps, and then to have crossed the plain to Mogador. Geographers cannot but feel grateful to Mr. Thomas Davidson, the traveller's brother, for allowing his rough notes of this novel route to be made public.

The Council of the Society have, in the course of the last year, appropriated the sum of 50*l*. from the funds of the Society towards a project, formed by a society of gentlemen interested in Egyptian and Æthiopian discovery, for sending out a native of Dongolah to explore the course and sources of the Bahr al Abyad, or the western and principal affluent of the Nile.

Explorations in the interior of Africa have been almost from their first commencement towards the middle of the last century, the exclusive inheritance of British ardour and enterprise, for Hornemann and Burckhardt we have long been accustomed to identify with our own countrymen; and the circumstance of the African Association having been fused into this Society, has given to this noble feature in the history of inland discoveries an additional value in our eyes. It was therefore with much gratification that we witnessed within these walls the discussions, in which Mr. Macqueen took an active part, upon the subject of the basin of the Chadda and its contributory or departing waters—that is, whether it be still water, or have an outlet. Major Denham had maintained the former hypothesis, and it is clear from his description and his own personal experience that the Shari does not flow out of it; but Captain W. Allen, who is so well known to you as the navigator of the Quorra, and of the Chadda as far as Fandah, and who has given much attention to this portion of African geography, is strongly of opinion that the Yeú River, mentioned by Denham as a clear, deep, and rapid stream of sweet water, communicating with the lake, and traced by Clapperton as far as Zangeia, 300 miles from the lake, and supposed to be the same with that which Lander crossed and recrossed on his return from Zaria to Danrorah, must either be identical with, or one of the principal affluents of, the Chadda, which passes by Jakóba and discharges itself into the Quorra a little below Fandah. From these data Captain Allen draws the natural conclusion that the Chadda, a larger river than the Quorra, is the outlet of the Lake Chad, drains the lofty Komri or mountains of the moon, and thus affords an uninterrupted water-communication perhaps to the very centre of the African continent.

Turning to the north-western coast, M. D'Avezac has published a critical analysis of one of the earlier journeys of the late Réné Caillié among the Moors of Beraknah, near the banks of the Senegal, for the details of which I refer you to the *Bulletin* of the Geographical Society of Paris; and in the same publication is to be found an interesting biography, by M. Jomard, of a man who, born with a true spirit for geographical discovery, if he had had the advantages of a more cultivated education, would have been entitled to a very distinguished place amongst the travellers of our days; though his journey to Tumbúktú, owing to the misfortunes which awaited him, has added but little to our knowledge of that part of this continent.

Dr. Edward Rüppell, of Frankfort-on-the-Main, is well known to you as one of the most distinguished travellers and naturalists of the present day. A few years ago we had the pleasure of seeing him at one of our meetings, when he gave us some valuable information respecting the interior of Africa. Dr. Rüppell has twice successfully explored the interior of that continent, and of the peninsula of Arabia. His former works need not on the present occasion be alluded to further than to say, that the Council had them also in their mind, when they awarded to him one of the royal medals at their disposal for the past year. Of his second journey Dr. Rüppell has hitherto given to the public the result in part of his geographical discoveries. These comprise a full view of the political and administrative state of Egypt under Mehemet Ali, seen by the eyes of a judicious and inquiring traveller, various notices on the productions of Lower Egypt, an excursion into Arabia Petræa, a journey from Cairo to Jiddah, and thence to Musawwa', the Island of Dahlak, and the ruins at Zula. Making Musawwa' his head-quarters, he visited Halai and Ategout in Abyssinia; from thence he gained the banks of the Tacazzi, near the village of Ber Agow. His route then lay through the province of Semen, and other parts of the region watered or rather penetrated by the Tacazzi, and the two great affluents of the Nile, the White and Blue Rivers. Besides the great mass of statistical and zoological information, the result of Dr. Rüppell's travels, they still owe their highest value to the great number of astronomical observations, by which he has fixed the latitudes and longitudes of every remarkable spot which he visited. These observations have long formed one of the most valuable divisions of the "Correspondance Astronomique," published at Genoa by the Baron Von Zach, to whose instructions Dr. Rüppell was chiefly indebted for his proficiency in the science, which he cultivated with such ardour and success throughout his African travels.

Dr. Rüppell's explorations in Kordofan have been followed up by

Mr. A. Holroyd, who has given us, very lately, the notes of his journey in the year 1837, from Wadi Halfah, at the second cataract of the Nile, to the ruins of Musawwerát, in the neighbourhood of Shendi, and to Khartúm, the station of the Pasha's government for the provinces S. of that cataract. Khartúm is situated upon the W. bank of the Blue River, about 1½ mile above its junction with the Bahr el Abyad, in 15° 34′ 40″ N. lat.; and, though but a small village a few years ago, has since risen rapidly into importance at the expense of Shendi and Sennár.

Mr. Holroyd observed in more than one place, in crossing the desert, many siliceous fossil trees, the superficial stratum of the ground being a coarse sandstone. Some of these trees were 51 feet in length, and 20 inches in diameter; and partially buried in the sand: splinters of them are used as fire-flints. They were apparently doom-trees, or the Thebaic palm-tree.

At Sennár, Mr. Holroyd describes the banks of the Blue River as between 40 and 50 feet high, the rise of the river being about 20 feet. Returning from Sennár to Wádí Medinah, he crossed from the Blue River at that point to Monkarah on the White River, a journey of 86 miles; and he gives a most favourable account of the capability of the ground for fertility, and for irrigation by canals, by which cotton, indigo, tobacco, sugar, and grain might be cultivated with success.

From Kajebi, where the thermometer was at 112° on the 19th March, Mr. Holroyd proceeded into Kordofan by the desert of Habshábeh. He found El Obeïd, the capital of Kordofan, with a population of 30,000 inhabitants, having doubled in the last ten years. Here we have another instance of the improvements likely to accrue to geographical knowledge from the progress of conquest, even in the hands of a power only in a comparatively advanced state of civilisation. The influence of the Pasha of Egypt has made countries now of easy access, which a few years ago were shut out from all but the traveller in disguise; and though the necessaries of life are so cheap, that their ordinary currency consisted of pieces of iron forged from the ore obtained in the neighbourhood at Wad Dha-s-Sákiyah, English gold had also its current value in the Bazár.

Mr. Holroyd returned from El Obeïd to Monkarah, on the White River, by the desert of Sakrah, a distance of 170 miles, and he decidedly prefers this route, as more abounding in villages, and better supplied with water, to that by Habshábeh. He has described the appearance and effect of two hurricanes of the desert, when the air was thickly charged with sand for two or three days; the thermometer at 102° and 104°. The hurricanes lasted about two hours.

Besides the services rendered to geography by Mr. Holroyd, I cannot omit making mention of one which he has rendered to humanity. He was instrumental in obtaining from the Pasha of Egypt a promise that he would put an end to one of the practices attendant upon the state of slavery, so long the curse of these unfortunate countries. In the dearth of other means of paying his troops, that chieftain had been in the habit of giving to them, in lieu of arrears of pay, one or more of the unfortunate beings whom the fortune of war, or what was still worse, predatory incursions, had thrown into his hands. Owing to Mr. Holroyd's intercession we have reason to hope that this new feature in the history of the horrors of the slave trade has been abandoned.

Dr. Bowring has also, within a few days, communicated to the Society the Journal of Mohammed 'Ali, dictated by himself, of his expedition from Cairo to Fazoglo during the past winter.

AMERICA.—I have said that the attainments, which geography is perpetually making in Asia, are rather to be described as recoveries than as discoveries. The latter term applies, on the contrary, in its fullest extent to all which we are doing in the New World; and I am happy to be able on this occasion to signalise as a triumph, wholly of this description, the expedition to the north coast of America, undertaken, at a great pecuniary sacrifice, by that high-spirited body, the Hudson's Bay Company. Messrs. Simpson and Dease have passed a second summer in carrying on their meritorious surveys of that portion of the continent; and we have been recently gratified with the announcement that the same gentlemen, after ascending Dease's River in the month of June last year, and crossing the portage between that and the Coppermine River in boats carried along the ice with a fair wind and hauling-ropes, descended the last-named river to the mouth, which they reached on the 1st of July. Here they were imprisoned in the ice till the 17th, whence, after encountering the greatest difficulties, they at length succeeded in reaching in their boats Point Turnagain, after making a circuit of 140 miles by Arctic Sound and Barry's Islands.

On the 9th of August they had attained a point 3 miles to the southward of Franklin's Furthest in 1821. The season was so much more unfavourable than on that occasion, that no hope was now left to them of making any further way by sea. The boats were therefore sent back on the 20th of August, and Mr. Simpson, with a small party, proceeded on foot, prepared for a limited journey of ten days. In this interval they reached on the 25th of August a point on the coast, on which they erected a pillar, in lat. 68° 43′ N., 106° 3′ W. long., and where the magnetic variation was 60° E. The compass had grown sluggish and uncertain in its movements as they advanced eastward, and frequently had to be shaken before it would traverse at all.

A long range of high lands had been seen out at sea to the N., capped with snow, which they called Victoria Land; and an archipelago of islands within 100 miles of coast were thus traced; and with what they could clearly descry to the E., when they were obliged to return, about 120 miles of continental discovery were effected. A large open sea was also seen to the eastward, from the farthest point reached; but there were no means of ascertaining whether it might extend to Ross's Pillar or to the estuary of Back's Great Fish River, though the trending of the most distant land in view rather seemed to favour the latter supposition.

There can be no doubt that these successful results of the first attempts of the Hudson's Bay Company to connect scientific pursuits with those in which they are more immediately concerned, will encourage them to persevere in the same honourable course; and it is only necessary to observe that they will add greatly to the services which they have already rendered to the cause of science, if they will avail themselves of any opportunities which may occur to instruct their servants in the arctic regions, and more particularly in the neighbourhood of Hudson's Bay, in reference to the statement made by Captain Sabine in his "Report on the Variations of Magnetic Intensity in different Parts of the World," to observe with the greatest care the intensity of the needle in those parts, the maximum of which Captain Sabine imagines will be found nearer to Hudson's Bay than to New York, at which latter place the highest force hitherto recorded in the northern hemisphere, namely 1·8, is stated to have been observed.

Dr. Richardson, the companion of Sir John Franklin in his Arctic journeys, has recently communicated to the Society a valuable paper, containing the discussion of all Sir Edward Parry's thermometric observations while in these regions, between the years 1826-8; and on this subject I must notice that, in accordance with the suggestion of Professor Baer, of St. Petersburg, the council of this Society has, through the kind assistance of the Hudson's Bay Company, caused twenty well-graduated thermometers to be distributed throughout their territory, in the hope of obtaining a series of observations, which may enable us to determine the curve of equal temperature throughout that extensive region.

A Society like ours, instituted for the purpose of promoting, mainly, physical geography, cannot have seen, without much gratification, that Her Majesty's government have decided upon an Antarctic voyage of magnetic research; and Captain James Ross, so well known as having borne his part in all the Arctic voyages since 1817, has been appointed to the command of an expedition, the main object of which is to establish in the island of St. Helena, at the Cape of Good Hope, and in Van Diemen's Land, sufficient stations for making regular hourly observations of the fluctuation of the three elements of variation, dip, and intensity, or their equivalents, with magnometers on the most approved construction, during a period of three years. The geographer cannot be insensible to the importance of accurately ascertaining the minutest properties of an instrument, namely, the compass, on which all his admeasurements so immediately depend. And it may not be out of place to remind you that nearly three years ago public attention was first called to the subject of an Antarctic expedition in a printed letter addressed to Sir John Barrow, then our president, and was more recently brought forward at the meeting of the British Association, by one who has proved himself not the least zealous of the Members of the Geographical Society; you will at once perceive that I allude to your Secretary, Captain Washington; and I am happy that this occasion has been afforded me, of declaring to you that the Royal Geographical Society owes a large share of its fame and success during the last two years to his indefatigable industry, to the extensive correspondence which he carries on with all the most intelligent geographers on the continent of Europe, to the suggestions which he offers to travellers for their guidance on their departure, and his readiness to assist in making the best use of their information on their return, to the courtesy with which he receives all who come to our rooms, and to the admirable manner in which, in conjunction with our learned Foreign Secretary, Mr. Renouard, he arranges the documents which are transmitted to us, and conducts the publication of them through the press. I am sure you will all warmly respond to the feelings which have dictated this very inadequate expression of the conviction of your departing President.

You will hail with much satisfaction the promised appearance of a work by Admiral Wrangel, on the Russian possessions in North America. In the capital of these countries, New Archangel, where is a population of nearly 10,000 persons, the Russian American Trading Company have already founded a library, schools, hospitals for the sick, churches, and chapels; an observatory is shortly to be erected, as well as a cabinet of natural history for the productions of the colony; and magnetic observations have been taken in an appropriate building, since 1833, in pursuance of the desire expressed to that effect by the Imperial Academy of St. Petersburg. The author observes that the inhabitants have as yet little, if anything, in common with other European settlements in America; they ought rather to be compared with the Danish settlements on the west coast of Greenland. The climate and soil are so little congenial to agriculture or pasturage, that the entire occupation of the colonists is confined to fishery, as was the case with the original inhabitants previous to the arrival of the Russians.

We are indebted to our zealous corresponding member, Mr. Worcester, of Cambridge, Massachusetts, for a table of corrected positions of places in the United States, many of them fixed by his own observations; and I may, perhaps, be permitted to express our regret that we cannot hear any account of the progress of the national survey of that vast territory, which we understood had been confided to the able direction of Mr. Hassler more than three years ago.

Passing to the southern division of this great continent, I have the satisfaction of announcing the publication of the Narrative of the surveying voyages of her Majesty's ships Adventure and Beagle, during the examination of the coasts of Patagonia, Chile, and Peru, within the years 1826 and 1836. As an abstract of these voyages, giving all the most important positions determined, and the chief practical results obtained, has been printed in our Journal, and I had occasion so recently from this chair, when presenting our gold medal for the year 1837 to the commander of this expedition, to state my opinion, and I believe I may add that of most geographers, of the value of these results, I need now only express my conviction, that henceforth the names of HUMBOLDT and FITZROY must be inseparably connected, as the chief authorities for our knowledge of the Geography of South America.

Of the four volumes now lying on your table, and for which we are indebted to Captain FitzRoy, the first contains the journals of Captain P. P. King, who began the survey of these shores in 1826, and prosecuted it for four years; the second and the fourth relate entirely to the voyage of the Beagle under command of Captain FitzRoy, during the continuation of the survey and circumnavigation of the globe, between the years 1831 and 1836; the third, written by Mr. Charles Darwin, is devoted to natural history, every branch of which, during this expedition, appears to have been explored by this distinguished naturalist.

When we consider the immense mass of materials contained in these volumes, the detailed tables of positions (which, though unheeded by the common reader, are invaluable to the geographer), the great extent of coast minutely examined and described; and, independently of the various official duties, as revising charts, writing sailing directions, &c., that the whole account has been given to the public in its present complete, and I may add, beautiful form, within little more than two years since the return of the Beagle to England, I am sure that I do but express the opinion of the greater part of this Society and of geographers in general, in saying that the 'Voyages of the Adventure and Beagle during the examination of the shores of Patagonia, Chile, and Peru,' have not their parallel in the annals of maritime surveying.

In addition to the numerous illustrations contained in this work, there is a general map of South America, of which I cannot deny myself the pleasure of saying a few words, were it only as a mere act of justice to Mr. John Arrowsmith, who has bestowed upon it a great portion of the labour of the last two years. Besides embodying the whole of the Survey, by Captains King and FitzRoy, of the coasts from the river Plata round to Guayaquil, reduced from the original charts, this map is corrected from the following documents; parts of Columbia, and the course of the Orinoco, from Baron Humboldt; the coast of Brazil from the survey of Admiral Roussin; Venezuela, from a large Spanish unpublished map; the Amazons, from observations by Lieut Smyth, R.N.; British Guayana, from the recent researches of Mr. Schomburgk; New Granada, from MSS. collected by mining companies; Equador, from a MS. sent home by General Miller, and received by him from the government of that State; Peru, from maps received also from General Miller and Colonel Belford Wilson, both of which officers have greatly exerted themselves to collect these materials; Bolivia, from Mr. Pentland's original observations, including the survey of the shores of the lake of Titicaca, and from some MS. maps received from him. The provinces of Chiquito, Moxos, and Cochabamba, with the Affluents of the Madeira, from a MS. brought home by Sir Woodbine Parish; Chile, from a MS. map of Dalbé, procured by Captain FitzRoy; the provinces of La Plata, chiefly from Sir Woodbine Parish's materials, for a detailed account of which, I may refer you to the introductory chapter of his recently published and excellent work on that country, in which he had profited by the opportunity afforded him as Chargé d'Affaires, to collect all the geographical documents that could be obtained, many, indeed, of the highest value.

The province of Rioja has been corrected from the map of Mr. French, which, with its accompanying memoir, has been presented to the Society: the direct route from Cordova to Mendoza is laid down from Captain Gosselman, of the Swedish navy; the northern frontier of the interior of Brazil, from original tracings furnished by Mr. Mornay; and the southern part, from St. Catherine's to the Banda Oriental and the Rio de la Plata, as also, Paraguay, from original MSS. brought to England by Sir W. Parish.

Upwards of 100 MS. maps, besides other written information, have been worked up with judicious criticism by Mr. Arrowsmith, in the construction of these two maps of South America and the provinces of La Plata; and they may safely be pronounced to give by far the best delineation yet produced of these countries, or likely to be expected for many years to come. To the work on the Provinces of La Plata I have already alluded, but I cannot refrain from recommending all who take an interest in the geography, the commercial prosperity, or the future prospects of those republics, to read carefully the faithful account of them here offered by an impartial and accurate observer.

Of some of the upper regions of the Andes we have received very interesting information from the distinguished geographer and naturalist, Mr. J. B. Pentland, Her Majesty's Consul-general at La Paz, then the capital of Bolivia. Near to the Guallillas-Pass, in crossing the western Cordilleras, at an elevation of 14,700 feet, Mr. Pentland examined the gigantic undertaking of some Englishmen, by which the whole stream of the Uchusuma, now flowing in a contrary direction, will be made to enter into a canal formed for the purpose, at about 14,000 feet above the level of the sea, the extent of the canal being 40,000 yards, before it reaches the culminating point of the Andes, over which it is to pass, on its way into the valley of Tacna, for the purposes of irrigating an extent of territory on the western slope, to which nothing has been denied by nature but a competent supply of water.

In the neighbourhood of Cuzco Mr. Pentland found vast remains of ancient structures of Cyclopean masonry, raised by the industry and and skill of the Peruvians; he visited and fixed the positions of some of those extraordinary natural phænomena of the country, where large rivers cut through the Andes, particularly where the Yucay passes from the basin of Ollantay-tambo and Urubamba, into that of Santa Ana, and where the Mapuri breaks the chain of the eastern Cordillera at the N.W. foot of the Nevado of Lachisani, a part of the gigantic mass of Ancuma. Mr. Pentland has also determined by astronomical observation the position of nearly 40 points, and, by barometrical means, their elevation above the sea, comprising every important feature of the eastern Cordillera between the parallels of 16 degrees and 13 degrees of south latitude. He has also, with the assistance of Mr. Bowring, the real author of the map lately published at Paris, completed the survey of the great Lake of Titicaca, where, at the height of 11,000 feet, the extensive ruins of the ancient Peruvians, especially of sepulchral monuments, indicate the former existence of a very dense and active population.

You have already heard, in the Report of the Council read to you this morning by your Secretary, that Mr. Schomburgk has now nearly completed the fourth year of his explorations into British Guayana and its contiguous provinces; the reports of his ascent of the river Essequibo, Berbice, and Corentyn, with some notices on the basins which they water, have already appeared in the sixth and seventh volumes of the Journal. During the two last seasons which he has spent in the interior of that country, during which he has frequently experienced the benefits of the continued courtesy and assistance of the present governor, Colonel Light, as he did before from the lamented Sir J. Carmichael Smyth, he has explored the source of the Essequibo, which he reached on the 27th Dec., 1837, in lat. 0° 41′ N., and long. 59½° W. He then crossed the Equator, and penetrated to 0° 12′ south lat. Returning thence to Pirará, he crossed the Brazilian frontier to Fort San Joaquim, where he experienced the greatest civility and assistance from the Brazilian commandant; he then descended about 30 miles of the Rio Branco, and explored the range of the Carumá mountains, on its eastern bank. Returning again to Pirará, Mr. S. has travelled in a N.W. direction through the valleys of the Paracaraima range, to the sources of the river Caruni, and to the remarkable lofty table-topped sandstone mountains of Rorima; and at the date of his last letter, Nov. 20th, 1838, he was about to proceed westward to the head waters of the Orinoco, with the intention of descending the river to Esmeralda, thence to cross over the range of mountains, which separates the basins of the Orinoco and the Amazons, descend the Padaviri, to the latter stream at Thomar; thence ascend again the Rio Branco, and thus by Pirará return to Demarara. Mr. Schomburgk and all his party had suffered much from fever and climate; but his zeal in the cause of discovery appears unabated; and should he reach the coast safely, with his historical and other collections, there is little doubt that in addition to his geographical information, he will bring with him a rich harvest in several departments of natural history.

Australia.—Colonization, and in good part geography, continue to make rapid strides in the great island of Australia. In New South Wales we have the account of three expeditions into the interior by Sir Thomas Mitchell, illustrated by beautiful sketches of the country and portraits of the natives. At Port Phillip, the town of Melbourne has already 3000 inhabitants; and an accurate survey of this fine bay has been completed by Lieutenants Symonds and Henry, R.N., and published at the Hydrographic Office.

At Adelaide, St. Vincent's Gulf, it is said that 10,000 persons are already settled in a colony, which only sprung into existence a few years ago. Several overland expeditions, driving large herds of cattle, have reached this place from Sydney; amongst others that of Captain Sturt himself, the original explorer of the River Murray, and another by Mr. Hawdon, the details of whose route have been communicated to the Society.

On the north coast of Australia a settlement named Victoria has been formed at Port Essington, which, from its very convenient site, and its immediate neighbourhood to a large but little known archipelago, bids fair to become a second Sincapore. Farther to the west the nautical survey by Captain Wickham has completed some of the portions left unfinished during the more extended examination of these coasts in the years 1820–3 by Captain P. P. King, R.N., and the account of which is still the chief authority for this portion of its shores, as the excellent work of Flinders is for the rest of this vast island. The survey of the Beagle has proved Dampier's Land, or Dampieria, not to be an island, by tracing Roebuck Bay and King's Sound to their termination, the latter affording an outlet to the largest river yet discovered on this part of the coast, and which, in honour of the late commander of the Beagle during her survey of the coasts of South America, has been named FitzRoy River.

Captain Wickham has since been employed in surveying the approaches to Swan River and Rottenest Island: and Sir James Stirling has just brought home a plan of Warnborough Sound, lying immediately to the southward of Cape Peron, which if the entrance be not obstructed by some hitherto unknown rocks, will afford one of the finest harbours on the western coast of Australia.

On the north-west coast, Lieutenants Grey and Lushington are the first Europeans who have ever penetrated into the interior. Starting from Hanover Bay on the 1st February, 1838, they proceeded in a general S.S.E. direction. At about 20 miles they came on the bank of a considerable river, which they named Glenelg, flowing through a fertile tract of country: continuing on in the same direction, they had great difficulty in getting their cattle over the sandstone hills; and when at a distance of 71 miles from the coast, want of pasture obliged them to return. By the latest accounts from Lieutenant Grey, dated Swan River, 22nd December last, we learn that during the three months he had passed there, he had made several short excursions into the interior, as far as any colonist had yet penetrated: he had lived much among the natives, studying their manners and customs; and had collected a vocabulary of their language, which he has just communicated to the Society. Whatever may be the result of this officer's explorations, we cannot but admire the zeal, activity, and perseverance, which appear to have animated him on all occasions, even under circumstances of no common trial.

I must not omit on this occasion also to mention that Mr. John Arrowsmith has lately published a general map of Australia, besides separate maps of South East Australia, the settlements at Adelaide and Port Phillip, and Western Australia, including all the recent discoveries.

Public attention has lately been much directed to New Zealand, and we hope that Captain Symonds, who has just sailed for that country, well supplied with instruments furnished by the Society, will obtain for us some useful geographical information.

Considerable expectations had been entertained that the voyage of discovery in the Antarctic Seas, undertaken by the French government, and under the command of Captain Durville, would have added to our knowledge of these inhospitable regions, and in which it is still hoped that a large field may yet be opened for scientific and commercial enterprise. The very unusual severity of the season during the South Polar summer of 1837-8, prevented Captain Durville from penetrating further south than the 64th degree of latitude. The same seas which had enabled Captain Weddell in 1822-4 to reach ten degrees nearer to the South Pole, presented to Captain Durville a compact and impenetrable barrier of ice; but the attempt to proceed southwards was not abandoned, until the ships had fruitlessly struggled during a whole month against the obstacles of nature.

The observations which I have ventured in the preceding pages to submit to your notice on the progress of geography, imperfect as they are, would perhaps by some be thought to be still more so, if I were to omit altogether, however familiar the subject is to us all, to allude to the great and rapid strides, which have been made within a very recent period, for the extension of steam navigation. This mighty power, which is spreading its influence over all the seas of Europe, and many of the waters of the other parts of the world, to the annihilation, we might almost say, of space and time, conveys the traveller as it does the merchant or the soldier, from one continent to another, with so many facilities as to make the communication between the most distant regions as safe and easy, and almost as rapid, as it formerly was between remote districts of the same empire. This has essentially contributed to make us better acquainted with each spot. The voyage to North America is by this mode of conveyance reduced from six weeks to about a fortnight, that to India from four to two months, and that to the farthest shores of the Black Sea is almost for the first time since the middle ages opened to the flag and commerce of Western Europe. The coasts of the Baltic may be visited from our own ports within the space of a short week: with a slight interval of overland conveyance, we may reach Constantinople with the same facilities by the Rhine and Danube.

The Indus, the Euphrates, the Tigris, the Ganges, and the Mississippi, are penetrated by the same means almost to their sources, with the same certainty, and thrice the rapidity, which have hitherto been experienced on the most ordinary canal for inland communication; and, we may confidently hope, that at no great distance of time the great rivers of South America, and even those of China, may be subjected to the same dominion of science and enterprise. In the mean time we have reason to apprehend that we may shortly be brought into rapid and regular communication with the coasts of Chile and Peru, by the establishment of steam-ships from our own shores to the Bay of Chagres, and from Panama to Valparaiso; for which we shall be principally indebted to the exertions of Mr. Scarlett, Mr. Wheelwright, and their fellow-labourers in this noble project.

Another remarkable isthmus, that of Suez, will also, in all probability, for the same or similar purposes, become again what it once was, the highway of nations; and the possessors of Egypt may again see themselves holding the key to lock or to unlock the most direct line of communication for the exchange of the commodities of the East and West.

And not only may we anticipate from the success of this undertaking large additions to our geographical knowledge of these distant portions of the world, but we may justly expect that such an instrument of intercourse between man and man will be the signal for rousing the intellectual energies which have long lain dormant in the two great continents of South America, and of the long-benighted Asia. This Asia, in which man has hitherto been almost always in a state of infancy, the infancy either of youth, or of premature and decrepid age, we hope to see at length throw off the chains, which have in all times held her whole social system in a state of unnatural servitude, checking the wholesome developement of the best faculties of our race by institutions hostile to all improvement, and to which the inhabitants are the more strongly attached, in proportion as they are more destructive of the means by which their happiness and their comforts might be increased. But the great engine to which I have alluded must inevitably teach them that there are other worlds besides their own, that the banks of the Ganges and of the Indus are not the only spots on the globe, on which Nature has been lavish of her choicest productions, that there are other enjoyments than those of listless indolence, and the observance of the idle or superstitious practices of their fathers; that they, as well as the other creatures of a benevolent Creator, are born inheritors of a cultivable mind, that they are not debarred from partaking with Europeans of the charms and advantage of Science; that the opportunity is now offered them of bearing an active part in shedding the blessings of civilization over the whole world, that they in short may, and must in due time, become navigators, astronomers, mechanicians, philosophers, geologists, and geographers.

If I were asked, what I conceive to have been the one mainspring, to which modern times owe their immense advantages in Science generally, I put out of the question any comparison between the two ages of the world in respect to literature, or to eloquence, or to moral and intellectual philosophy, poetry, or the fine arts; but speaking strictly of Science, to what one faculty, or rather tendency, of the human mind, are we most indebted for our superiority over the ancients? I should say, it was the spirit of geographical and nautical discovery, which developed itself in the middle of the fifteenth century. This spirit, which guided Diaz, in 1486, to double the Cape of Good Hope, and Vasco di Gama to penetrate to India, Columbus to discover the New World, Magellan, Drake, Schouten and Le Maire to find their way into the Pacific, has forced every succeeding generation to cultivate astronomy, navigation, chemistry, metallurgy, botany, and every science which could contribute to dive into the depths of the elements of our own globe, and of those which govern the great phænomena of the system of which we form a part. Nature herself has here presented larger and more varied laboratories, in which these phænomena, and all her various monuments may be studied. Without this accession to our knowledge, without this enlargement of the sphere of our observations and experience, without this excitement to improve the machinery of our minds, we should probably have long remained immured in the darkness, in which our ancestors groped their way in the pursuit of knowledge; we should have lived on the ill-digested remnants which the antients had left us; we might have guessed, but might never have been convinced of the form of our globe; we should have been satisfied with hugging the coast in our most distant voyages; marine architecture would not have advanced beyond a galley or a trireme; our eyes would not have been opened to the splendid natural productions of a tropical climate; the precious metals would have become so scarce in Europe, that capital would have been unknown, commercial adventure would have slept its long sleep, distant colonization could not have been dreamed of, and there would have been no escape from domestic tyranny, from religious persecution; no opening for daring adventure. But the spirit of discovery was the happy spell which changed the destiny of nations; and Europe is not insensible or blind to the advantages she has derived from it. New societies are annually springing up in the principal cities, to promote and encourage it; each forming, as it were, a nucleus for the reception and promulgation of the results of the observations of travellers and navigators. We rejoice in the increasing number of our fellow-labourers in this rich and inexhaustible field. The contest we are engaged in is one in which all are proud of their success, all are sure to win: we play into each other's hands, we are all partners, and all rivals; the stake we contend for partakes of the infinite divisibility of nature, with this difference in our favour, that the greater the number of divisions, the larger is the portion which each of the parties may appropriate to himself.

ROYAL GEOGRAPHICAL SOCIETY.

Patron.

THE QUEEN.

Vice-Patron.

His Royal Highness the DUKE of SUSSEX.

COUNCIL, ELECTED MAY 27, 1839.

President.

G. B. GREENOUGH, Esq., F.R.S.

Vice-Presidents.

Sir John BARROW, Bart., F.R.S.
The Hon. MOUNTSTUART ELPHINSTONE
W. R. HAMILTON, Esq., F.R.S.
Francis BAILY, Esq., F.R S.

Treasurer.

John BIDDULPH, Esq.

Trustees.

Sir George T. STAUNTON, Bart., F.R.S.
Francis BAILY, Esq., F.R.S.
John BIDDULPH, Esq.

Secretaries.

Captain WASHINGTON, R.N.
Rev. G. C. RENOUARD, Foreign and Hon.

Council.

Captain Sir George BACK, R.N.
John BACKHOUSE, Esq.
Captain BEAUFORT, R.N., F.R.S.
Captain WENTWORTH BULLER, R.N.
Lieut. Col. CHESNEY, R.A., F.R.S.
Lord COLCHESTER.
Maj. Gen. Sir A. DICKSON, R.A., G.C.B., F.S.A.
Colonel FOX.
J. BAILLIE FRASER, Esq.
Bartholomew FRERE, Esq.
W. I. HAMILTON, Esq.
Right Hon. Sir J. C. HOBHOUSE, M.P.
Colonel JACKSON.
Lieut.-Colonel LEAKE, F.R.S.
George LONG, Esq.
James MEEK, Esq.
James MORIER, Esq., F.R.S.
R. I. MURCHISON, Esq., F.R.S.
Sir WOODBINE PARISH, F.R.S.
Lord PRUDHOE.
Rt. Hon. Sir Charles VAUGHAN.

FOREIGN HONORARY MEMBERS.

His Imperial Highness the Archduke JOHN OF AUSTRIA
Chevalier BALBI . . Vienna
Mons. C. F. BEAUTEMPS-BEAUPRE, Mem. Inst. . . . . . Paris
Leopold VON BUCH, For. M.R.S. L.S. and G.S., Mem. Acad. Berl. . Berlin
General CAMPANA . . . Milan
Chevalier CASSALEGNO . Turin
General CLARKE . United States
Captain DUMONT D'URVILLE . Paris
Admiral DUPERREY . . Paris
C. G. EHRENBERG, For. M.R. and L.S., Mem. Acad. Berl. . . Berlin
Carl FALKENSTEIN, Corr. Mem. Acad. Berl. . . . . Dresden
Captain L. DE FREYCINET, Mem. Inst. Fr. Paris
Professor HANSTEEN . Christiania
Baron Alex. VON HUMBOLDT, For. M.R.S., L.S. and G.S., Mem. Inst. Fr., Mem. Acad. Berl., &c. . . Berlin
Rev^d. Padre G. INGHIRAMI Florence
Mons. E. F. JOMARD, Mem. Inst. France, Corr. Acad. Berl. . . Paris
Admiral VON KRUSENSTERN, Corr. Mem. Acad. Berl. . St. Petersburg
Count Alexandre L. J. LABORDE, Mem. Inst. Fr. . . . Paris
Admiral F. B. LUTKE . St. Petersburg
A. J. LETRONNE, Mem. Inst. Fr. . Paris
Dr. Charles VON MARTIUS, For. M.L.S. Corr. Inst. Fr. and Acad. Berl. Munich
Conrad MANNERT . . Munich
Count MEYENDORF . St. Petersburg
Professor Carl RITTER, For. M.R.A.S. Mem. Acad. Berl. . . Berlin
Dr. E. RUPPELL, For. M.L.S. Frankfort
G. S. SCHOOLCRAFT . United States
Professor J. F. SCHOUW . Copenhagen
Colonel Ferdinand VISCONTI . Naples
Dr. George WAHLENBERG, For. M.L.S., Corr. Mem. Acad. Berl. . Upsala
Baron C. A. WALCKENAER, Mem. Inst. Fr. . . . . . Paris
Captain ZAHRTMAN . Copenhagen
Augustus ZEUNE . . Berlin

1836.

Count GRÅBERG AF HEMSÖ, Corr. Inst. Fr., For. Mem. R.A.S. . Florence
Baron HUGEL . . . Vienna

1837.

Professor Heinrich BERGHAUS . Berlin
Colonel FORSELL . . Stockholm
Professor Finn MAGNUSSEN Copenhagen
General PELET, Chef du Dépôt de la Guerre . . . . Paris
Mons. P. VANDERMAELEN . Brussels

CORRESPONDING MEMBERS.

Col. Don José ARENALES Buenos Ayres
Mons. DAUSSY . . . Paris
Dr. DUPONCEAU . . Philadelphia
William HILHOUSE, Esq. . Demerara
T. A. KJAR, Esq. . St. Thomas
M. KUPFFER, Mem. Ac. Sc., St. Petersb.
R. H. SCHOMBURGK, Esq. . Demerara
Mr. TANNER . . . Philadelphia
Dr. WOERL . . Freiburg
Don José URCULLU . . Oporto
Mr. WORCESTER . Cambridge, U. S.

1836.

Mons. D'AVEZAC . . . Paris
Counsr. J. J. da Costa DE MACEDO, Sec. Roy. Acad. Scien. . . Lisbon
Gen. Don Juan ORBEGOSO . Mexico

1837.

Prof. Adolph ERMAN . . Berlin
Don M. F. DE NAVARRETE . Madrid
Baron WRANGELL . St. Petersburg
Professor Paul CHAIX . Geneva

1838.

Colonel LAPIE . . Paris
Don Pedro DE ANGELIS Buenos Ayres
Colonel OBERREIT . . Dresden
Colonel SKRIBANECK . . Vienna

1839.

Professor C. C. RAFN . Copenhagen
Dr. VON SCHLIEBEN . Dresden
Capt. Don Eduardo CARRASCO Lima

VII.—*Notes on a Journey to Kordofán, in* 1836-7. By Arthur T. Holroyd, Esq.

[Read 25th February, 1839.]

It was not until I had arrived at Wádí Ḥalfah, at the second or great cataract of the Nile, in lat. 22° N., that I finally resolved to penetrate into the Beléd-es-Súdán (countries of the Blacks), and even then I did not anticipate prosecuting my journey beyond the ruins of Muṣawwerát,* in the neighbourhood of Shendí, or at the utmost to Khartúm (Proboscis), the seat of the Páshá's government for the provinces to the south of the second cataract. The sequel will show that I visited Sennár, and then proceeded to the west to El 'Obeïd,† the capital of Kordofán.

In the afternoon of December 5th, 1836, having engaged camels at 35 piastres each, I left Wádí Ḥalfáh, accompanied by Ḥájí Suleïmán my interpreter, and 'Alí an inferior domestic; and, having crossed the Nile, determined to take the road on its west bank to New Dongola (Donḳolah). Our route lay through Abú-ṣír, Taḥtí, Kagmí, Semneh, Askur, Melik-en-naṣr, and 'Oḳmeh, all small villages consisting of only a few huts: at this latter place there is a hot saline spring close to the river; it is about 4 miles S. from the village. A small quadrangular burnt-brick ruin, without either door or window, but with an opening at the top, now nearly closed by the drifted sand, shows that this was, in all probability, anciently used as a bath. At the back of this building is the principal spring, which is so close to the edge of the river, that it must be concealed during the inundation: The temperature of this water at its source is 130° of Fahrenheit. Smaller and more insignificant springs ooze through the ground on each side of the principal one to the distance of 250 paces, and the saline matter left by the evaporation gave the bank a frosty appearance. The peasants of 'Oḳmeh have an idea that the hot spring comes under ground from the Oasis of Selímah, 70 miles to the W., where there is abundance of muriate of soda in a state of great purity; and I afterwards asked an intelligent camel-driver, who had made many excursions to Selímah, at what point he thought he should come upon the Nile if he travelled due E. from Selímah, and he replied at 'Okmeh. This is strongly corroborative of the position of Selimah laid down by Mr. W. G. Browne, and of its position relative to 'Okmeh as given in Mr. John Arrowsmith's map. From 'Oḳmeh we passed through Dál, Sáḳiyeṭ el Abṭ, Deïr Ḥamíd, Sedaënga, Dóshe, Ṣoleïb, Kóyeh, Ṭináreh, Gurgód, Saesa, Marakol, Ḥannek, and

* Muṣawwerát, "the figured (buildings)," is misspelt, as is often the case, in M. Cailliaud's table of proper names.—F. S.

† Vulgarly pronounced L'obeyet: it means "the little slave."—F. S.

Hafír, to New Donkolah, which I reached at noon on the 22nd of December. The journey is usually performed in ten days, but I stopped on the road to examine all the antiquarian remains.

U'rdeh, A'gí,* Marákah,† or New Donkolah (for it is known by all these names), has sprung into a place of importance within the last twelve years. The population is estimated at 6000, including 800 troops, their wives and families, an estimate which appeared to me much too high. The number of Copts is about 100. The bázár is daily increasing, and is supplied from Cairo; the principal articles brought from thence being shoes, printed cottons, calicoes, sugar, rice, cloth, hardware, &c. The duties upon goods entering New Donkolah are heavy, so that articles of consumption are dear, in many instances treble or quadruple what they are in Cairo. Donkolah can boast of a coffee-house. The government is at present building baths, and there is already a large manufactory for indigo. The thermometer on Christmas-day stood in the shade at 2 P.M. at 86°, and at 8 P.M. at 80°. The position of New Donkolah, as stated by M. Linant, is 19° 7′ 30″ N. lat. and 29° 54′ 35″ long. E. of Greenwich.‡ The town is placed on the edge of the river, the banks of which, when the water is lowest, are about 25 feet high, and I should think that the Nile rises here about 18 or 20 feet.

On the 31st of December, having procured a boat, I left New Donkolah, and on the afternoon of the 3rd January, 1837, I reached Old Donkolah. The town is in ruins, and does not contain a population of more than 300. The most striking object here is a mosque on rather an elevated site, from the top of which there is an extensive prospect of the arid Desert and meandering Nile. The sand is of a very bright yellow colour, and has accumulated in such quantities as in many places to conceal the houses; its surface being level with their roofs, and the entrance to the apartments being through the ceiling of the rooms. There is no land capable of cultivation near the ruined town; in short, almost all the east bank of the river, between Old and New Donkolah, is covered with drifted sand, and rare is it to see even a few feet of cultivable soil. Our direction from Old Donkolah was S.E., then due E. until near Ambukol, whence we travelled a few miles a point to the N. of E.

The only object of interest in the neighbourhood of Ambukol is a portion of desert near the river, about 8 miles W. of the village. This waste is called *Haagbarlak.* The superficial stratum here is a coarse sandstone, curious and interesting from its

* The court, a Turkish term.—F.S.

† Capital of one of the ancient ecclesiastical divisions of Nubia (Quatremère, Mémoires sur l'Egypte et la Nubie).—F. S.

‡ Dr. Ruppell gives its lat. 19° 10′ 19″, long. 30° 22′ 15″ E.

containing many siliceous fossil trees. I observed five or six, the largest of which, situated 20 minutes' walk from the river, is 51 feet in length and 20 inches in diameter at its largest extremity. It is partially buried in the sand. The peasantry splinter off fragments and use them for gun-flints, and to strike a light. It appeared to me that these fossils were dóm-trees (Cucifera Thebaïca).

On the 8th of January I arrived at Ambukol, and a severe attack of fever arrested my progress until the 24th, when, being sufficiently recovered, I resolved to cross the desert of Bayúḍah* to Khartúm, a route, I believe, not previously passed by any European traveller. Thermometer 75°. I engaged camels at 25 piastres each from Ambukol to Khartúm. My first day's journey was short, as first day's journeys always are in the East; and I rested for the night, after travelling 2¾ hours, at a place to which the Arabs have given the name of Berj el ḳurán (Koran tower).

On the 25th I travelled 7½ hours, and rested at Khór-el-ghaním. On the 26th 9 hours to Abú Ṣamúd.

On the 27th, in 3 hours we reached the wells of Bayúḍah. Many wells have been sunk, but we found water only in three, and in very small quantities. The water is good, free from any brackish taste, thick, and of a deep yellow colour, deriving these latter properties from the soil through which it passes. The water which we had brought from the Nile was putrid and nauseous, and we were glad to avail ourselves of the opportunity of procuring a fresh supply. We were delayed several hours in taking a small quantity, the wells having been visited during the morning before we arrived by Arabs, who had drawn largely upon the springs for their goats, sheep, asses, and camels. These are the only wells between Ambukol and El Ḥajír; the latter the point at which we met the Nile again. In the afternoon we resumed our journey, crossed a small ascent of about 70 feet, and rested for the night, after 3 hours, at Khór-el-Laban. The distance on the 27th, six hours (18 miles). On the 28th, 8¼ hours, to Kús Abú Deluah. On the 29th, 7 hours to El Atherleh. On the 30th, 8 hours to El Gharah.

On the 31st, in 3½ hours we arrived at the Nile, leaving the Desert between two rocky hills, El Ḥajír, about 150 feet high, on the right, and Jebal-el-Róyán (situated on an island of the same name), about 250 feet high, on our left.

I crossed the desert of Bayúḍah, as I have shown, between Ambukol and El Ḥajír, and the journey is usually performed in seven days, though it might with ease be made in six. It is flat, with few hills, and those small and of easy ascent. After leaving Ambukol, the surface is alternately sand and gravel, but the sub-

* بيوضة from its whiteness.—F. S.

stratum is sandstone, and continues to be such till within an hour of reaching Abú Ṣamúd, where we came upon grey granite, furrowed by veins of primitive quartz. The gravel, both to the N. and S. of Abú Ṣamúd, contains quartz pebbles in great abundance; and I noticed also portions of siliceous fossil trees, similar to those observed at Ḥaagbarlak. At Abú Ṣamúd, the wells of Bayúdah, and in short as far as El Ḥajír, sandstone occurs with only one exception, where a vein of red granite crossed from W. to E. between these two latter places. I observed occasionally stones of a yellowish grey colour, spotted with black points, resembling those which are thrown up by Vesuvius during its eruptions. I also found abundance of fragments of hard sandstone, containing quartzose pebbles, which were in patches, and at no great distance from each other. The patches of stones of a grey colour, which appeared volcanic, were at least 40 or 50 miles from the hills; and though they appeared to have been deposited after some volcanic convulsion, we were unable to discover any extinct volcano; the sand does not drift much in this desert. I also occasionally found hard black cinders. It is, however, remarkable that these isolated patches are found at so great a distance from the mountains. At 8 hours S. of Ambukol we came upon patches where water had collected during the periodical rains; and upon looking at the map I observed that Mr. Arrowsmith had laid down their northern limits very correctly.

The desert of Bayúḍah abounds in herbaceous plants, and one or two varieties of the mimosa. It also possesses many objects of great interest in the zoological kingdom, especially the leopard, oryx, and gazelle. Its few inhabitants belong to the tribe of Kabábísh Arabs.*

In 2 hours and 40 minutes we halted for the night at Jagjoke.

Feb. 1. We travelled 8 hours and 40 minutes to Kerrerí. On the following day, in 4 hours, we arrived at the northern point of the White Nile (Níl el Abyaḍ), on which we embarked, and in less than an hour arrived at Khartúm.

The direction from Ambukol to El Ḥajír was from N.W. to S.E., and the distance occupied us 52 hours' march, which, at 2½ miles per hour, would give us 130 miles; from El Ḥajír to Khartúm, from N. to S., following the course of the river, 16 hours 20 minutes, or about 42 miles.

The position of the junction of the Bahr el Abyaḍ, or White River, with the Bahr-el-Azrak, or Blue River, in the observations given me by M. Linant, is 15° 34′ 40″ N., and 32° 11′ 25″ E. of Gr. These observations do not agree with those already made by the same gentleman, and published in a memoir on the Bahr el Abyaḍ, in the second volume of the London Geographical

* Plural of Kabbásh, "a shepherd."—F.S.

Journal; for M. Linant there gives the position of the junction of the two rivers as 15° 34′ N. lat., and 32° 30′ 58″ E. long.* Mr. Perring, a civil engineer in the employment of Mohammed 'Alí Páshá, who kindly undertook the construction of my map, pointed out the discrepancy between these observations; and we immediately saw M. Linant on the subject, when he told us that he considered the list furnished to me, from which I took all the fixed points, to be the most deserving of reliance, because he carefully looked over and corrected his observations after his return to Caïro, and subsequently to his sending his papers to the African Association.

Khartúm (the Proboscis) is situated on the W. bank of the Blue River, about 1½ mile from its junction with the Bahr el Abyad. It is the seat of government of the Beled-es-súdán, and the present governor is Khúrshíd Páshá. It was a small village when Mohammed 'Alí subdued the kingdom, but has risen rapidly into importance at the expense of Shendí and Sennár, and is now a place of considerable trade, being convenient as a rendezvous for the slave-caravans from Abyssinia, Sennár, and Kordofán. It contains 15,000 inhabitants, including about 1600 soldiers and their families. Parts of it are regularly built. Many of the houses are large and isolated, enclosed by a garden. They, as well as Khúrshíd Páshá's palace, which is by no means striking, are built of sun-burnt bricks. The bázárs are irregular, small, and confined, and, when the troops are absent, extremely ill-supplied. The principal articles of consumption are shoes, calicoes, printed goods, sugar, rice, broad cloth, pistol-belts, saddles, worked saddle-cloths, a little tea, crockery, hardware, *mishmish*,† *kamar-ad-dín*,‡ &c., and a considerable trade in slaves is carried on both by auction and private contract. Just outside the bázár is the market for vegetables, bread, fruit, sugar-cane, butter, dates, grain, straw, grease for the head, &c. &c. And here stands the frame to which criminals are suspended when executed, complete with rings and accommodation for three persons.

The land upon which Khartúm is built, as well as that which surrounds the town, is rich alluvial soil, similar to the deposit of the Nile. The ground is flat and without trees. The banks of the river at the season of low water are about 30 feet in height, and I should think, from their appearance, that the average rise of the river here is about 20 feet. During this, my first stay, at Khartúm, the mercury in the thermometer never stood in the daytime below 75°, but on the 10th of February it rose to 90°.

Feb. 11. I left Khartúm to proceed up the Blue River, Khúrshíd Páshá having provided me with an excellent boat

* Vol. ii. p. 171.—Long. 32° 41′ is adopted in the map from Bruce's observations at Halfayeh.

† Dried apricots.

‡ The pulp of apricots spread out and dried.

for that purpose, and on the 15th I arrived at Abú Kharráz, on the E. bank, and at the junction of the Rahad, an eastern tributary stream about half a mile wide, of blue water and rapid current. Here is a village of considerable size, and a station of 300 Moghrebí cavalry. Most of the houses are of straw, of a circular form, and resembling corn-stacks. There are only one or two, of sun-burnt brick, and the largest of these is that of the káshif. There is a market here daily; the wares are scanty, in small quantity, and of inferior quality. On the same day I crossed the river to Wád Medínah, and in an hour reached a military post, where one battalion of 800 men is stationed. Here is a daily market, small, and ill supplied with articles of a very coarse description.

There are two large bázárs held in the neighbourhood of Wád Medínah weekly; that to the northward is at Sálemíyah, the other, to the S.W., at Sorríbah. To the latter I went on the 16th. The village is situated about 6 miles from the river, and the road to it is over a rich plain. Dhurrah (millet or sorghum) is cultivated round it in patches after the kharíf or rainy season is over, but for want of capital and labour by far the greater proportion of the soil lies idle. This market was well attended, principally by Bedowíns, and the wares exposed for sale much better than those at Abú Kharráz and Wád Medínah. The banks of the river at Wád Medínah are about the same height as those of Khartúm, and I should think that the rise of the Nile is in the same proportion. Thermometer at Wád Medínah, Feb. 16, 88°.

On the 17th I left Wád Medínah, and at 4 P.M., on the 21st, I had moored my boat immediately below Sennár. I took up my quarters in a house in the barracks, provided for me by the military commandant; and I was hardly settled in my new habitation before the peasantry flocked in numbers with their manufactures to endeavour to seduce me into becoming a purchaser. Among their wares were mats with beautiful devices made of split dóm (Cucifera Thebaïca) leaves, and dyed of various colours, conical straw covers for plates in elegant patterns, silver zerfs (stands for coffee-cups) in filagree, warlike weapons, spears, knives, &c., and lastly, hegabart, or amulets for security from every kind of disease and casualty, including the fatal fever of the country and the voracious jaws of the crocodile.

A small bázár is held daily at Sennár, and the principal articles for sale are butcher's meat, grease, oil, tobacco, dhurrah, bilbil, and merísah;* but on Mondays and Thursdays a market attended by persons for many miles round is held at Kadero, a village to the S. of Sennár, about 1½ mile from it, and between it and Jebel Mowíl. That held on Mondays is the largest and best supplied. Near the place where the daily bázár of Sennár is held is a mosque,

* Bilbil and merísah are kinds of beer made from fermented dhurrah.

adjoining to which may be traced the foundations of the palace (now no longer existing) of the last sultán, and near it is a coffee-house, the favourite resort of the military officers and fashionables of this capital.

The banks of the river at Sennár are between 40 and 50 feet high. The Nile rises about 20 feet. The country around is dull, uninteresting, and flat, with the exception of Jebel Mowíl, a hill about 800 feet high, six miles S.W. by S. of the town. Scarcely a tree is to be seen except in the distance towards the S. To the N., S., and E., are villages about 2½ miles off.

The inhabitants of Sennár are of a dark-brown colour. The women are lighter than the men, and both are handsome. The latter wear drawers and a cotton shirt reaching to their ankles, or a fold of cotton round the waist, and afterwards thrown loosely over the shoulder. Most of them wear rosaries of black seeds, of ebony, or other wooden beads, round their necks. They almost all have a purse suspended from the neck, hegabart, or amulets, on the right arm above the elbow, and a knife on the left in the same situation. Some wear leather ornaments round the ankles. They generally allow their hair to grow, and do not wear the Ṭáḳiyeh, or white skull-cap.

The women wear a fold of cotton round their bodies, thrown afterwards over the shoulders, and sometimes covering the head also. The unmarried girls and slaves usually wear only a ráhat, or leather fringe, round the waist. Their hair is plaited into tresses about the thickness of rats' tails; and this operation of plaiting is preceded by the destruction of the parasites which abound in these localities; a massacre which they carry on against those insects every fortnight or three weeks. The head is then dressed, and loaded with three or four pounds of mutton-suet or camels' fat. Some of the women wear silver ear-rings; all, necklaces of glass beads; and most of them an immense bunch of hegabart suspended from the neck. They also ornament themselves with bracelets of silver, of rings of horn or ivory, plain or slightly adorned with black spots; leather ornaments on the wrists and ankles; and I noticed a few who had a ring or beads passed through a perforation in the right nostril; but this was by no means common. Both men and women have good teeth. In many the gums are black or brown, deriving this peculiarity from their admixture with the negroes. Many of the women allow the nails on the fingers of the left hand to grow to an extraordinary length, an inch or more beyond the tip of the finger, similar to the women and grandees in China. They, however, pare the nails of the right hand, as the not doing so would interfere, they said, with their custom of eating with their fingers. Most of the lower orders possess one or two slaves.

The neighbourhood of Sennár is interesting to the naturalist. A very great variety of animals and birds are to be found during the kharíf or rainy season, and when the dhurrah is ripe. At all times the crested crane abounds, as well as several other species of cranes, storks, eagles, and vultures. I observed great numbers of wild guinea-fowls, and many varieties of ducks and geese. Elephants, hippopotami, and crocodiles are in great abundance.

During my stay at Sennár the thermometer never stood lower than 88°; and on the 27th and 28th of February it rose to 94°.

March 2.—At 5 A.M. I took my departure from Sennár by the Nile; at 10 A.M., on the 5th, I arrived again at Wád Medínah; and in the afternoon of the 9th I left it in order to proceed to Monḳarah, near Wád Shellaï, on the White Nile. In 1 hour and 40 minutes from Wád Medínah we reached the village of El Bessátnér, and in 1 hour and 10 minutes more we arrived at El Leïweh, where we rested for the night.

March 10.—In 2 hours to Heglígah, and in 3 hours more to Abút. This tract is a dead level, with a soil of the richest quality; but from want of capital, and paucity of inhabitants, little of it is cultivated, and that only during the rainy season. The peasantry are content to raise a crop of dhurrah sufficient to answer the demands of the government, and to maintain themselves during the year. The water of Abút is good and wholesome. Thermometer 90°.

March 11.—In 5 hours we arrived at Monáḳil, and took a meal of sour milk, pastry, and honey, with a Turkish soldier, káïm-maḳám, or governor of the village. A market, held here every Sunday, is better attended than any other in the neighbourhood. The water of Monáḳil is slightly impregnated with iron, but by no means unpalatable. In 5 hours from hence we arrived at the small village of El Fákirkír; ½ a mile from which are wells of good water. Our journey to-day was 10 hours.

March 12.—In 1 hour and 20 minutes from El Fákirkír we arrived at the wells of Emm-dakkat, where the water is slightly brackish, though not so much so as to be despised. 3¾ hours more brought us to Egarín, a village consisting of several huts, but only three inhabited. The water here also is brackish. We left Egarín in the evening, and, after travelling 5 hours, halted for the night.

March 13.—I resumed my journey at an early hour, and in 6 hours 15 minutes arrived at Monkárah, with the thermometer at 93°. Our general direction, from Wád Medínah to Monḳárah, was W., and the distance 86 miles. The price we paid for each of our camels, 12½ piastres [2*s.* 6*d.*].

From Abút to Monáḳil the country is not cultivated; but I am of opinion that in proper hands it might be made available for agricultural purposes. From Monákil to Monkárah the land is of excellent quality, with the exception of a small strip, about 7 miles distance from the latter place. After having made several excursions into this desert, and crossed from Wád Medínah to Monkárah, I have come to the conclusion, with respect to the country from Jebel Mowíl, near Sennár, to Al-leïs, westward, and to Khartúm northward, that the greatest portion, if not the whole, of this triangle, is formed by the alluvial deposit of the Nile. The soil of this desert—if desert it may be called—is precisely similar to that near the banks of the river; its surface is nearly a dead level; there are no hills or mountains; and the Nile, even now, rises during its increase to within 2 or 3 feet of the top of its banks, on the Blue and White Rivers; and in many parts of the latter, where no banks exist, it spreads itself over a large tract of country. If a canal were cut from Wád Medínah to Monkárah, with branches N. and S., almost all the land might be used for the production of cotton, indigo, tobacco, sugar, grain, &c. Much might be accomplished by means of tanks for collecting the rain-water during the kharíf, and by sinking wells. But still this would not suffice to irrigate the whole of this rich and valuable land. Between Wád Medínah and Monkárah the country is beautifully studded with prickly acacias. Monkárah is one of the Páshá's boat-building stations. There is abundance of wood in the neighbourhood; but little of it is used, as larger and better timber is found near Al-leïs, or in the country of the Shillúks. About thirty boats are built here annually. I observed eighteen or twenty pairs of sawyers, and five or six boats on the stocks. The sawyers are negro slaves and convicts. The whole population may be 100 persons.

In travelling in this country it is absolutely necessary to prevent the skin being exposed to the scorching rays of the sun, or it is immediately attacked with swelling and inflammation. Between Abút and Monkárah I was so imprudent as to ride with my legs bare from the knees downward, and when I arrived at the latter village they were so swollen, painful, and inflamed, that I was confined to my bed several days before I was able to proceed to Kordofán.

On the morning of the 15th I crossed over from Monkárah to Kajebí, a small settlement of Hasaníyeh Arabs, on the W. bank of the Nile, and N.W. of the former place. Here were also the tents of Suleïmán Káschif, who was collecting the contributions from the villages in the neighbourhood. During my stay at Kajebí a party of Shillúks arrived, accompanied by the son of their sultán. This prince was dressed in a blue cotton shirt, similar to those worn by the felláhs (labourers) in Egypt. They were on their way to Khartúm, for the purpose of endeavouring to make terms of peace with Khúrshíd Páshá, and adding their dominions to those already subject to the Viceroy of Egypt; a policy they thought it better to adopt than to expose themselves to the annual or biennial incursions of his troops for the purpose of seizing them as slaves. I heard that, subsequently to my leaving the Beléd-es-Súdán, Khúrshíd Páshá had concluded treaties with them, and that they are now subject to Mohammed 'Alí. These Shillúks were fine men; none of them less than six feet high, and many of them several inches taller. They were clumsily formed, their legs being too short for the size of the trunk. The heads of some were shaved: the hair of those unshorn was curled and woolly. Their countenances were harsh and savage, their cheek-bones high, and noses narrow near the root, but broad and flattened towards the nostrils. Like the inhabitants of Denkah, the incisors of their lower jaw had been extracted. The only weapons they had with them were sticks, shields, and spears of a rude construction. Some of them wore a single ring of ivory above the elbow of the right arm. Their prince was distinguished from the others by two large rings of solid silver, which he wore as bracelets on his left wrist. They indulge in smoking, but not to excess. They amused us one evening with several choruses and catches, which they sang in a very pleasing manner, keeping excellent time. They are very expert sportsmen, killing the crocodile and hippopotamus with the same spears which they use in self-defence. At Kajebí, in a double tent, the thermometer reached 112° on the 19th March.

The west bank of the White River is occupied to the N. and S. of Kajebí by Hasaníyeh Arabs, whose only other settlement which I observed was opposite Berber. They are generally fine men, a shade lighter than the Sennárese; and their females not so dark as the males. They are social, lively, and gay, and their females particularly fond of their national dance, which they accompany by clapping of hands and singing.

March 22.—At ½ past 7 A.M., having procured camels at 25 piastres each, I left Kajebí to proceed to Kordofán, by the desert of Habshábeh. Our guide lost his way and took us in a S.W. direction to Túrah, where we arrived in 6½ hours. The track is over land some portions only of which are tilled; much more might be brought into cultivation, though not all, for some of it is light and too sandy. I noticed the impressions of hippopotami in several places, the first I had seen: they were as much as two miles distant from the present channel, though at the height of the inundation the river passes through temporary channels close to the place where I observed the foot-marks, and there is little doubt but that the animals had been basking in some of these shallow streams which at this season of the year are dry. Túrah is a village consisting of a few circular straw huts; there is good water in four wells. In the evening I left Túrah, our direction due W., to Abú ghárát, where we halted for the night. The distance travelled to-day 8 hours 40 minutes.

The next morning a caravan arrived at Abú ghárát, and I soon ascertained that it was that of Abú Medyán, half-brother to the present Sulṭán of Dár Fúr, who was on his way to Kharṭúm, intending, with Khúrshíd Páshá's permission, to go to Cairo to endeavour to persuade Moḥammed 'Alí to furnish him with troops to proceed against his brother, and, if possible, make Dár Fúr tributary to the Egyptian viceroy. A gentleman at Kharṭúm had given me a letter of introduction to this sable prince, and I lost no time in presenting myself at his díván. I found him about 25 years of age, of the most complete jet black, with a countenance frank, ingenuous, and pleasing, and features partaking little of those characteristic of the negro; in stature, short and inclined to corpulency. From him I learnt that the present Sulṭán would detain all Europeans who should enter his kingdom, as prisoners—under the impression that they had visited it to make surveys with the idea of returning to Europe and bringing an army to subdue and overthrow him: that such prisoners he would treat with courtesy, furnish them with a house, good provisions, horses, slaves, a harem, &c., but that a guard would always be stationed at the door, and the state-prisoners would never be allowed even to ride or walk out unless accompanied by an escort. The Sulṭán resides at El Fásher, which is called also Tendeltí; he has an army equipped with swords, spears, and shields: fire-arms are unknown amongst them. The inhabitants of Dár Fúr, as well as Dár Márah, to the S. of it, are Mohammedans.

Abú Medyán informed me that the distance from El-'Obeïd, the capital of Kordofán, to El Fásher, was 13 days of camel-journeys. From Ḳobbeh to Kubkabíyah 3 days, and from Ḳobbeh to Debbah, nearly opposite to Old Donḳolah, 25 days.

In the evening I resumed my journey in a S.W. direction, and in 3 hours arrived at El 'Adáyír. The water at Abú-ghárát is good, while that at El 'Adáyír is brackish. This latter village is situated on the edge of an uninhabited waste, and, as it was a bright moonlight night, I determined to enter it immediately. In 1 hour and 50 minutes we were obliged to stop, as the Ḥájí was taken suddenly ill. Our journey to-day was 4 hours 50 minutes. Thermometer 96°.

March 24.—I set off early, and in 5½ hours arrived at a single isolated tree, the only one between El 'Adáyír and Habshábeh. In 2¼ hours from this spot we passed Jebel esh-shawáyír, a hill about 250 feet high, a little to the left of our route; and 4¾ hours beyond halted for the night, after having travelled this day 12 hours 30 minutes.

March 25.—Started at ½ past 6 a.m., and in 3½ hours arrived at the village of El Habshábeh. In passing this 'Aḳabah or desert-ascent, it is necessary to carry water, as none can be procured between El 'Adáyír and El Habshábeh. Before leaving Monḳarah or Kajebí, the traveller should see that each camel is provided with two water-skins to be filled at the wells of Abú-ghárát, in preference to those at El 'Adáyír. The camel-drivers will provide the skins, and suspend one on each side of each camel under the baggage. The water at El Habshábeh is good and plentiful. The desert between it and El 'Adáyír is uninhabited; but we observed occasionally Kabábísh Arabs pasturing their camels upon a plant to which they give the name of Askanít,* and which the camels eat with avidity. The plants are covered with fine thorns, which readily enter the skin and give great inconvenience and annoyance to the traveller.

In the evening we travelled 3¼ hours before stopping for the night. The distance to-day 6 hours 40 minutes. Therm. 99°.

March 26.—In 8 hours from our resting-place we reached El Kowermát, the water of which is good. Here I met with two pilgrims from Bornú, who were on their way to Mekka. They had come by way of Dár Fúr and Kordofán, and had already been twelve months on their journey. They had a very vague idea of their country, and all the information I could obtain from them was that it was situated upon a river called Shárí, which they believed to be one of the tributary streams of the White River, and that the name of their present Sulṭán is Moḥammed.

Between El Habshábeh and Kowermát innumerable trees are decayed and laid prostrate, the work of those destructive insects the white ants, which abound in this desert. The soil between these two places is light and sandy; and very little is at present under cultivation. Encamped for the night after a day's journey of 12 hours 20 minutes. Thermometer 103°.

March 27.—In 50 minutes we arrived at Ed-du'amá, and, our camels and asses being fatigued, stopped here for the day: between El Kowermát and Ed-du'amá the soil is sandy, and few patches of it are capable of cultivation. The only grain produced is dhukhn (sorghum), which appears to thrive well on a light sandy soil. The inhabitants of all these villages are almost wholly uneducated, and have very primitive ideas. A boy asked me at Ed-du'amá how many days there were in an hour?

In the afternoon I left Ed-du'amá, and after 4 hours and 20 minutes stopped for the night. Thermometer 104°.

March 28.—We started this morning early, and in 4 hours arrived at Wád Desakkí [Dhá-s-sáḳiyeh?]. After having killed a lamb which I procured in the village, I observed that our

* Pennisatum dichotomum?—F. S.

camel-drivers, who were Ḥasaníyeh Arabs, ate the tripe and intestines raw; and, upon inquiring if such was the custom of these Arabs, they said that they never eat those delicacies otherwise. From Ed-du'amá to Wád Dhá-s-sáḳiyeh much dhukhn is produced. The soil is light and sandy.

In the neighbourhood of Wád Dhá-s-sáḳiyeh iron ore, yielding 25 per cent. of metal, exists in considerable quantities. It is found within from 3 to 6 feet of the surface; and the metal is extracted by placing the ore in a heap with wood and charcoal, and covering it with sand: the fire is then kindled, and by means of bellows a white heat is produced, and the iron runs off. The inhabitants work the ore at their own expense, and make the hasshahshah, or iron money of El-'Obeïd, and send also presents of the metal to the shipbuilding station at Monḳarah.

The water from the wells of Wád Dhá-s-sáḳíyeh is good. I left the village late in the afternoon, and proceeded on my journey for 4 hours, when I rested for the night. Our distance to-day 8 hours: thermometer 101°.

March 29.—In ½ hour we arrived at Korsí. The soil between it and Wád Dhá-s-sáḳiyeh is sandy, and annually produces a considerable quantity of dhukhn. Korsí is the largest village between the White River and El-'Obeïd, and contains about 500 inhabitants. A Káschif resides here; a sheïkh, a military commandant, and a few troops. A market is held daily, but poorly attended: on Mondays and Thursdays are the best markets, and these are frequented by the Bedowíns and Arabs from the neighbouring villages. The water from the wells is good and abundant. In the afternoon I left Korsí, and after travelling 6 hours and 20 minutes halted in the desert. It is necessary to carry water from Korsí to El-'Obeïd, as there are no wells on the road, the villages being at some distance from the caravan track: thermometer 99°.

March 30.—We continued our march early this morning, and in 3½ hours arrived at El-'Obeïd. The road from Korsí for the first 4 hours is over sand; afterwards the soil is firmer; and, lastly, the sand appears again on approaching El-'Obeïd. Our general direction from El 'Adáyír was S.W., and the distance from Kajebí 180 miles; from Ḳorbáj, a village about 2 miles to the eastward, the descent is very gradual to El-'Obeïd; a little to the N. of the village we passed an isolated hill, about 150 feet high, called Jebel Ḳorbáj.

El-'Obeïd, vulgarly pronounced L'Obeyet, the capital of Kordofán, is situated in a plain: it contains 30,000 inhabitants, having increased 15,000 since 1828. The town is straggling, extending about 2 miles from N. to S., and 1 mile from E. to W. The houses are principally built of dhukhn reeds or straw, in the form of English corn-stacks, and two or three of these houses frequently belong to the same family, and stand within a small enclosure. The house of the governor Moḥammed Bey, of Selím Káschif, of the military commandant, and of the Europeans in the service of the Páshá, with one or two more, are built of sun-burnt bricks.

The inhabitants of Kordofán belong to several tribes. The most numerous, called Gúnjárah, consists of adherents of Sulṭán Faḍl; the second is called Meserbát, and were subject to Háshim, formerly Sultán of Kordofán; the third, El Fúng, belonging to Sulṭán Idrís Ibn 'A'dlám, called also Fúngaráwí, probably originally from Jebel Funjí; and the fourth, Iddellaglí, is a tribe of Donḳolah. The Meserbát is the tribe properly belonging to Kordofán. In choosing a wife a man is not obliged to marry a woman of Kordofán, and *vice versâ.* In many Arab tribes, if a man marries a woman of another tribe and loses his wife, he is not afterwards allowed to marry one of his own tribe. The tribe of Auláḍ Ṣeïd, for example, which inhabits Jebel Ṭór, in the peninsula of Mount Sinaï, does not permit a man who has once married out of his tribe to take a wife from it afterwards: but in Kordofán, if a man of Gúnjárah marries a woman of the Meserbát, and either divorces her or loses her by death, he can marry into the tribe to which he himself belongs, or into any other, as he chooses. Kordofán has been subject to Moḥammed 'Alí Páshá about sixteen years. Before he obtained possession of it, it was a province of Dár-Fúr; and when subdued by the Defterdár Bey was governed by Ṭowáshí Emm-sellam, a black eunuch, who, with the assistance of Ibráhím Idwír, a sulṭán of Dár-Fúr, maintained a contest against the Turks for an hour, when, both he and the sulṭán being killed, his people gave way, and submitted to the government of the viceroy of Egypt.

A market is held every afternoon at 4 o'clock, where provisions of the most ordinary kind are offered for sale, principally the produce of the country. Grease, for anointing the hair and skin, is in the greatest demand; and in April and May, just before the periodical rains, good water fetches a high price: for, with the exception of the well attached to the governor's house, and one or two others in the town, all the water obtained from the public wells is so brackish as to be unpalatable even to the natives of the place. The Páshá's money, Spanish dollars, and English gold, are current in the bázár; but as every article of life is so remarkably cheap, and small change very scarce, the peasantry in Kordofán make an iron money from the ore obtained from the neighbourhood of Wád Dhá-s-sáḳiyeh, and to which they have given the name of Hassháshah. This money resembles a section of a mushroom, is made without any reference to weight, and each piece passes for one párah, forty being equal to one Turkish piastre, which, according to the present rate of exchange, is equal to about $2\frac{1}{2}d.$ sterling. All money transactions are made in reals, an imaginary coin, equal in value to 15 piastres.

The troops stationed in Kordofán were marched annually after the kharíf to Jebel Núbah, for the purpose of capturing slaves from these mountains. These expeditions were called Ghazíyeh,* and when I arrived at El-'Obeïd the troops had just returned with the produce of such an expedition. The handsome women were sold for the harems of the Turks and Arabs; the able-bodied men were placed in the ranks; the decrepit of both sexes, the pregnant females, and young children, were allotted to the soldiers in lieu of money to the amount of a moiety of their arrears. I once witnessed this distribution; and a more heart-rending scene cannot be imagined: for, though these blacks had been seized two or three months, and had been deprived of their liberty, they felt severely the final separation from their friends and families. As the soldiers were many months in arrear, they were obliged to receive the slaves considerably above their value, and sell them again at a great sacrifice; and many were compelled to turn their slaves into money, in order to relieve their immediate embarrassments. A slave, therefore, who had been received by two soldiers in lieu of 300 piastres, was sold in the bázár for little more than half that sum; and many were daily hawked about and disposed of by public auction.

The slaves are of various prices, a child 4 or 5 years old is worth 50 or 60 piastres (10*s.* or 12*s.*); an adult sells from 4*l.* to 6*l.* Beautiful Dár-fúwí girls are in great request; and an unexceptionable looking one will fetch from 15*l.* to 20*l.* Abyssinian women are much in demand, and sell from 10*l.* to 20*l.* a-head. But, for domestic purposes, slaves from Dár-fúr are preferred.

Up to the period of my visiting Kordofán, the bodies of negroes and criminals were never interred, but thrown outside the town to be devoured by birds and beasts of prey. When Moḥammed Bey succeeded Muṣṭafá Bey as governor, one of his first orders was to collect and burn all the human bones in the vicinity of El-'Obeïd, and to direct that slaves, in future, should be interred according to the Moḥammedan rites of sepulture.

But the most distressing circumstance resulting from the slave trade, and one more than any other calculated to excite our sympathy, is the system of emasculation carried on at El-'Obeïd. This

* It is gratifying to add that, owing to representations I made to Dr. Bowring, when he was in Egypt, he succeeded in obtaining a promise from Mohammed 'Alí Páshá to abolish the Ghazíyehs in the Beléd-es-Súdán, to discontinue paying the troops in slaves, and to punish slave-dealing amongst the soldiers. And there seems reason to believe, from what I have since heard, that his orders have been carried into effect.

operation is principally performed by Melik Tamar, brother to the late Sulṭán of Dár-fúr. He belongs to the tribe of Gúnjárah, which tribe is remarkable for having black nails, caused by the natural deposit of the *rete mucosum.* For many years he enjoyed an exclusive monopoly of this brutal practice, but, as his success was great, other persons, finding they could realize a larger profit by making their captives eunuchs than by selling them as ordinary slaves, adopted a similar profession. As Sulṭán Tamar is a man of considerable celebrity, I visited him; I found him infirm, about seventy years of age, with a haggard countenance, of a jet black hue, and having a scanty white beard, which he had dyed red with hinná (Lawsonia inermis). When he spoke he held his hand before his mouth, anxiously concealing his nearly toothless jaws. He said that he first came to Kordofán 36 years ago, and remained there till the Defterdár subdued it, when he returned to Dár-fúr, but, having received a hint that his presence was not agreeable, and fearing assassination, he removed to Sennár, and when the new government was finally arranged returned to El-'Obeïd, where he now enjoys a pension of 600 piastres (6*l.*) a-month from Moḥammed 'Alí Páshá. He emasculates from 100 to 150 slaves annually, and the same number are mutilated by the other operators in Kordofán. Sulṭán Tamar is always in difficulties, and (which is remarkable) he is very charitable, keeps open house, and his consumption of grain amounts to about 500 ardebs (2500 bushels) a-year. He inhabits a curiously-arranged straw-built house, and possesses 200 slaves. His slaves, whenever they bring him water or sherbet, fall upon their knees, present the cup, and do not rise again till they have left their mighty master's presence. The subjects selected for emasculation are boys from 7 to 11 years old. All the organs are removed,* and the operation rarely proves fatal, not more than 5 per cent. There is a military hospital at Kordofán; a powder-magazine enclosed in a fortification of mud walls, upon which are mounted two pieces of cannon; a mosque recently erected; and about a mile to the east of the Governor's house is a large water-tank, made by Rostan Bey when he was governor, on the edge of which he built a summer residence, now in ruins, where he passed much of his time in the society of his women.

The men of El-'Obeïd are well made, spare, and tall. Their features are handsome, and their skin of a dark-brown colour. Their hair is slightly curled, and they generally wear it long and plaited. Their dress consists only of a piece of cotton cloth over the pubis, with a cord round the hips to keep it in its place, or of

---

* The Arabs have a peculiar term for this complete mode of castration: former observers however have found the mortality in such cases to be very great.—F. S.

a cotton shirt and drawers. Some wear the táḳíyeh, and a very few the tarbúsh and turban: many have *hegabart* above the elbow on the right arm; and almost all carry a knife upon the left, and frequently a pair of tweezers are thrust into the sheath of the knife, which the peasants use in removing the thorns of the askanít.

The women are generally very beautiful, and are a shade or two lighter than the men. They wear their hair long, plaited, and loaded with grease. Their dress consists of a piece of cotton cloth round the waist, generally passed over the shoulders; many of them wear silver ornaments round their necks, bracelets, earrings of a tremendous size, ponderous nose-rings, and ornaments for the ankles. Some have a few beads of gold, mixed with glass or other beads, round their necks; and those who cannot afford ornaments of gold or silver wear them of horn, leather, wood, or iron.

The female slaves and girls wear the ráhat or leather fringe round the waist, sometimes ornamented with cornelians and silver balls. The children of both sexes generally go naked till the age of 6 or 8 years.

Throughout the Beled-es-Súdán, but more in Kordofán than in any other part of it which I visited, many of the men, and almost all the women, have three or four perpendicular gashes on each cheek. This disfigurement is considered by them as a great beauty. I observed, also, some of the women cut on the temples, shoulders, fore-arms, breasts, and back, on one and sometimes on all these parts. This operation is performed with a razor, and the parts cut are then rubbed with wheaten flour and water to prevent the edges of the skin from uniting, and to cause the parts to heal by granulation; for the higher the skin is raised after this process, the more beautiful is the effect produced. Both sexes perfume themselves once or twice every month. The women are fond of dancing, and perform on the darabúkah or drum. The men amuse themselves with playing on the zomárah or flute, made of a reed; while both sexes perform on the harebábah, or lyre of five strings.

The diet of these people is simple: their fare consists of dhukhn, a grain which grows well on a sandy soil, made into porridge or bread; bámiyah, a mucilaginous vegetable,* sour milk, and the seed of the askanít, a plant abounding in the desert, but occasioning a violent diarrhœa to those who are unaccustomed to eat it. They also drink arrack,† a spirit distilled from dates,

* Okra of the West Indians: the *Hibiscus esculentus*, "le plus glutineux des légumes," says M. Olivier.—F. S.

† Arakí. The word here used is a Greek word adopted by the Turks.—F. S.

bilbil, and merísah. The women are more frugal in their diet than the men, and rarely eat animal food.

The prevailing diseases are fever, dysentery, and small-pox, the latter being very destructive. The plague is unknown here, and ophthalmia a disease of very rare occurrence. Senna (*Cassia Senna*) is indigenous and very abundant in Kordofán. Round the town there are twenty or twenty-five very large trees, called Tebeldír or Ḥamrá: I have no doubt they are the monkey-bread (*Adansonia digitata*), as they bear a fruit of an oval form, something like the vegetable marrow, and when dried, the natives eat the inner substance, which is white and farinaceous, surrounding irregularly-formed brown seeds. The flavour of this fruit is a pleasant acid. I measured the circumference of one of these trees at the base, and found it to be 48 feet. White ants (termites) abound at El 'Obeïd, and commit great devastation. There are a few dogs which belong to nobody, but not so many as at Kharṭúm and Sennár: hyænas are frequently seen in the neighbourhood. Within a few days' journey from El 'Obeïd the camelopard is found, and Muṣṭafá Bey had procured six the year before I was there, but from want of proper management all but one had died. There is a beautiful variety of the goat at Kordofán: it is not indigenous, but was originally brought from Jebel Núbah, where it is wild, but when taken is easily domesticated: the same species is found in Jebel Fungí, to the south of Sennár.

The exports, with the exception of slaves, are entirely monopolised by the government: they consist of gold, silver, hides, ivory, and gum Arabic. Slaves are allowed to be exported from Kordofán on the payment of a heavy duty.

On the 14th of April Mr. Rusegger, a German mineralogist, employed by Moḥammed 'Alí, arrived at El 'Obeïd on his way to Sháhún to inspect the gold-mines.

Seven miles to the S.E. of El 'Obeïd is the village of Milbess, beautifully situated, and the summer residence of the principal persons in authority. The land here is rich and productive; good water is found in shallow wells, of which the inhabitants avail themselves for artificial irrigation, raising it by manual labour. Half a mile to the W. of Milbess are some hills about 140 feet high, the formation of which is primitive quartz, and from the summit of any of them there is an extensive view of the desert, and of Jebel Daïr,* about 25 miles to the S., a small chain of mountains which, though only one day's journey from El 'Obeïd, is not yet subject to Moḥammed 'Alí Páshá, and is the favourite resort of slaves who have absconded from the capital.

* The Dyre and Tuggala Mountains of Bruce.—F. S.

Six or seven days' journey to the S.W. of El 'Obeïd is Jebel Núbah, whence the Páshá used to carry off slaves annually. The inhabitants of this chain are, generally speaking, handsome, and, if they are not frequently sold from hand to hand, turn out good and faithful domestics. They are dark, though not black, with much less of the negro feature than the Shillúks, Denkáwís, or inhabitants of Dár-fúr. I observed many of the women from Jebel Núbah with perforations in their ears and chin, into which small pieces of wood had been introduced; and the females of Jebel Minmin, an offset of Jebel Núbah, extract the incisors of the lower jaw, a practice adopted by the Shillúks and people of Denkah: the men do not observe the same custom. The Núbáwís salute each other with the two first fingers of the right hand, snapping them twice with those of their friend. The thermometer ranged from 90° to 97° Fahr. at 2 P.M. in the shade during my stay at El' Obeïd.

April 15.—After sixteen days spent at El 'Obeïd I quitted it at half-past 4 P.M. on my return to Korsí, and arrived there at a quarter to 7 A.M., on the 16th, having rested only 1½ hour on the road during the night. Having determined to proceed by the desert of Sakrah, I took a guide from hence to the village of that name. In 2 hours from Korsí I arrived at Umm-ḥájir, and was obliged to stop, as the Hájí was suffering from bruises which he had received in a fall from his dromedary on our leaving Korsí. The people of Umm-ḥájir were kind, willing, and obliging: they gave me, immediately upon my arrival, the best house in the village, and provided me without delay with a kid, fowls, milk, straw, dhukhn-bread, &c.

April 17.—I left Umm-ḥájir at 10 minutes to 5 P.M. In 1 hour we arrived at the village of Rokab Ardam; in 1 hour 30 minutes at Gamrah Hamdasíd; in 50 minutes, after passing El Fásher, we reached Wád Berrí; and in 2 hours 5 minutes from thence we arrived at Umm-bálaghí, where we rested for the night. Here we found comfortable quarters and procured provisions without difficulty as soon as the inhabitants perceived that it was my intention to pay for everything I required. The soil from Korsí to Umm-bálaghí is very light and sandy: it is adapted for dhukhn, and a great deal is grown. After leaving Korsí the best water is found at Umm-ḥájir and Rokab Ardam. In all the other villages the water is brackish, and it is slightly so at Umm-bálaghí: our journey to-day was 5 hours 25 minutes.

April 18.—At 5 P.M. we left Umm-bálaghí and in 1 hour arrived at El Kaú; here iron-ore is found close to the surface of the ground, and is excavated and wrought for the government. Two hours farther we passed El Margler, a populous village for this part of the country, where I observed a schoolmaster teaching twenty or thirty boys by the light of an enormous wood-fire. One hour beyond we arrived at 'Abd-al'ír, where we halted, having found a comfortable straw hut, and without difficulty obtained everything in the way of provisions, &c., I required. Good water is found at all three of these villages: thermometer 96°. Our journey to-day 4 hours: for the last two days we travelled slowly, in consequence of the indisposition of the Ḥájí.

April 19.—At 4 hours and 20 minutes P.M. we left 'Abd-al'ír, and in 1 hour we passed Aúlád Marghah; 2 hours thence we reached Aṣṣereághah; 1 hour beyond El Serreh; and in 3 hours farther we arrived at Umm-ḥájir: its present local governor is Sheïkh Bakít, a man of some importance, inasmuch as he commands and collects the revenue of sixteen villages. He is an excellent fellow, ready, obliging, and attentive. He brought us abundance of provisions, and a little dhukhn, though there was a scarcity in the village, the poorer people being obliged to subsist on bread made from the askanít. Good water is found at all these places with the exception of El Serreh, where it is brackish. Our distance to-day 7 hours.

April 20.—Left Umm-ḥájir at 5 P.M., and in 2 hours arrived at Esh-shaṭíb, and in 1 hour from thence at Almána. Thus far Sheïkh Bakít accompanied my caravan to provide a guide to conduct it to Abú Ghárát. My conductor from Korsí was discharged, and a new arrangement was quickly made with the son of the Sheïkh of the village, to pilot us through the desert of Sakrah. During the delay occasioned by coming to terms, the Sheïkh regaled us with fresh milk, sour-milk, and dhukhn-bread: quitting this place, in 3 hours and 20 minutes we arrived at the village of Sákrah. The Sheïkh had retired to rest, but we soon aroused him, and, after a little grumbling, he provided us with comfortable quarters. Our journey to-day, 6 hours and 20 minutes. The water good throughout—thermometer 102°.

April 21.—We left Sakrah at 5 P.M., and, after proceeding 6 hours and 30 minutes, we halted in the desert. It is necessary to take water from Sakrah, as none drinkable is found between it and Abú Ghárát. Thermometer 104°.

April 22.—At 3 A.M. we were *en route*, and in 5 hours and 30 minutes arrived at Bír el Ḥelbah. This well was dug a year ago by the Jowámát Ash-shenít Arabs, a tribe who inhabit a portion of the desert near Wád Dhá-sákiyeh, and who are under the immediate command of Sheïkh Hayásín of Korsí. These Arabs informed Muṣṭafá Bey of their intention of sinking a well, and said that if they succeeded in finding good water, they would settle here, but, after digging to a very considerable depth, though they found water in great abundance, it was so bitter and brackish as to be totally unfit for use. They therefore gave up all idea of settling at this place. Adjoining the well is a small straw hut built by these Arabs.

Just before we arrived at Bír el Ḥelbah, to the N.W. of the well, and on our left, we passed a small hill named Jebel el Ḥelbah, and to the N.E. of it is a hill somewhat larger, called Jebel el Mináwerát. Each is isolated, and they are the only two hills I observed in this neighbourhood.

At 5 P.M. we resumed our journey, and, after travelling 5½ hours, rested in the desert. Shortly after leaving Bír el Ḥelbah, we met with a great many ostriches and a herd of oryx: the former are found all along this road from Korsí to Abú Ghárát, but especially between the latter place and Sakrah. Our journey to-day was 11 hours. Thermometer 107°.

April 23.—We started again at 4 A.M., and in 5 hours and 45 minutes arrived at Abú Ghárát. I went to my old quarters, the residence of Abdallah the Ḳáïm-maḳám. He was absent: but his household attended to our wants. I observed in the village a number of women dancing their national dance to the music of the darabúkah, singing and clapping of hands, and, upon inquiry, I found that they were showing this demonstration of feeling out of respect to one recently deceased. I had now passed the two routes from Abú Ghárát to Korsí, the one by the desert of Habsliábeh, the other by the desert of Sakrah, and I decidedly give the preference to the latter. In the first place there are abundance of lice which attack the person most furiously on the Habshábeh route—none on the Sakrah. There are more villages and better water on the latter route than on the former. Few caravans pass by the desert of Sakrah, and the inhabitants, from not being plundered by the Turkish and Arab soldiers, are more willing and obliging. The only inconvenience on this route is the greater quantity of askanít, and the annoyance of the thorns on the trees, which in many places overhang the track.

April 24.—At 4 A.M. we left Abú Ghárát, and in 7 hours arrived at the river opposite Monḳárah: we took the road direct between these two places without going round by El 'Adáyir and Túrah. The distance therefore from El 'Obeïd to the White River, opposite Monḳárah, is 67 hours 45 minutes, which, at 2½ miles an hour, gives about 170 miles. The general direction is N.E.

I shall abstain at present from any observations relative to the White River, beyond mentioning that I left Monḳárah on the 25th of April in a boat belonging to the government, arrived at the junction of the Bahr el Abyaḍ with the Baḥr el Azreḳ on the 29th, and on the following morning reached Kharṭúm. During my second visit there were two remarkable hurricanes of wind and sand—the first on the 2nd of May, at 2 P.M., and the other on the 14th of the same month, at 4 in the afternoon. That on the 2nd commenced more gradually than that on the 14th. In both these hurricanes the wind was from the S.E. On the 2nd, after blowing tremendously for twenty minutes, the atmosphere was of a blood-red colour, succeeded by total darkness: this lasted for a quarter of an hour, when it began to be lighter, and, at the end of two hours from the commencement, subsided a little, though the wind blew hard all the afternoon and during the night. The air was charged with fine sand for two or three days afterwards. The thermometer during the hurricane was 102°.

On the 14th, while riding near the river, I saw the second storm approach, and it appeared as if an immense chain of sandstone rocks was driven before the wind. I had just time to reach the residence of Signor Boreani, when it overtook me, and darkness came on instantaneously. This lasted for ten minutes, when the atmosphere assumed a deep-red colour, becoming gradually lighter. It was nearly two hours before the hurricane subsided, and it was followed by a slight shower of rain: on the following day the air was loaded with sand. The thermometer at 2 P.M. on this day was 104°. These winds occur two or three times in every spring, and the inhabitants call them *dohr*. The Arabs assured me that these storms had not visited Khartúm for more than three years.

May 18.—I left Khartúm; on the 22nd arrived at El Metemmeh; and on the day following I crossed over to Shendí. Here I remained until the 2nd of June, and during the interval made an excursion to the temples of Musawwerát, El Auweh Tayyib, and Kenísat el Kerbekán. On the 11th I reached Berber, having stopped a few days at El Bekráuwiyáh, to visit the Pyramids of Baï and El Jelásafrát, to which M. Linant has given the name of Gabina.

The greater portion of the land between Khartúm and Berber is uncultivated, though it might be advantageously used for the production of grain, tobacco, cotton, and indigo. The inhabitants have been compelled to serve as soldiers, or have absconded into the desert, in order to avoid it, and there is great room for colonization. They are in the lowest state of degradation and oppression, and are addicted to lying to an inconceivable extent. They entertain the greatest fear of a person wearing the tarbúsh, or red cap, mistaking every one with this badge for a Turk; and whatever a traveller requires he must get done by compulsion. Their services are not to be obtained by promises of money or by kindness: they have been so often deceived by their superiors, that they expect the same from travellers; and I never, but with one solitary exception, found that the word of a native of the Beled-es-Súdán was to be depended upon.

Berber is a place of some importance. From its size, I should think that it contained 8000 or 9000 inhabitants. The houses are generally built of sun-burnt brick, of a square form; and many are enclosed in a small court-yard. It is the rendezvous of the slave-merchants from Sennár and Kharṭúm, on their way to Cairo by the Great Nubian Desert; and a considerable traffic in slaves takes place there. There is a daily market: its present governor is 'Ábbás Aghá, a very severe disciplinarian. Berber is a place of great resort of the Bishárí Arabs. 'Abbás Aghá has built a mosque, adorned with a lofty minaret. He also built some shops in the bázár, with the profits arising from the rent of which he pays the people employed about the mosque. The thermometer at Berber varied from 100° to 106°.

Before I left Berber I was obliged to provide myself with skins to carry water from hence to Jebel Berkel, as it is not the custom of the country for the camel-drivers to furnish them. Fortunately there are two magazines, one of new and the other of old skins: from the latter I selected those required for my journey.

June 24.—Early in the morning I crossed from Berber to El Wobsh, a village situated on the opposite side of the Nile; and at 6 P.M., on the same day, I set off for Meraweh; and in 3½ hours I stopped for the night.

June 25.—We proceeded at 5 A.M.; and in 1 hour passed some Bedowín tents. The part of the desert which they inhabit is called Abú Kharráz. In 2 hours from this place I rested until the evening, when I resumed my journey, and halted for the night after proceeding 3 hours farther. Our journey to-day 6 hours.

June 26.—We set off at 5 A.M. and in 2 hours 20 minutes arrived at Korrobí. Here is a small natural reservoir in the mountains, which are of a coarse grey granite. We found water left from the last year's rains, sufficient to fill 7 skins, and, though it had a slightly earthy flavour, it was by no means unpalatable. The main feature of the country between the Nile and Korrobí is its flatness. After leaving the river we passed over alternate strata of sand and coarse gravel. At 6 hours' distance from the river we crossed a stratum of coarse grey granite, and afterwards a fine thin slate, containing abundance of mica. Both the granite and slate are furrowed by large veins of primitive quartz. Afterwards sand and broken quartz cover the desert as far as Korrobí, where, as I have already mentioned, there are mountains of coarse grey granite, broken in a rude, wild, fantastic manner; the masses, large and small, having assumed, from some cause or other, round and oval forms. In the afternoon I left Korrobí, and stopped for the night, after travelling 3 hours.

June 27.—We proceeded at 5 A.M., and in 3 hours passed some hills of granite on our right, to the N. of the road, and about 1½ mile from it. One is of a conical form, higher and more pointed than the rest, and is called Naṣbel el Ḥosán; and this hill, our guide informed us, is reckoned by the Arabs to be half-way between the Nile and the Wells of Shimáíl: it may reach 800 feet above the plain. The surface of the desert from Ḳorrobí is generally sand; but after leaving Ḳorrobí we crossed a vein of red granite. Jebel Naṣbel el Ḥosán, as well as the hills around it, is of grey granite; and in many places veins of the same rock appear through the sandy surface. Half an hour beyond Jebel Naṣbel el Ḥosan I rested until the afternoon; when proceeding for 3 hours and 20 minutes farther, we halted in a portion of the desert, where, on our left (½ mile south of the road), is a small granite hill named Jebel Naṣbel el Melḥ.

June 28.—We travelled 7 hours to-day, and rested at the foot of a hill on our left, called Jebel el Medáyet, where we found a solitary Bedowín tent. From its inhabitants we procured some water, which they had brought this morning from Bír Shimáíl; and a great luxury it was, for our own was putrid in the highest degree, and I was suffering from a severe relapse of dysentery, and my thirst was insatiable.

June 29.—We moved on again at 6 A.M., and in 1¼ hour arrived at a steep ascent of 25 feet, which having ascended, we passed over a plain strewed with broken stones, very much resembling in appearance fragments of lava. After passing over this small plain we wound round the S. side of Jebel es-Safrah, the hill being on our right, and then arrived at a spot surrounded by hills, where is situated Bír Shimáíl. The water of these wells is good, free from saline taste, and possessing only a slight earthy flavour. Between Berber and these wells I noticed many gazelles and large coveys of rock-pigeons. The latter abound also in the immediate neighbourhood of the wells. Our journey to-day 3 hours. As I was still suffering severely from dysentery, I determined to stay a couple of days at these wells. Thermometer 100°.

July 1.—At ½ past 4 P.M. I left the wells of Shimáíl, and in 50 minutes we passed on our right what I at first supposed to be an ancient ruined village; but upon inquiry the guide informed me that it was a cemetery of the Manáṣír* Arabs. Upon the spot where each body was deposited a stone wall was built, in form resembling the gable end of a house. Each wall was about 5 feet high in the centre, and 7 feet in length, and constructed of small stones, with some neatness, and not unlike the way in which the stone walls are built as fences to the fields in many parts of Yorkshire. This cemetery is called El 'Arfah.

---

* Plural of Manṣúr.

In $1\frac{1}{2}$ hour from the cemetery we reached Bír Abú-ḳaráët. These wells are situated in a small grove of dóm-trees. I had taken sufficient water from Bír Shimáïl not to require to replenish my skins from these, as the guide informed me that though the water was good it was not always in abundance. I observed, however, that the water was springing freely, and that the depth of each well did not exceed 20 feet. At 40 min. beyond I halted for the night, having travelled 3 hours from Bír Shimáïl.

Our tract to-day was exceedingly rough. First, we passed a plain covered with broken stones, followed by a smooth hard surface. Afterwards, we made an ascent, gradual and tedious, and passed over a second plain covered with broken stones, bearing a strong resemblance to the lava of Vesuvius: in fact, the whole plain appeared of volcanic formation. I noticed in many places the lava in a state of decomposition, presenting a similar appearance to that seen at the foot of Vesuvius, though, from the intense heat of a tropical sun, and the small quantity of rain which falls annually in these latitudes, the decomposition was proceeding very slowly. I could not find any craters, or extinct volcanoes.

The hills on each side of our route are distant from each other about $1\frac{1}{2}$ mile. They are numerous, some isolated and others forming small continuous chains. Some are formed of grey granite of a coarse texture, whilst others appear to be composed of freestone. There is no regularity in the disposition of these formations, each appearing when and where one least expected to find them. Of all the scenery I ever witnessed this was the wildest and most savage. There is no verdure, and rarely a single shrub. The desolation is complete.

July 2.—After 4 h. 45 min. travelling, we arrived at Bír Sarniyah, where we remained for the remainder of the day. Before arriving at the wells, we passed a spot where the Arabs turn up the surface of the desert, and at the depth of from 8 to 12 inches they find a saline earth containing impure muriate of soda in large quantities, and from which they make salt for domestic purposes.

Bír Sarniyah is picturesquely situated in the midst of mimosas and dóms. The latter were particularly fine trees, and the ripe fruit hung in clusters upon the branches. There is only one well. It is large, and its sides substantially built of stone. The water is plentiful, and quite free from any brackish flavour, though it is slightly tainted with a putrid taste, which disappears after being a few hours in the skins.

The guide mentioned that the Manáṣír Arabs, who inhabit this part of the desert, sometimes attack and pillage caravans whilst reposing at the wells. I therefore deemed it prudent to put on a watch during the night: we were not, however, molested.

July 3.—We left Bír Sarniyah at a ¼ past 6 A.M., and shortly afterwards passed a cemetery of the Manásír Arabs. The sun was so hot and oppressive, that, after travelling 2 h. 50 min., we were obliged to stop until the afternoon, when we proceeded for 2 h. 50 min. more. Thermometer 102°. We travelled for 6 h. 30 min. on the following day.

July 5.—We resumed our journey at 20 min. to 5 A.M., and in 2 h. 50 min. arrived at Ed-duḳáyet, a village situated on the Nile. Soon after leaving Bír Sarniyah, the stratum is coarse grey granite, with veins of quartz running from S.W. to N.E., and this structure continues to the Nile.

I was delighted once more to arrive at the river, more especially as passing this desert is very severe travelling, and doubly so to me, suffering as I was from a violent attack of dysentery, which would not give way to any remedy that I adopted, though, for the benefit of travellers in Africa, I may state that I found it universally yield in other cases in the course of a few hours to the internal use of ipecacuanha, in about the dose of 3 or 4 grains every 4 hours. With me this remedy produced no effect, but, on the contrary, aggravated the symptoms, and I consequently tried laudanum, and partially succeeded in controlling it until I arrived at Cairo. From Berber to Ed-duḳáyet scarcely a tree is to be found along the route capable of affording the least shade, if we except those in the vicinity of the wells: the trees in fact are little more than bushes. Fortunately, for the first 7 or 8 days, the sun was slighly obscured by clouds, or the heat would have been insufferable. Of all the journeys I have made, this desert is the most wild, arid, comfortless, and uninteresting; and I should strongly recommend future travellers to take the line from El Metemmeh to Jebel Berkel in preference to that from Berber.

The whole of this desert is thinly peopled: from Berber to Bír Shimáïl it is inhabited by the Hasáníyeh Arabs; from Bír Shimáïl for two-thirds of the way to the river, by Manásír Arabs; and the last third, towards the Nile, bŷ Seïḳíyahs.

At Ed-duḳáyet the placidity of the stream of the Nile is broken by rocks; and here is a slight rapid. At 1 day's journey from hence, higher up the river, is the fourth cataract.

At ½ past 4 P.M. I left Ed-duḳáyet: in 1 hour 30 minutes we passed the village of El Bellal on our right, and in 30 minutes afterwards arrived at the Pyramids of Núrí. In 10 minutes from the Pyramids we reached El Jeráyib, and I repaired to the house of the Sheïkh, where I remained all night.

The land about Ed-duḳáyet is rich, but only a small portion of it is under cultivation. It continues good to El Ḳeráyib. Our journey to-day was 6 hours.

July 6.—At El Ḳeráyib the senna-plant is indigenous, very luxuriant, and in great abundance. If properly cultivated, it might afford a great profit to the grower.

At 5 P.M. I resumed my journey, and in 3 hours 20 minutes arrived at Abdúm, a small village situated opposite Meraweh. The distance from Berber to Ed-duḳáyet was 54 hours 25 mins., or about 137 miles, in a general W.N.W. direction. From Ed-duḳáyet to Abdúm, S.W., 14 miles. The expense of each camel from Berber to Meraweh was 15 piastres [3*s.*].

July 7.—At daybreak I crossed the Nile to Meraweh, which may have a population of about 700. I remained here until the evening of the 14th, when, having procured a boat, I returned to Donkolah, where I arrived in the morning of the 22nd. On the 25th I took my departure from New Donḳolah, and, after a detention for camels of a few days at Ḥafír, I arrived at Wádí Ḥalfah on the 16th of August, 1837.

To the foregoing paper I beg to append a few observations on the practicability of exploring the sources of the White Nile with a small steam-vessel. If an expedition left Caïro in the month of July in a steamer with a draught of water not exceeding 2 feet, it might, with a little care, pass all the cataracts between Caïro and Khartúm. At high Nile the cataract of Essuan disappears and becomes a rapid. The second cataract is a succession of rapids from Wádí Halfah to the third cataract at Ḥanneḳ, which latter is most improperly called a cataract, since even at low Nile there is hardly any fall, and not so much as to attract notice. The fourth and fifth cataracts I have not seen; but, from inquiries I made, I learnt that they would form no obstacle; and the sixth is passed without the least difficulty.

In passing all the cataracts it would be judicious to have the steamer preceded by a pilot-boat, sounding as it proceeded, to prevent any accident to the former. I mentioned the subject to Mr. Perring, an eminent civil engineer in the employment of Moḥammed 'Alí Páshá, and from his knowledge of the Nile and its peculiarities, he very kindly made a drawing of a steamer which he calculated to be adapted for a steam expedition. He recommended a boat of light sheet-iron, 70 feet long on the water-line, 16 feet beam, and 8 feet deep, including the keel, and which would not, with all requisite stores on board, draw more than 2 feet of water. The power to be two twelve-horse oscillating high-pressure engines, and the fuel to be used wood or charcoal.

If a steam expedition left Caïro in July, it would have time to get to Berber by September, where it should remain until the termination of the tropical rains, which generally takes place there during that month. An express might be sent from thence to Khartúm to order provisions to be prepared and ready against the arrival of the expedition, so that no delay might be occasioned. It would then proceed up the Bahr el Abyad, and I think six months would be ample time to make surveys of both branches of the river. The expedition might then return to Berber, and when the Nile had risen sufficiently high to pass the cataracts in the following year, it should proceed immediately to Cairo.

The expedition might be composed of the following persons:—

Commander.	4 Artillerymen.	
Naval Officer, as Surveyor.	4 Seamen.	
Surgeon.	Dragoman or Interpreter.	Arabs.
Naturalist and Geologist.	Cook.	Arabs.
2 Engineers.*	4 Sailors.	Arabs.
Serjeant of Artillery.		

Just before my departure from Khartúm, I inquired of Khúrshíd Páshá if he thought it possible to send an expedition with safety to endeavour to discover the source of the White Nile. His ideas of navigation were extremely limited; and being unacquainted with the powers of steam, was unable to give an opinion as to the utility of steam-vessels in such an expedition. He said that he had been 21 days above Al-leïs, on the Bahr Abyad, in the boats of the country, and took with him 700 soldiers. He reached Denkah, having passed the country of the Shillúks. He had several times been a considerable distance above Al-leïs. The expeditions were sometimes attacked at night, but never during the day; and he found it necessary to moor the boats at night off the islands, in preference to the banks of the river. The Shillúks were armed with spears and shields, being ignorant of the use of other weapons of war. The pashá was able to procure dhurrah occasionally from the natives; though he said he should recommend an exploring expedition to take biscuits and all its provisions from Khartúm, and provide sufficient for its return. He thought it would be necessary to take 400 soldiers. He mentioned that the river divided a considerable distance from Khartúm, but he was not prepared to say which was the direction of the White Nile, properly so called. He was of opinion that an expedition could not go and return to Khartúm in the boats of the country in less than twelve months; but when I informed him that a steamer went twice or thrice as fast as a country-boat, against the stream and against the wind, he thought that it might be accomplished in six. He said there were a few rocks above Al-leïs. Between Wád Shelláï and Khartúm the river is clear of these incumbrances to navigation.

---

* Who would put the vessel together in Egypt.

The advantages of an expedition by steam for this survey must be obvious. First, the tribes to be passed through are hostile, not only to Turks and Europeans, but frequently to each other; and a sailing expedition would require a large number of troops for protection, a steam expedition very few, because it would have no difficulty in moving quickly away in case of emergency. Secondly, if troops were taken, provisions must also be taken for them, and this would greatly increase the size and expense of the expedition. Thirdly, a sailing expedition would require more time for making the survey—in all probability a year or more; and, if so, would be obliged to remain in the country during the rainy season, which is so hurtful to Europeans not acclimatised: whereas, in a steamer, all might be accomplished in six or seven months from Berber, and twelve or fourteen months from Cairo.

The probable expense of such an expedition would not exceed 5000*l.*, and if assisted by government with men and stores, considerably less; doubtless, also, volunteers would be found who would gladly serve in a cause which must excite the greatest interest in all geographers.

Annexed is a brief vocabulary of the inhabitants of Jebel Nubah, which is nearly the same as the Koldagí (Rüppell, *Nubien*, p. 372):—

Man	Iddli	Goat	Ogninin	Make haste	Murko do sheï
Woman	Iddoniu	Ostrich	Tigambin	Come here	Tweï
Boy	Nirttando	Pig	Kigan	Go away	Ishur
Girl	Tergo	Rat	Komanin	Yes	Oliuko
Father	Annuggan	Flesh	Kwaji	One	Ber
Mother	Annanneng	Knife	Kwatwar	Two	Ora
Brother	Anningtang	Milk	Idju	Three	Toju
King	Kuju	Porridge	Kalju	Four	Kinju
Head	Undiu	Dhurrah	Windin	Five	Tishu
Hair	Tilgin	Water	Otú	Six	Kwasú
Mouth	Aljo	Rain	Harikojú	Seven	Kwalut
Hand	Osigi	Cold	Kedí	Eight	Iddu
Leg	Koórdo	Stone	Kwakandin	Nine	Weddu
Skin	Dortonin	Black	Wurrindin	Ten	Bore
Ass	Ondonin	White	Horinnin	Eleven	Bore-ber-ku
Camel	Calenin	Red	Kelindin	Twelve	Bore-ora-ku
Cow	Tí	Wood	Oronin	Thirteen	Bore-toju-ku
Dog	Boldin	Cannon	Motting	&c.	&c.
Horse	Komjin	As-salám Aleïkom	Kendi Ar-krá	Twenty	Tarbu
Fowl	Kokondin				

[Mr. Holroyd has also communicated to the Society a list of seventy-seven positions of places on the Nile, between Philæ and Sennar, from observations by M. Linant. Many of these have been used in the construction of the annexed map; but as some of the others differ considerably in longitude from the determinations by Dr. Rüppell, their publication is delayed till we can hear from M. Linant on what *data* his positions depend.]

VIII.—*On the Ethnography of High Asia.* By JAMES COWLES PRICHARD, M.D., F.R.S.

SINCE the time of Visdelou and De Guignes, the sanguine hope, once entertained, of illustrating the ancient history of Northern Asia, from resources hidden in the Chinese language and in the archives of the Celestial Empire, seems to have been gradually abandoned. Those writers traced in Chinese books obscure notices of an infinity of barbarous tribes, whose names—and it often happened that their names only were recorded—disguised by a Chinese orthography, could not be recognised as having belonged to any people known to the learned of Europe. Attempts made to identify the races, of whom more ample accounts were delivered in the historical works of the Chinese, with nations known to western geographers and historians, were particularly unsuccessful. They were founded chiefly on some slight resemblances of names, or on accidental synchronisms in the accounts of migratory movements, or on passages of history too imperfectly recorded to admit of a comparison leading to results. It was thus that De Guignes was led into the mistake of identifying the Hiong-nú of the Chinese traditions with the Huns, so well known in the history of the West. That this identification was erroneous, was less apparent to most readers than the circumstance, more easily perceived, that it rested on no shadow of proof, and it was generally inferred that no data were to be found, in the voluminous works which De Guignes examined, that might have rendered possible a more accurate research. But since Abel-Remusat and Julius von Klaproth applied themselves to the investigation, that unfavourable opinion has been gradually changed; new lights have been struck out which are likely to elucidate dark passages in the history of those great nations who performed so conspicuous a part on the theatre of human affairs during the middle ages, and some reason appears for looking to the same quarter with expectation of further success. Klaproth and Abel-Remusat had not access to the oldest and most authentic documents of Chinese history. They have principally consulted abstracts and compilations of a later era; and some important documents have come into the possession of Europeans since the death of those distinguished writers, from which we may expect further contributions to the ethnography of Northern Asia.*

* It seems that Abel-Remusat and Klaproth had never consulted the original annals of the elder Han, the celebrated dynasty who ruled over China from B. C. 163 to 196 after Christ, and in whose time the principal wars between the Hiong-nú and China were carried on. They were only acquainted with the contents of this important work through the great historical cyclopædia of Ma-tuanlin, or the Wen-hian-thoung-khao, compiled A. D. 1321, and in other comparatively later documents. The original annals of the Han have been translated from the Chinese

In the mean time some knowledge has been obtained which is fitted to promote this investigation. It is well known that the high region of Central Asia, reaching through the whole longitude of that continent from the Euxine to the Sea of Japan, is occupied by tribes chiefly nomadic, belonging to three great races of men. These are the Turkish, or the improperly called Tartar race, whose chief country is between the Caspian and the Blue Lake, or Koko-Nor, viz., Western and Eastern Turkestan; the Mongolian race, somewhat farther towards the E. and N.; and the Tungusian or Mandschurian race in Daouria, and what is called Mantschu-Tartary. Since the races of mankind have been distributed with reference to their physical organisation and especially to the form of the skull, it has been a prevalent opinion that the two latter of these races, who nearly resemble each other in the shapes of their heads, belong to one great stock, which is termed Mongolian, and is supposed to have had its origin and abode for many ages in the remote East, and probably beyond Lake Baikal. The Turkish race, on the other hand, is set down without hesitation as a Caucasian stock, or a people akin to the western nations, and originating in the same region of the world as the inhabitants of Europe. The Turks of the race of Osmanli, subjects of the Sultan, and the Tartars in European Russia and the Siberian towns, have the form of the head which is termed Caucasian, and this is generally supposed to be the primitive type of the Turkish race. The great nomadic nations of this race, the Nogays, Kirghises, Turkomans, as well as the remote offsets of the same stock in the distant parts of Asia, as the Jakutes, are known to have a different organisation, approaching nearly to the Mongolian and Tungusian character. It is common to refer this deviation from the form assumed to be the original type of the race to mixture with the Mongolians, whose intermarriages are supposed to have modified the true Turkish form, and to have given rise to the comparative ugliness of the nomadic Turks, according to the European idea of beauty. The fact that many of the nomadic races speak pure Turkish dialects, and display few or no traces of intercourse with Mongolians, must be allowed to be an objection of some weight against this assumed intermixture in stock. But this difficulty has been overlooked. Now, it is an essential part of this hypothesis that the local position of the original Turks was in the same region with that of other Caucasian tribes, and far remote from the cradle of the Mongolian race. All the nations of the West have nearly one type; those of the north-eastern parts of Asia display another

by Father Hyacinth, and published at St. Petersburg in 1829. A part of this version has been translated from the Russian by Dr. Schott, and inserted by Ritter in his excellent Erdkunde von Asien. Th. 5.

form, peculiar from immemorial times to all the various races beyond the Altai. As there is not one instance of a nation that bore originally the Mongolian type originating from or having their abiding place, from the first ages of the world, in the western parts, or amidst the group of Caucasian nations, so it is not to be expected that one particular race from the Mongolian centre should be found to bear, as its genuine and original character, the Caucasian type. Here the historical information derived from Chinese sources comes to our aid.

That the Turkish nations are descended from the Hiong-nú, whose early history is preserved in Chinese records, has been clearly established by Abel-Remusat and Klaproth. The Chinese records are, in the opinion of the former of these writers, fully worthy of confidence, and sources of authentic history as far back as the accession of the dynasty of Han. Klaproth deduces from them a series of events much more remote in the history of the Hiong-nú. According to that learned writer, the Turkish race, soon after the great deluge, came down from the lofty and now perpetually snow-clad mountains of Tang-nú and the Great Altai, and soon spread themselves towards the S.W. and S.E. It seems that in very early times they took possession of the country on the southern declivity of the highest steppes of Mongolia, to the northwards of the Chinese provinces of Shan-si and Shen-si, and particularly of the region of In-shan, where the mountain of that name, near the northward bending of the Hoang-ho, forms the continuation of the second great hill-system of Central Asia, that of the Thian-shan, or the Mountains of Heaven. This people were, according to Klaproth, styled Hiun-yue, under the dynasty of Shang, which reigned over China from 1766 to 1134 B.C.; under the Tsin and Han, from 256 B.C. to A.D. 263 they obtained the name of Hiong-nú. The time of their greatest power, when they were formidable enemies of the Chinese, and waged frequent and bloody wars with the generals of the Celestial Empire, was before the middle of the second century of our era. Their power was then broken by various accidents, by severe famines, by internal dissensions; the consequence of which was a division of their race into the northern and southern Hiong-nú. The southern tribes allied themselves to the Chinese, and by these the northern hordes were expelled from the ancient domains of the Hiong-nú race, situated between the upper Amur, the Selinga, and the mountains of Altai. The dispersion of the northern Hiong-nú is supposed by Klaproth to have given occasion to the first great movement among the nomadic nations of Asia towards the west.

The enemies of the Hiong-nú increased in power in subsequent times; and, in the first quarter of the third century, the remains of that celebrated people were overcome and finally expelled from the country which the southern tribes had continued to possess between the desert of Gobi and the northern boundary of China. This is the date of the dispersion and of the wandering march of a great part of the Hiong-nú or Turkish race towards the western parts of Tartary, and of the final occupation of the desert of Gobi and the northern provinces of the Chinese empire, by the tribes who have since possessed the country, and who over China itself have raised several imperial dynasties.

After the destruction of the empire of the Hiong-nú, the Turkish race, aboriginal in the region of Asia which lies between the Amur and the Hoang-ho, abandoned the vast steppes which border on the desert of Gobi and reach to In-shan and the north of China, and dispersed themselves over various parts of Northern Asia. The main body of them found a new country, which became their second home or permanent abode, in the high plains now included in Chinese Turkestan, viz., in the mountainous region of Thian-shan, around Turfan, and reaching eastward to Chamil or Hami, and westward to Karaschar, and northward to Uran-tschi or Bischbalig. The Chinese annals in subsequent ages contain accounts of several barbaric dynasties, founded by tribes of the same race, who held a temporary and limited dominion in the countries farther westward, whither the Hiong-nú had retired. One of them was the empire of the Thu-khiue: a third dynasty of much later origin was that of the Hoei-hou.

It was in this region that the Thu-khiue became known to the Byzantines and to Europe in general, under the name of Turks. An embassy was sent by the younger Justinian to the court of the sovereign of the Turks, in 596, who reigned in Ektag, or the Golden mountain, over the greater part of Central Asia. It was on this occasion that the people of Constantinople first heard the name of their future conquerors. It seems that the Turks had at this period conquered the greater part of the country between Mount Altai and the eastern shores of the Caspian.

In the middle of the eighth century, according to Klaproth, the empire of the Thu-khiue was destroyed by another Turkish people, likewise descended from the Hiong-nú, who came from the countries situated to the southward of lake Baikal. These people were called Hoei-hou, and by some writers Hakas: they ruled for 100 years over the Turks of Altai, but were partly extirpated and partly expelled by the Chinese. A remnant of the Hoei settled in Tangut, to the northward of the Koko-Nor or Blue Lake. At length, in 1257, the Hoei were conquered by the overwhelming armies commanded by Tschengis and his Mongoles. A part of them retired farther westward and settled in the towns which are to the southward of the chain called the Mountains of

Heaven, where they became amalgamated with the Ouigours, the earlier inhabitants of the same districts, whose language and origin were the same as their own.

I have mentioned the Hoei-hou first, in order to distinguish them the more clearly from the Ouigours, with whom they are frequently confounded. The Chinese term the latter people Wei-ou-eul (ŭy-ŭ-ŭl), which answers in Chinese orthography to Ou-ig-ŭr. The Ouigours are the ancient inhabitants of the plains of Chinese Turkestan, where they had dwelt for many centuries before they were conquered by the nomadic people of the same race, termed Hoei-hou. They came in ancient times from a high country to the N.E. of the wilderness of Gobi, near the sources of the Orghon and the Selinga, where they were followed by the Mongoles, who occupied their former abodes, to the plains of Turfan and Chamil or Hami, and between Lake Lop and the river Ili. Here at an unknown period they laid aside their nomadic habits and became agriculturists and the inhabitants of towns, among which were those of Turfan, Chamil, Aksou, Kashgar. To the northward of these places was the more celebrated state of Khotan, which was for centuries the principal town of High Turkestan, and the history of which Abel-Remusat has elucidated from Chinese sources. It was in the vicinity of Khotan or Iu-thian that the celebrated traveller Lao-tseu is said to have erected a temple in honour of Buddha in the sixth century before the Christian era. Khotan became afterwards the metropolis of Buddhism or the religion of Fo, in Tartary, and the inhabitants of Central Asia had adopted that religion, and with it a certain degree of civilization, many centuries before it reached the Mongoles or the Mandschures. According to the historians of the Wei, the religion of Fo was widely spread in these countries in the fourth century B.C., and they were overspread with temples, towns, and monasteries of persons of both sexes, devoted to the monasticism of the Lamaite superstition. In times long anterior at least to the era of Islam, Khotan was the emporium of trade between China, Persia, and India, and it was the medium by which the religious doctrines brought into it from India, through Kashmir, were disseminated over China and the whole N.E. of Asia. It retained its independence till the extension of the Mongolian power, and subsequently, with the rest of Eastern Turkestan, came under the dominion of China. Khotan has been visited by two Europeans: by Marco Polo about 1280 A.D., and by Benedict Goez, the discoverer of China, in 1604. The former found the people already Mohammedans, and their towns abounding to superfluity with all the supports of human life—with cotton, flax, corn, and wine: the inhabitants were devoted to agriculture and manufactures, but were bad soldiers.

We thus discover the patriarchal stem of the great Turkish race at that era when the light of history begins to dawn upon the East, still planted as one member in a group of nations, isolated in a remote region of Asia, on the borders of China, or between them and the lofty desert of Gobi which sends its streams on one side to water the Chinese empire, and on the other to the Frozen Ocean. These nations are far removed from the centre of the western or Caucasian race, whose physical type is so strongly contrasted with the Mongolian. Without assuming that there is a close relation between the physical organisation of human races and the regions of which they have been the immemorial inhabitants, it yet seems extremely improbable that a stock so situated was originally of the Caucasian type. The question is one of more interest in regard to the physical history of our species than it at first appears to be. It is fitted to excite us to further inquiry into the relation between the Turkish race and the other nations, who by their local positions belonged originally to the same group. Philological researches are alone calculated to throw satisfactory light upon this subject. A comparison of the language of the Turkish, Mongolian, and Tungusian races is not, as may be supposed, altogether new or unattempted. It has suggested itself to several writers since the time when Adelung and Vater first laid open and marked out the great field of philological research; but Abel-Remusat, until very lately, was the only writer who had entered far into this subject. The philology and literature of the Turkish nations was in particular his investigation. By exploring the history of the Ouigours or eastern Turks, and of their curious language, the first of the Scythian dialects that was reduced to writing, he discovered an important link in the chain of these languages which connects the western Turkish with the idioms of the Mongols and Mandshurians. Some later writers have followed up this investigation; among them the most distinguished is Dr. Schott, who has seen further into the structure and affinities of the Scythian languages than any former philologist. As this inquiry affords the only probable way of discovering what original relations existed between the three great races of High Asia, I shall endeavour to collect and lay before the reader, in as short a compass as possible, the results of all the inquiries yet instituted.*

Adelung considered the Turks, Mongolians, and Tungusians, to be three distinct races of men, each having a peculiar and original language; and he supposed these languages to be quite uncon-

* See Adelung, Mithridates Th. 1.; Scherer in Fried. Adelung's Nachträge zu dem ersten Theil des Mithridates; Abel-Remusat, Recherches sur les Langues Tartares; Julius von Klaproth, Asia Polyglotta; Dr. Schott, Versuch über den Tatarischen Sprachen; F. Ritter von Xylander, Das Sprachgeschlecht der Titanen; Gyamathi, Affinitas Linguæ Hungaricæ, &c.

nected with each other, and with all other human idioms. It is so much the more remarkable, in his opinion, that the Tungusian, though confined to the eastern extreme of the ancient continent, contains some words common to it with languages spoken in Europe, such are the following:—

TUNGUSIAN.	EUROPEAN.
Sengui	Sanguis (Latin).
Ura (hinder-part)	οὐρά (Greek).
Tschop (top of a hill)	Zopf, top, tuft (Germ. Eng.)
Non (virgin)	Nonne, nun (German).
Heren (master)	Herr, Herrn (German).
Kiesun (to talk)	Kosen (German).
Hife (oats)	Hafer (German).
Farsche	Part.
Fialhou	Foul.
Furu	Fury.
Fourdan (a way)	Fordh (Welsh).
Latu	Late.
Malu	Malleus.
Morin (a horse)	Mähre (German).

In the time of Adelung there were no accessible sources of information respecting these languages that were sufficient for enabling him to form a correct opinion as to their affinity or diversity. Of late years materials have been acquired which lessen this deficiency, and an opinion has gained ground which is opposed to that maintained by Adelung. The first writer who appears to have been strongly impressed by the analogy which really exists between the different idioms of Tartary was Scherer, a librarian at Münich. He observed that a comparison of corresponding sentences in these languages indicates a striking resemblance both in the structure of phrases and in particular words. Scherer's remarks are founded on certain passages in versions of the Lord's Prayer in Turkish or Tartar dialects, in the Mandschu dialect of the Tungusian language, and in the Kalmuk dialect of the Mongolian, and they likewise contain one or two references to an extract from a Kalmuk romance in Benj. Bergmann's Nomadische Streifereyen. The following are Scherer's specimens. Octorgai-du (Kalmuk), i. e. *Heaven in.* In Turkish the same construction, *Tangri-de.*

Tani neretani, Kalm. i. e. *vestri nomen vestrum.* Turkish, Senin adin, i. e. *tui nomen tuum.*

Abga-de thege meni ama—Mandschu. Compare Gock-de degen benin atam—Turk. i. e. *Heaven-in dwelling my Father.*

The expressions, "Bayassuk-sani dula," K., i. e. "*the rejoicing for,*" in German, "*Zu erfreuen um,*" and "oengoeroel-duktu adali," i. e. *the forgiving like,* or "*dem Vergeben gleich,*" also

Kalmuk phrases, and "eget-schi," *evil-from,* in Mandschu, are constructions completely analogous to those of the Tartar or Turkish language.

Scherer subjoins a collection of words which strikingly resemble each other. In this point of view we shall hereafter have occasion to compare the languages of the three nations. He also first pointed out the analogy which is to be recognised in the elements and composition of words. These elements, especially the guttural and nasal consonants, are strikingly alike in the Mongolian, Mandschu, and Turkish. This correspondence is displayed in the fact that the same alphabet has been found well adapted to all these idioms. The Ouigours, as it is well known, were the first Turkish nation who learnt the art of writing. They were taught the use of letters by Nestorian or other Syrian missionaries, and by them the art was communicated to the Mongoles and the Mandschures. These Ouigourian characters, says M. Abel-Remusat, are found to express the vocables of the Mongolian, Mandschurian, and Turkish languages, just as texts drawn from the Neskhi, Cufic, or Mauretanic, are spelled with nearly equal facility in the ordinary Arabic characters.

Scherer also noticed in the structure of Mandschu and Mongolian words, and in the arrangement of the elements of which they are composed, that remarkable peculiarity which Viguier pointed out in his Turkish grammar as a characteristic of some Turkish dialects, and which he termed the Quadruple Harmony of Vowels. It consists in the rule that a given vowel occurring in one syllable of a word, or in the root, requires an analogous vowel—that is, a vowel belonging to the same set, of which sets there are in the Turkish four—in the following syllables of the same word, or in the particles appended to it, which therefore alter their vowels accordingly.

This law in the formation of words constitutes a very remarkable feature of analogy in the languages of Tartary or of Central Asia. The rule which exacts the harmony of vowels, as above described, pervades the original component materials of these dialects, and it therefore seems necessarily to carry us back for the period of its origination to the age of their first developement. The languages in which this law prevails are not only the Turkish, Mongolian, and Mandschu, but likewise some others, all of which have been observed to display, in various particulars, marks of relationship more or less decided with the same class of human idioms. The Hungarian language, which belongs to the class of the Finnish or Uralian dialects, displays the influence of this same principle of formation in the most extensive degree. In that language, the vowels of the primitive word or root have a predominant influence over the vowels of the adjunct or servile syllables. In the other languages above mentioned, the vowels of the suffixes, or of the additional syllables which are appended to words, and perform the office assigned in other languages to inflexions or the terminations of case and number, take different vowels, according to the vowels of the word to which they are appended. Thus, *sza* and *ta*, which are signs of the plural in Mandschu, become *sze* and *te*, when they are suffixed to words containing *e*, or a vowel analogous to it. An attention to this harmony of vowels is likewise perceptible in the entire structure of polysyllabic words; such in Mandschu are *sarapa, angara, ele, eme.* Words similar to αγκυρα, χαριζομαι, ελωρ, would not be tolerated in this language.

Another peculiarity prevails through the formation of words in the Turkish and Mongolian languages, in which certain consonants can only be pronounced in juxtaposition with certain vowels: some consonants require to be joined with *a, o, u;* others admit into connection with them only *e*, or the analogous diphthongs *oe* and *ue*.

These analogies in the structure of sentences, and still more such as are found in the composition of words themselves, are very remarkable. It seems difficult to account for them satisfactorily, either by referring them to accidental coincidence, or on the supposition that peculiarities so deeply inlaid in the component materials of languages can have been acquired or adopted through the imitation of foreign idioms; and this leads us to inquire whether confirmatory proofs can be discovered of a common origin, either in the grammatical framework of these languages or in examining the primitive words of which they are composed.

On comparing the Mongolian, the Mandschu, and Turkish languages, in relation to their grammatical structure, a series of very remarkable analogies is discovered. The resemblance of the Mongolian and the Mandschu is much closer than between either of them and the western dialects of the Turkish language. These dialects, especially that of the Osmanli, have been subjected to a foreign influence and culture, the result of intercourse with Persians and Arabians, and the introduction of Mohammedan literature among the Turks. But in the most essential points in which the western Turks differ in the grammatical forms of their language from the more remote nations of Turan, the eastern Turks or the Ouigours, tribes of the same race who dwell within the Chinese empire, and have never emigrated, and have therefore much less associated with people foreign to their race, actually differ from the Osmanli and approach to the Mongolians and Mandschures. The Ouigours themselves have indeed cultivated a peculiar literature from a remote period, and it is therefore likely that their idiom should be found more improved, both by the development of its native resources, and by embellishments from foreign intercourse, than the Mongolian or Mandschu.

Yet it retains much of its apparently original simplicity, and certainly some very remarkable traits which are common to it and to those eastern languages. Some of these characteristics are also discovered in the idioms of the Finnish and Hungarian nations.

The principal features of this grammatical resemblance may be comprised under the following heads :—

1. Words have in these languages no inflexions which can be so termed in the strictest sense; they admit no formative prefixes, allow no modification in the constituent elements of roots, nor any change, generally speaking, in the endings; they express the relations of nouns only by suffixed particles, of which they have a sufficient variety; the modifications of meaning in verbs are denoted likewise by suffixes; all these are joined for the most part to the unaltered root; to it they rather become adherent than are really compounded with it. There is a juxtaposition or aggregation of such auxiliary words, and no real cohesion.

Thus the plural number of nouns is marked by additional particles which do not form a part of the words, and may sometimes be written separately. Separate words, indicative of plurality or multitude, are added in the Mongolian and Mandschu, as also in Chinese. Among the separate words indicative of plurality one is common, as Dr. Schott has observed, to the three principal languages of Turan; *chamuk*, in Mongolian, is *qamuk*, or *qamu*, in Turkish; in Mandschu, *gemu*. Compare ὁμου and γεμει.

It is a peculiarity of the Mandschu that the only nouns which have plurals are significant of things which have life; all other substantives are indefinite as to number. In the Ouigour dialect of the Turkish, the particle which in other dialects denotes the plural is never appended to nouns, which are therefore found in the same indefinite state; yet this particle exists in the Ouigour language, and is used for forming a plural in pronouns.

The derivation of these pluralising particles is unknown; but it is apparent, as Dr. Schott has observed, that those used in the different languages are of cognate origin. The Mandschu and Mongolian have only what may be termed a dialectic difference, and even the Turkish and Finnish are plainly allied. Thus the Mandschu plural particles *sza, sze, szi, ta, te, ri,* are analogous to the Mongolian *sz, d, od, nar,* and the Mongolian *nar* to the *lar,* or *ler,* which is the Turkish plural ending, since *n* and *l* are in these languages interchangeable consonants, as it has been abundantly proved by Dr. Schott. In the Finnish dialects, properly so termed, the plural is formed by adding *t,* which in the Laponic, as well as in the Magyar, is replaced by *ch* or *k.*

Cases are likewise formed exclusively in these languages by appended syllables, or suffix particles, if they may be so termed.

These particles display unequivocal marks of a common derivation in the several languages. Thus, the signs of the genitive case, or of possession, are as follows:—

Mandschu—ni, i. Turkish—ning, üñ. Mongol—ũn, ũ, yin.

The sign of the ablative is in Mandschu the particle ZI, that is, TSCHI, postfixed; in Mongolian, ETZE: these are, as Schott remarks, nearly related among themselves, and with the Finnish ST or STA. The Turkish has DEN, a form nearly parallel to the Greek θεν, and used precisely in the same manner. The Turkish *den* is a modification of the locative and dative particle *de*—compare θί. It is remarkable that *da, de,* or *du,* is the particle answering to *in,* in the Mandschu, Mongolian, and Turkish.

Abel-Remusat has remarked that the termination *tschi,* after a verb, becomes the formative of the agent noun in each of the three great Tartar languages. Another characteristic feature in all the High Asiatic languages, including the Mandschu, the Mongolian, and Tartar dialects, and the Finnish and Hungarian, as well as the Chinese, is, that the nouns, both substantive and adjective, have no gender; they are in form neither masculine nor feminine. When, in speaking of living creatures, it is necessary to distinguish sexes, this object is attained by appending words meaning male or female. These words are, in Mandschu, *khakha* and *khekhe;* in Mongolian, *ere* and *eme;* the Mongolian *ere* has been compared with the old Scythian word for *man,* *οιορ,* and with the Greek *ἄρρην.* In the Turkish *er* and *erkhek,* meaning male, correspond to the Mongolian *ere,* and *kas* or *kys* to *eme* or *khekhe.*

2. A want of inflexion in the forms of words always renders it necessary to observe strict rules as to their collocation, in order that their mutual dependencies and relations may not be mistaken. But in the Turanian languages very peculiar laws prevail as to the precise order of words and their arrangement in sentences. Dr. Schott observes that every word which influences the meaning of another, and denotes any circumstance, or defines any mode in its relations, must always be prefixed to it. Thus, adjectives uniformly precede nouns, adverbs verbs, the possessive pronoun the thing possessed, and clauses dependent on a relative precede the relative; nouns affected by a preposition always precede the preposition or the word which has the force of a preposition. It is difficult to bring this last fact under the meaning of the rule above stated. The circumstance, however, that prepositions, or rather particles used instead of prepositions, always follow nouns, is one of the most striking characters of the Turanian languages. In all other instances, prepositions are appended or suffixed in a manner precisely analogous to that in which they appear as substitutes for the terminations of cases, and it is indeed somewhat

difficult to point out a distinction between these modes of using servile or subsidiary particles.

In the Turkish and Finnish dialects, although the same laws prevail in general, yet some occasional deviation takes place from the rigid law of collocation established in the Scythian languages. This appears chiefly in the Osmanli and the Magyar, and is attributed to the influence of Persian and Arabic on the one, and of the European languages on the other. In general, however, the Turkish dialects have in this particular the same habitude which is characteristic of the High Asiatic languages:—

"Un trait commun à tous les dialects Turcs," says M. Abel-Remusat, "sans excepter le Turc oriental (*i. e.* the Ouigour), c'est l'inversion perpetuelle si contraire à nos habitudes, il semble même qu'on peut dire, si contraire à la nature. Ici, comme en Mandschou et en Mongol, le mot qui régit se place toujours après celui qui est régi, et le verbe principal, auquel viennent ressortir directement ou indirectement tous les mots d'une phrase, doit toujours être mis à la fin. Les mots composés, les noms en rapport, les particules, les phrases incidentals, tout est soumis à la même règle; et si dans les textes Ouigours on trouve des cas où elle semble negligée, on s'apperçoit aisément que ces exceptions sont l'effet immédiat et palpable d'une influence étrangère."

3. The preceding observations relate chiefly, though not exclusively, to nouns, and the method of collocation which the peculiarity of their structure makes necessary. The following remarks refer principally to verbs:—

Circumstances or modes in action which other languages express by means of adverbs, by separate pronouns, or by auxiliary verbs dependant on the principal verb, are expressed in the Scythian languages by means of particles or particular syllables brought into immediate connexion with the verbal root, and serving to denote all such modifications in its meaning.

With one single exception, which is that of the Osmanli Turkish, a mixed dialect disguised by a peculiar culture under the influence of a foreign literature, all the languages of Eastern Asia display the greatest simplicity in the formation of verbs. The root of the verb is always the imperative. By the addition of particles to the imperative, all the modifications of which these verbs are susceptible are induced. It has even been proposed by some writers to characterise all these languages by a term derived from this circumstance. The Osmanli, as I have observed, forms an exception to the simplicity or poverty of the other languages comprehended in the same class. It employs a verb substantive as an auxiliary verb, a thing unknown in the cognate idioms. It forms a great variety of compound tenses and moods, and has complex derivative conjugations, such as those called transitive, co-operative, reciprocal, and inceptive verbs. Of all these no vestiges have been discovered in the Ouigour or

Eastern Turkish, which, on the contrary, has all the simplicity of the other Turanian languages. It employs no auxiliary verb, has no compound tenses. By means of post-fixed particles it forms a present and a past tense, and it has been found to possess no other modification of the verb indicative of time.*

4. Most of these languages are scantily provided with conjunctions, but rich in gerunds, which actually contain conjunctions, and render separate and distinct particles unnecessary; even when the gerunds, or the participles which supply their place, are less numerous, they are still frequently used. Thus, as Dr. Schott observes, are formed long-winded and singularly involved periods, especially in the Turkish language, which it is quite impossible to translate, their construction being preserved.† From this want of conjunctions and tendency to form involved periods, the infinitives and participles assume the character of verbal nouns, and are brought more frequently and with greater boldness into connexion with pronominal suffixes, and with the terminations of cases belonging to nouns, than in any other language. Hence an apparent copiousness in verbal inflexion which is in fact a simplicity and poverty of structure. Some instances of this kind may be seen in the short citation above made from Scherer, in which the same construction appears in all the languages compared. Verbal nouns, which resemble in form the Greek *θελημα*, become in construction with pronouns real verbs.‡ With that form indeed the infinitive mood ending in *me* or *ma* in Mandschu, and in Tartar or Turkish in *mek* or *mak*, may be compared. In this want of analysis in construction, which is so striking a characteristic of the High Asiatic languages, the Turkish fully participates. It is remarkable that the Osmanli, which possesses a greater variety of forms than the pure Turkish or Ouigour, scarcely derives any advantage from them, as if it had not been able to shake off the yoke originally impressed upon it, and to accustom itself to the liberty which it has acquired.

"L'usage des temps simples et impersonnels," says M. Abel-Remusat, "viennent souvent y obscurcir les idées, que les temps composés exprimeraient avec netteté et précision." "Les Ouigours ont évité les principaux inconvéniens de ce système, en suivant un marche simple et naturel, qui empêche d'être élégant, mais permet d'être clair. Les autres Turcs, qui, sans renoncer à l'emploi fréquent des participes, ont voulu porter dans leurs compositions un style plus orné, et construire leurs phrases d'après un plan plus compliqué, n'ont réussi qu'à rendre la construction embarrassée. Chez eux une longue periode, imparfaitement soutenue par le rétour fréquent du gérond ou du participe, conduit souvent le lecteur au bout d'une page, sans lui offrir le verbe d'où

* Abel-Remusat. Recherches sur les Langues Tartares.
† Versuch über die Tatarischen Sprachen.
‡ "Ritter F. Von Xylander." Die Sprache der Titanen.

dépend le sens de la phrase entière. C'est de l'aveu des plus habiles dans ce genre de littérature, ce qui fait que la lecture des ouvrages Turcs est toujours difficile et fatigante."

These general remarks are sufficient to point out the nature of that analogy in genius and structure which exists between the languages of the Tartarian nations. I shall now add some particular grammatical forms, which exemplify their connexion.

There is no class of words in which the oldest forms of languages are so well preserved as in pronouns, as any one is aware who has any accurate knowledge of the classical languages.

The personal pronouns are as follows:

	I.	Thou.	He.
Mongolian	Bi	Tschi	Tere
Mandschu	Bi	Si	Tere, I
Turkish	Ben	Sen	Ol, O
Ouigour	Man	San	Ol
Finnish	Ma	Sa	
Esthonian	Ma, Minna	Sa, Sinna	Ta, temma
Lapponic	Mon	Don	Son
Hungarian	En	Te	Oe

N.B. The third person is in reality wanting in most of these (as in Greek), and the place is supplied by a demonstrative.

Genitive.	Of me or mine.	Of thee or thine.	Of him, his.
Mongolian	mini	tschini	ta
Mandschu	mini	sini	tereï
Turkish	benüm	sening	aning
Ouigour	maning	sangge	aning
Dative.	**To me.**	**To thee.**	**To him.**
Mongolian	mendou	tschimdou	
Mandschu	minde	sinde	
Turkish	bange	sange	ange
Ouigour	mangge	saning	angge
Plural.	**We.**	**Ye.**	**They.**
Mongolian	bida	ta	te-det
Mandschu	be	souwe	te-set
Turkish	biz	siz	an-lar
Ouigour	bis-lar	sis	o-lar
Finnish	me	te	
Esthonian	meie	teie	neet, nummad
Lapponic	mi, mige	ti, tiye	si
Hungarian	mi, miyuk	ti, tiyed	oe-k
Genitive Plural.	**Of us, our.**	**Of you, your.**	**Of them, their.**
Mongolian	minu	tschinu	
Mandschu	mini	sini	ini
Turkish	bizím	sizing	anlarong
Ouigour			alar-din

Dative Plural.	To us.	To you.	To them.
Mongolian	mendou	tschendou	tedendou
Mandschu	mindeu	sinde	terede
Turkish	bize	size	anlare
Ouigourian	bis-ke	sis-ke	alar-ke.

It must here be observed that the pronoun of the third person is in many instances defective in several of these languages, and made up of demonstratives and of other substitutes. Hence there is great variety in the forms apparently belonging to this personal pronoun. But those of the first and second display the most striking resemblances in all the above languages; the differences may be considered as merely dialectical; and the pronouns of the third person correspond when they are extant. Even the irregularities of one language correspond with those of another.

We must not omit to observe that the Mandschu has another form answering to the plural of the 1st person, viz. *mouse*, regularly formed as a plural by addition of the pluralising particle from *mou*. Compare the Lithuanian *més*, genitive *músû*, dat. *mús*. This word is used when the speaker includes the person whom he addresses together with himself in the same *we;* a variety in the number and conception of personal pronouns, which is well known in the languages of the Algonquin and other American nations, and has been considered as peculiar to them.

The Mandschu, Mongolian, and Turkish languages have, according to the peculiar genius of those idioms, syllables which, suffixed to the attributes of the subject, form a sort of possessive or attributive participle or adjective, and answer the purpose of relatives governing clauses. This requires explanation.

In Mandschu, *aracha*-ngge means *written-having*, or "quiscripsit;" minde-buche-ngge, to me given-having, "qui mihi dedit." The particle *ngge* means possession, *belonging to;* as nijalma-ingge, *i.e.* "menschen-wo," "what belongs to man," and it thence becomes the sign of the possessive case. This *ningge*, or *ngge*, of the Mandschu is likewise found in Turkish, in the particle indicative of the possessive case, which is *ning*. The Mongolians have a particle, *ki*, *gi*, and *kei*, which they use just as the Mandschures use *ngge* for a sort of relative suffix. The Turkish alone has a separate relative pronoun, *kim* and *ki*, which is even found in the Ouigour, or pure Eastern Turkish dialect, and this may be prefixed and used as is the relative pronoun in the European languages. The use of this is, as M. Abel-Remusat observes, foreign to the grammatical structure of these languages, and the Ouigour has the means of substituting for it a suffix particle like the other idioms of Northern Asia.

From the Mandschu *ningge*, or *ngge*, compounded with the possessive pronoun, or rather with the possessive case of the personal pronoun, and from corresponding words in Mongol and Turkish, arises a sort of abstract relative which is very remarkable from its almost exact identity with the German *meinige*. Thus, mini, sini, &c., make in Mandschu, *miningge, siningge, iningge* (literally mi-ni-ngge, viz. "mine what is" or "mine being," in German, "das meinige," "das deinige," "das seinige." In Mongolian the corresponding form is minüge, tschinüge, ekonüge; in Turkish, minningki, seningki, aningki.

The verb-substantive is another part of speech which retains very old forms in many languages. The Mandschu has two verbs used in the sense of *to be;* these are *bi,* meaning properly *to hold,* or abide, as in the infinitive *bime,* pres. indic. bimbe; this resembles and *may be* cognate with the Sanskrit, bhu; Welsh, bû; be, bin, buwain, φυω, fuo, of European languages. Another verb, more strictly a verb substantive, in Mandschu, is *ome* (εμεν). The Mongolian has bü-kü (φῦναι), and present bui (ich bin) and acho (esse), amui (I am). The Turkish has only ol-maq (properly, fieri): with this we must compare the Finnish infinitive olla, Esthon, ollema, in the pres. tense, 1. olek, 2. olet, 3. on; pl. 1. olemme, 2. olette, 3. owat; and the Lapponic corresponding form, 1. lem, 2. lek, 3. la; pl. 1. lep, 2. lepped, 3. lak.

The preceding instances answer the double purpose of showing resemblances in vocabulary, and also in grammatical forms between the different Scythian languages. It is impossible to doubt of the original affinity of the pronouns in all these idioms, or of those of the verb-substantive. What is more surprising is the resemblance which displays itself, without being sought, between these Scythian forms and those of the Indo-European languages. The consideration of this last subject would be foreign to my present undertaking, and I shall merely remark, with respect to it, that the resemblances in particular grammatical elements, as in the pronouns especially, and also those which may be pointed out in radical words—of these a short specimen has already been given from Scherer, which has been greatly extended by Klaproth—between even the most western European languages and the Mongolian and Mandschu, spoken in the extreme east of Asia, are certainly too strong and decided to be attributed to mere accidental coincidence, while, on the other hand, it is impossible to account for these phenomena by referring them to occasional intercourse, a thing which cannot be imagined between nations so widely remote from each other. If we attempt to resort to the only explanation that remains, namely, the hypothesis of a common origin, we seem to be carried back beyond the period open to historical or even ethnographical research.

We come now to the inquiry whether, and to what extent, there is a connexion in regard to the vocabulary and the stock of primitive words between the Turanian languages. It may be observed that, if no such affinity is found, we shall consider them to be one class of languages, a class strongly marked, and the members of which are nearly related to each other by such analogies as constitute a class, but we shall not venture to declare that a family relation exists between them, unless it be allowed that resemblance in grammatical construction, where it amounts to a certain degree, constitutes by itself this relation. Many have thought so, and they have reckoned the polysynthetic idioms of America, and the monosyllabic idioms of the Chinese and Indo-Chinese as languages respectively of one kindred stock. On this ground, the languages of the Turanian nations would be considered as one family of languages, even if no roots should be found common to them.

Father Gerbillon, who travelled in Chinese Tartary, in the suite of an expedition commanded by the Emperor of China, and whose "Elementa Linguæ Tartaricæ" was the first work that appeared in Europe on the Mandschu language, was of opinion that only seven or eight words in that idiom were similar to the Mongolian. It has been generally supposed that there is an almost equal diversity between the latter and the Turkish, notwithstanding the tradition collected by Abulghazi Khan of the common original of the two races of people speaking these idioms. M. Abel-Remusat says that Gerbillon was greatly mistaken; both he and Klaproth affirm that a large number of similar words exist in these languages, though they do not thence infer a common origin. M. Abel-Remusat makes a distinction in reference to this point, which supports a very strong argument, and cannot easily be set aside if the fact is exactly as he considered it to be. He divides words into different classes; one class he terms words of the first necessity and simplicity, and thinks common to all nations, springing from the same stock; another set are words of a merely secondary kind, such as simple nations adopt from their neighbours; a third class denote ideas which indicate refinement. The first class contains such expressions as those of kindred, father, mother, husband, wife, &c., and words denoting parts of the body, hand, head, &c., striking external objects, sun, moon, star, tree, river, and numbers up to ten. Secondary words are terms for domestic animals, metals, fruits, esculent plants, instruments of agriculture, of war, and other arts. The third class contains names for offices, dignities, &c. M. Abel-Remusat says, that of words belonging to the second class a great proportion are common to the Mongolian and Mandschu particularly, but that the terms included in the first class are distinct in each of these idioms. He therefore considers the Turanian nations as separate and different races.

A late writer, who has investigated this subject with great accuracy, has called in question inferences drawn by Abel-Remusat, as well as the data from which they are deduced. His work, to which he has given the modest title of "Versuch über die Tatarischen Sprachen," indicates a deeper insight into the genius of the Scythian languages than any former writer has displayed. Dr. Schott begins by observing, that even in idioms between which a near affinity has been completely established and is universally allowed, a multitude of words radically distinct from each other may be found to express the most universal ideas and objects of the first necessity. What resemblance is there, for example, between our German word sohn, a son, and the Greek υἱος, or between sohn and tochter, and filius and filia? Who would connect bruder, or brother, with ἀδελφος; sister with αδελφη; frau with woman, femina, or γυνη; man with vir; gattin or gemahlin, wife, with uxor and ἀλοχος? How is himmel, heaven, related to cœlum, οὐρανος, and the Russian niebo? How earth, terra, γη, and the Persian zemin? How moon, luna, and the Sanskrit tschandra? our sun and the Persian churshid, mihr, and afitāb? kopf, head, and the Persian szer? hand, manus, χειρ, and deszt? mund, os, dehān, στομα, and the Russian rott? baum, tree, arbor, δενδρον? vogel, bird, avis, ορνις, and the Russian ptitza? stone, saxum, πετρα, and Russian kamen? All these, and very many other words, in languages known to belong to one family, express ideas of the most simple class, and are yet totally diverse. Similar instances are afforded by a comparison of the Semitic languages, whose relation to each other is in other respects scarcely more distant than that of dialects of one speech. The moon is in Hebrew yārēach, in Arabic qamar; a hill, Hebrew, hōr, Arabic, jebel; a tree, Hebrew, ētz, Arabic, shedsher; a stone, Hebrew, ebhen, Arabic, hajar. Even in the very same language, words are often found expressive of necessary objects, which, though exactly synonymous, are totally different in several provincial dialects.

These instances are quite sufficient to prove that a considerable number of different words, even though expressive of ideas of the first necessity, do not disprove a family relation between languages. In the Scythian languages as in the Indo-European the same roots are often discovered, with some deviation in their meaning. It must also be observed that words themselves undergo in the Scythian dialects modification from the interchanges of particular consonants and vowels. In this way many words derived certainly from the same origin are so disguised that, without attention to the laws which govern this interchange, and which

Dr. Schott has been the first to explore with respect to the Scythian languages, their real affinity could hardly be recognised.

The following is a short selection of examples in which words of the same origin are found in several of these languages, expressive of ideas nearly related to each other.

In Turkish *gol* or *goel* means a lake: in Mongolian *ghool*, a river: *golo* in Mandschu is the bed of a river. The sea is in Turkish *dengiz* or *deñiz:* in the Magyar or Hungarian, *tenger: r* and *z* are shown to be frequently interchanged. In Mongolian *tenggisz,* and in Mandschu *tenggir,* means a *great lake.* In Mandschu *alin* (Mong. *aghola, oola*), a mountain: in Hungarian *alom*, a hill. In Turkish *qaya,* a rock: Hung., *koe, köv;* and in the Finnish languages *kii, ku,* a stone. The Mongolian *tsilagh-on,* a stone, resembles the Hungarian *szikla,* in the word *kö-szikla,* rocks, which seems to be compounded of two synonyms.

For *ice* the Finns have the word *yeg, yegna:* the Hungarians, *yég.* To this word the root of the Mongolian word *yik-ekün,* cold, frost, corresponds, while the Mandschu *juche* again means *ice.* With *juche* the Turkish *szuq,* or *saghouq,* cold, is closely connected.

The Mongolian *aldar,* and the Mandschu *elder,* mean shine, splendour: in Turkish we have *ilder-im,* or *yilder-im,* lightning, and *yeldiz* (for *yeldir*), star. The sun is in Mongolian *nar-an:* summer, the sunny season, is in Hungarian *nyár:* in Turkish *yaz,* written for *yar,* by the interchange of consonants. In German and English the names for sun and summer seem to be in like manner related.

The Turkish for heaven, *gök, gökler,* pl., does not occur in that sense in the other Scythian languages; but in the meaning of *blue,* which it has not lost in Turkish, we recognise it in the Hungarian *kék,* the Mongolian *köke,* and the Mandschu *kuku.*

Boi in Turkish, form, stature, is related to *beye* in Mongolian and Mandschu, meaning *bodies.* The Mandschu *udju,* head, seems isolated; but in Turkish *üsz,* or *üz,* means the upper part of anything, as in *uz-re,* upon, above. The root of the Turkish *qul-aq,* ear, is found in the Finnish *caul-en,* I hear: Hungar. *halla-ni,* to hear. The Turkish *göz,* eye, is connected with the Mongolian *üze* (*üze-kü*), to see, from which the Mongolian forms the word *üzel,* sight, and the Turkish *güzel, beautiful, spectabilis.*

These instances have been adduced by Dr. Schott as indicative of the fact, that, when the same roots are not detected in different Scythian languages in corresponding terms for the same objects or ideas, they are often to be found in use in a somewhat modified sense, in several of these idioms. There is also a considerable number of words bearing precisely the same meaning, both radical and derivative, which are either identical or very similar in the Turkish, Mongolian, and Mandschu languages. In some instances these words have been borrowed by one people from another. It is difficult to determine when this is the fact, and when they are a part of the original stock of vocables belonging to each language; but sometimes this may be done by noticing the form of such words, and whether their formative terminations belong to one idiom or to another; whether such words are in one dialect isolated and in another derivable from known roots, and associated with cognate words. When words nearly similar or identical exist in several languages, connected with ideas of the first necessity, we ought not, without proof, to conclude that they were derived by one people from another. Several examples illustrative of these remarks are cited in the work to which I refer. They afford additional evidence in support of the conclusion already suggested.

The preceding examples of analogy have been pointed out by Dr. Schott; they are few in number, but on comparing carefully the vocabularies of the Northern Asiatic languages given by Klaproth and others, I have found a correspondence equally decided, comprising a large proportion of words belonging to that class which Abel-Rémusat designates as terms of the first necessity. The collection is too long for insertion in these pages: I intend to avail myself of it on some future occasion. I must now give a few specimens of the interchanges of consonants and vowels, discovered by Dr. Schott in comparing the vocabularies of the High Asiatic languages, a subject which has been first elucidated by that writer. The following are some of the leading facts which he has observed:—

The final *n* in Mandschu nouns is frequently elided, and this is the only change produced in roots by grammatical construction: this consonant is in fact only a formal termination. *Morin*, horse, makes *mori-sa* in the plural. The Mongoles also omit *n*; as for *Khan-t*, plural of *Khan*, king, *Kcha-t*. Both these languages often drop the *n*, ad libitum, even without construction. Turkish nouns never drop *n*, grammatically or in construction, but often want *n* or *un*, in instances where the other languages have it as the usual ending. This syllable must therefore be cut off when we compare Turkish with Mongolian or Mantschu words. Examples :—

Mongolian,	kütz-ün	Turkish,	gütsch (strength).
„	mesz-ün	„	muz, buz (ice).
„	toghoz-un	„	toz (dust).

The Turkish avoids *n* at the beginning of words by omitting it or changing it into *j* or *d*,* as—

* *d* and *n* are interchangeable in the Celtic, *d* and *j* in many languages.

Mongolian,	neng	Turkish,	eng, (very).
„	nasz-un	„	jasz or jasch (age, period).
Mandschu,	nasz'ch-un (fortunate time).		
Mongolian,	nögür	Turkish,	ögür (amicus, consuetus).
Mandschu,	nadan	„	jadi, jedi (seven).

In the Hungarian or Magyar, *ny* takes the place of *d*.

*Mongolian,	neile-kü } to open.	Turkish,	del-mek (to bore a hole).
Hungarian,	nyil-ni }		
„	nyelv	„	dil, (tongue).
„	nyel-ni	„	dile-mek (to swallow).
„	negy (four)	Mandschu,	dechi (forty).
„	nyar	Turkish,	jaz for jar, (summer).
Compare Mongolian,	nar-an (sun).		
	nyol-ni	„	jal-maq (to lick).

The Turkish affects the medial or soft mutes and avoids aspirates and even tenues at the beginning of words. Thus it changes all labials, including *f* and *v*, into *b*; it drops *f* entirely in some instances.

The Mandschu, like the Chinese, avoids *r* by changing it for *l*. The Turkish often changes it into a soft and scarcely audible *z*. Examples of the above changes:—

Mandschu	falga	Turkish,	barq (a race).
Mongolian	ghar	Mandschu,	gala; Turkish, gol (hand).
Hungarian	tenger	Turkish,	dengiz (sea).
„	nyar	„	jaz (summer).
„	ökür	„	öküz (ox).
„	terd	„	tiz (knee).
Mongolian	mörū	„	omuz (shoulder).
Mandschu, ara:	Hungarian, îr	„	jaz (to write).

Words are in like manner disguised by the interchanges of gutturals and sibilants, and by the occasional omission of the former, phenomena which are observable in the dialects of most other languages.

The Turks and Mongoles change the hard *k* for a guttural *gh*, and the softer *k* for *q* and *j*. The Turkish final *k* or *q* is often a mere formative ending and is liable to be dropped: *qamuq*, Turkish, all, is in Osmanli *qamu*, in Mongolian *chamu*. The *maq* or *mek* of the infinitive in Turkish corresponds with *me* in Mandschu. *Gh* is dropped from the middle of words between two vowels, as *szighir*, a bullock, in Osmanli Turkish, becomes *szir* in Eastern Turkish, *schir* in Mongolian. So also *taghosz-un* (dropping also *un* as above) *toz*, Turkish; *chaghorai*, Mongolian, i. e. *chōrai*, in Turkish *qoru*, dry; *chabar*, Mongolian, a

* *Kü* in Mongolian, *ni* in Magyar, and *meq* in Turkish, are only the signs of the infinitive.

nose, dropping the initial guttural and changing the mute into its aspirate and adding a vowel, which the Mandschu requires uniformly at the end of words, becomes very near the Mandschu *oforo,* in Turkish *bur-un.* The common root is *var* or *vor;* *olcho-me* Mandschu, is *qorq-maq,* Turkish.

Omissions of the initial guttural and changes of *z* and *j* produce such differences as the following:—*göz,* eye, Turkish; *üzekü* (i. e. *üze*), Mongolian, to see; *üz-el,* sight, Mongolian; *güzel,* Turkish, spectabilis; *jasz-a,* Mandschu, an eye.

Mongolian,	zai	space, room	Turkish,	jai-maq (to extend).
„	zacha	border	„	jaqa.
„	zali	flame, cunning	„	jaling (flame) jalan (deceit).
„	zekü-kü	to draw	„	jük (a burden).
„	dschimesz	barley	„	jemisch.
„	zol-gha-cho	to meet	„	jol (a way).

Examples of the same interchange in other Turanian languages:—

Mandschu,	dsche-me	to eat	Turkish,	je-mek.
„	dschulergi	before	„	jileru or ilerü (ilergu).
„	dschaman	quarrel	„	jaman (bad).
„	tschala-me	to err	„	jalan (false).
„	botscho	colour	„	boja.
Hungarian,	szel	wind	„	jel.
Mandschu,	dschuche	ice	„	szouq (cold).
			Finnish,	jeg (ice).

By a similar comparison the author has shown that although the numerals differ considerably in all these languages, a sufficient analogy is discoverable between them to indicate an ancient though now obscure relation.

On comparing the phenomena traced in the preceding pages, it appears unquestionably to result that an extensive analogy of structure prevails through the four principal groups of languages compared—I mean the dialects belonging to the Turkish, the Mongolian, the Tungusian, and the Finnish, or Finno-Uralian families of languages, but I refer more particularly to the three former. They are all formed according to the same general laws. In the simplicity of their structure, and the want of real inflections, of which the place is supplied by juxta-position of particles, they approach in some degree to the character of the monosyllabic idioms spoken by nations who inhabit a contiguous region of the earth. They form a distinct class of languages, both from the Semitic, which inflects its dissyllabic roots and abounds in prepositions and conjunctions, and from the Indo-European idioms, which make so extensive a use both of inflection and composition, which affect changes at the beginnings of words, and have their endings for the most part without addition. The languages of all these classes of nations last mentioned display the influence of that active fancy which peopled the universe with sentient souls, and ascribed life to all the objects of nature. Groves and fountains, rivers and trees, even stocks and stones, are in all their idioms either male or female. But the rude inhabitant of cold and arid steppes, rendered dull and phlegmatic by the monotonous aspect of nature and the changeless manner of his existence, gave no play to his imagination; he affixed different terms to his bulls, cows, his horses, and mares, and to creatures of which he made different uses, but all other objects were to him of one sex; he never compared inanimate with living things. It is, however, impossible to explain the common construction of the Turanian nations by reference to physical or moral circumstances; they display one type and method of formation; all question on this point seems to be silenced by the discovery of so many particular grammatical forms as we have traced through all of them, by their having the same pronouns, verbs substantive, and resembling particles. If we go still further back and examine the very structure of words, we find the inference confirmed; the law of harmonic vowels, found as far as I know in no other languages, shows that the inventors of words themselves had their attention directed to one principle, or were governed by a similar habit. Even the idiom or style in the composition of sentences sometimes displays similar analogies, and this was the fact which struck the attention of Scherer, and perhaps first suggested a further examination. Lastly, in the vocabulary itself, or the material of the several languages, there is a considerable extent of analogy; perhaps this would be thought of itself scarcely sufficient to lead us further than it led Klaproth and Abel-Rémusat, viz. to the opinion that frequent and ancient intercourse between the Mongolian, Tungusian, and Tartar tribes occasioned the adoption by each of common terms from the vocabulary of others. But such intercourse could only have produced an effect similar to that which the mixture of Normans and Saxons has effected in our own dialect; the adopted words would be distinguishable as entirely foreign; they would not be found naturalized by such interchanges of the particular elements of articulation as we have traced.

On the whole, there seems to be sufficient evidence to constitute the languages of Northern Asia as not only a particular class of human dialects, but as belonging to one great family of languages, of which the different members, though more remote from each other than the idioms of the Indo-European class, yet bear and they are curiously contrasted with the African languages, traits that cannot be mistaken of a very ancient affinity. That the different nations who speak these languages, though they have been separated and scattered over interminable wildernesses from immemorial times, were yet allied in origin, or sprang from one primitive stock, is a further inference which it is difficult to avoid.

If we join the evidence thus deducible from a comparison of languages to the conclusion which historical testimony establishes respecting the original position of the Turkish race, and their early connection with nations of the remote East, we must give up entirely the notion that they were a Caucasian tribe, or nearly related in the first ages to the races of men who peopled Europe and the western parts of Asia. It has been observed, that while some Turkish nations, as the Russian Tartars and the Osmanli, have nearly the features of Europeans, other Turkish races display the type termed Mongolian. The latter are all the great nomadic races of Turkish extraction, in the central parts of Asia. The fact that the dialects of most of these nomadic nations are pure Turkish, without any considerable mixture of Mongolian words, strongly opposes the often-maintained opinion that their resemblance in features to the Mongoles has arisen from the blending of races, a supposition which the small numbers of the Mongoles at all times, in comparison of the great Turkish races, renders extremely improbable. This is not the proper place for a discussion of the question, to what other causes may be assigned such diversity in tribes descended from one race. I shall only remark, that it is not without parallel instances in the history of the Asiatic and European nations. The Finns and Lappes, for example, are allowed to be nations of one stock; yet they differ physically. The skull of the Lapp has the broad-faced Turanian form, while that of the Finn is entirely European, or of the type termed Caucasian.

IX.—*Notes on a Journey from Constantinople, by Heraclea, to Angora, in the Autumn of* 1838. By William Ainsworth, Esq., in charge of an expedition to Kurdistán.

Angora, 24th December, 1838.

In consequence of plague on board the steamer plying between Constantinople and Trebizond, and of the temporary interruption in the navigation of that part of the Black Sea, our party was reduced to the necessity of taking the land route through Anatolia, yet we still determined to fulfil as far as possible the instructions of the Geographical Society, as well as those of the Society for Promoting Christian Knowledge, who have so liberally supplied us with the means of travelling, to obtain a better acquaintance with the course of the Halys, and more particularly to explore the almost unknown districts between Ereglí and Angora. With these intentions, and in order to follow a less-beaten track, we endeavoured to procure horses for the purpose of proceeding by the site of the temple of Jupiter and the promontory of Kirpeh, to the mouth of the Sangarius, but the forests and the rocks on the sea-shore were declared by our guides to be impassable, and it was not till after we had arrived at Dúzchah that we were enabled to cross the mountains named Yáïlí Ṭágh, and reach the sea-coast, in the neighbourhood of the river of Uskúb.

Our party, consisting of Mr. Russell, Mr. Rasám, and myself, left Scútari on the morning of Tuesday, September 18th, 1838, and on the same evening reached Ḳartál, a village on the shores of the Sea of Mármora; passing on the road Mount Aïdós and the hills of Yaḳájiḳ, composed of sandstone and superimposed limestone, and associated at Scútari with rocks of the felspatho-pyroxenic series. Ḳartál is surrounded by gardens, which supply the market of the capital with bámiyahs (hibiscus esculentus), tomatos (solanum lycopersicum), capsicums, onions, and other vegetables, besides grapes and peaches.

Sept. 19.—From Ḳartál to Pendík (Pantichium), a fishing village, chiefly inhabited by Greeks. Near it are the remains of a castle, in part constructed out of the ruins of a still more ancient edifice. Geïbúzeh (Geybúzeh), the ancient Dacibysa (?), is now a large village, with a handsome mosque, surrounded by cypress groves. It is built upon a hill, composed to the west of trap rocks, to the east of cretaceous limestones. At the khán or resting-place, a sarcophagus in white marble serves as a reservoir for water, and numerous other fragmentary remains denote the antiquity of the site. The cretaceous rocks constitute a hilly country, from hence to Ṭavshánjiḳ, a pretty village surrounded by gardens and groves of cherry-trees or vineyards, which extend along the shores to Harakah, a khán and ferry to the opposite side of the gulf of Astacus. The ruins of a castle upon an adjacent hill, and solid walls near the sea-side, attest an ancient site. The ferry on the gulf of Astacus existed in former times between Libyssa and Pronectus. Libyssa, in Rennell identified with Cshebissa (*i. e.*, Jebísah, a name not now known by the natives)*, is in Colonel Leake's map identified with Maldysem. At present there are two villages on the N. side of the gulf, from whence the passage across is made; the one Tavshánjiḳ, the other Harakah, which latter appears from this circumstance, and from its ruins, to be the ancient Libyssa.

Sept. 20.—The hilly country from Harakah, as far as to Izmíd,† is composed of sandstones and limestones. On this road a mound of crumbled ruins, an ancient cornice, and two columns, converted into a gateway, are all that remains of Brunga of the Jerusalem Itinerary. The road from Scútari (Uskudár) to Izmíd, where we arrived this evening, is now measured off as far as Itinerary distances are concerned; and painted posts are put up at every $\frac{1}{4}$ of an hour of a horse's walking pace, or about 3 miles to an hour. The distance from Ḳartál to Geïbúzeh is 20 posts or 5 hours; from Geïbúzeh to Harakah, 18 posts or $4\frac{1}{2}$ hours, and from Harakah to Izmíd, 20 posts or 5 hours, making a total of $18\frac{1}{2}$ hours or $55\frac{1}{2}$ British miles, by the circuitous road, which is about the mean of the three ancient sets of Itinerary or 62 Roman miles, and not far from Rennell's deduction from the reports of seven different travellers compared with one another, which gave as a result 57 ordinary miles.

Sept. 21.—Nicomedia, so long a capital city, has been truly said to occupy a most imperial situation, both with respect to the scenery about it and its political and commercial advantages. Pliny the younger was prefect at Nicomedia, and the city of the kings of Bithynia was the residence of Diocletian and the historian. Its modern condition has been too often described to need any details here, but some travellers have almost denied the existence of any remains of ancient times, yet there is still a castellated building, and the tombstone carvers' yards are filled with fragments of antiquity. Lat. by two mer. alts. of the sun 40° 47′ 40″ N.; long. by chronometers, by three sets of altitudes, 29° 53′ 30″ E.

Sept. 22.—The country to the N. and N.E. of Izmíd, and between it and the Black Sea, is hilly and covered with wood, forming part of the forests designated by the Turks as the Agháj Denizí,‡ or Ocean of trees. It is said to be crossed in that di-

* Jebíseh is probably the Arabic name of Geybúzeh (pronounced Geïbizéh), and answers to the ancient Dacibyza.—F. S.

† Izmíd is a colloquial contraction for Iznikmíd (from Εἰς Νικομήδειαν).—F. S.

‡ Jihán numá, p. 666.—F. S.

rection by no great road, which information must be received with great caution, but the súrují's (muleteers) would not proceed in that line. The mountains of the sky (Gök Ṭágh), composed of limestone reposing on, and associated with, mica and chlorite schists, gneiss, and quartz, bound the Gulf of Astacus to the S., and extend eastward by Ṣabánjah to the Karám 'Alí Ṭágh, S. of Khandaḳ 7 miles, and by the latter to the Bólí Ṭágh, S. of the plain of Dúzchah; constituting portions of the Bithynian Olympus; they are uniformly covered in their northern divisions with forests, with the exception of one culminating point, bearing S. 23° E. of Khandaḳ, and rather in the rear of the Karám 'Alí Ṭágh, whose bare summit is said to expose the ruins of an old castle. The hills of Khandaḳ unite Olympus with the southern ranges of hills named the Yaïlá Ṭágh,* and these are also covered with forest-trees from their northern acclivities to their abutments on the Black Sea.

The country at the head of the Gulf of Astacus and between the Gök Ṭágh and the southern hilly districts, is at first low and level, watered by the Ḳizil Irmáḳ, and cultivated with rice and melons. Further inland are pastures diversified by hedges covered with wild vines, hops, and virgin's bower (clematis cirrhosa), the luxuriant creepers of these climates. On approaching the lake of Ṣabánjah, the northern and southern hills which enclose it prolong their rocky declivities into the plain, which is thus raised above its ordinary level, and is covered at first with a low and shrubby vegetation of evergreen oaks, &c., which soon, however, attain the magnitude and growth of forest trees. I have been thus minute in describing the features of this portion of Bithynia, because projects have often been made to construct a canal between that lake and the Gulf of Astacus. The younger Pliny, in a letter to the Emperor Trajan, proposed to convey a canal along this line, where he said there were already indications of a previous attempt to dig one; but there are at present no remains of such a canal. Plans for the same purpose, of the same nature, have been formed by the Turks in modern times; one in 1490 is noticed by Rennell. (Vol. ii. p. 104.)† Pliny reports that the difference of the levels between the lake Sophon (Ṣabánjah) and the Gulf of Nicomedia was 40 cubits, or about 60 feet, and the Turkish account is 30 lirás (cubits), also about 60 feet, and the lake was said to be 35 feet above the Sangarius.

The lake of Ṣabánjah, the ancient Sophon, is upwards of 8 miles in length, and of an oval form. At some seasons of the year it is said to overflow, and to pour its waters into the Gulf of

* Summer-quarters mountains.

† Others in 1505 and 1563 are mentioned by the Turkish historians Hammer, Reise nach Bressa, s. 171.—F. S.

Astacus; but there is a constant communication between it and the Sangarius by a rivulet called Killis. The lake has few pretensions to beauty, the hills to the north are low, there is little woodland, no villages, nor even any rocky scenery; but to the S. the woods are of noble growth and extent, and rise to the summits of the mountains, at least 1000 feet above the lake.

Ṣabánjah is a mere travelling station, full of coffee-houses and stables, of which the inhabitants vary every day, with about 500 houses and two mosques. The Greek Patriarch of Nicomedia claimed for this spot (which has now only a few broken columns and scattered fragments of ancient times) the name of Helenopolis. Colonel Leake marks it as Sophon, and Rennell as Lateæ, which latter in the Theodosian tables is placed 24 m. p. from Nicomedia, corresponding with the distance of Sabánjah from the same place.

Sept. 23.—The Theodosian or Peutingerian tables have a site marked as Demetriu or Demetrius, 13 m. p. from Lateæ. There are in the present day, at a corresponding distance from Ṣabánjah, the almost perfect remains of a handsome bridge of seven arches, 1087 feet in length, and carried over an old bed of the Sangarius, from which a small stream still finds its way along the same valley to the N. A plan of this bridge was accurately sketched by Mr. Russell; its name would appear to connect it with that of the son of Antigonus. As a remote proof that this was the old bed of the Sangarius, Mr. Rasám obtained from the natives a fragment of a tradition in verse, which relates that a dervish, or holy man, coming to the bridge, was required to pay, which he refused to do, alleging that his avocations forbade his carrying money about. The passage of the bridge was not however allowed, and in his anger the dervish prayed that God would change the bed of the river, that toll might not any longer be collected at the bridge; and it appears that his prayers were heard. The bridge is called by the people of the country Mahámah, and a road passing over it takes a southerly direction, being probably one of the Roman roads to Ancyra.

From the old bridge we proceeded to the modern one, a wooden structure carried over the river Saḳáriyyah, where it is 372 feet wide, with an average depth of 2 feet, and a rate of about 2 miles an hour. From the Saḳáriyyah, the road is carried over marshy land, which I knew by experience to be scarcely passable at some seasons of the year, but now tolerably dry: a wooden causeway, like an American corduroy, is carried for nearly a mile over this marsh. At a guard-house further on we found the residents suffering severely from malaria. Beyond this point the country improved, and low hills of trap-rocks led the way to the wooded hills of Khandaḳ, which we reached the same evening.

Sept. 24.—Khandaḳ * is a posting village in the forest, and contains about 200 houses. For the supply of the post, Scútari has 100 horses, Ṣabánjah 150, and Khandaḳ 200. There are few remains of antiquity, but occasionally fragments of columns and of hewn stones are seen, more particularly in the burial-ground of the town. Khandaḳ is identified by Colonel Leake with Latania. There was much thunder and rain during this day.

Sept. 25.—Left Khandaḳ by a circuitous route through forests of beech and oak growing upon rocks. After a ride of nearly 4 hours we came to the open plain of Dúzchah, remarkable for its picturesque beauty. It is itself very level, traversed in an irregular manner by the Mílán† river, which forms the lake of Ak-Tení Gölí,‡ in the S.E. corner of the plain. It is surrounded by wooded mountains, of which the Bólí Ṭágh, to the S., attains an elevation by trigonometrical admeasurement of 1490 feet above the plain, and the Yáïlá Ṭagh, to the N., is by barometer 1500 feet. To the W. are two openings with lower hills, and to the N. above Uskúb, vulgo Eskí Bágh, are some hills of minor elevation. The plain is everywhere verdant with green sward, trees, or shrubs, while the surrounding heights are covered from foot to summit with continuous forests, the different shades of which have a very pleasing effect; add to this that the whole is a plain of about 12 miles in length and 8 in width, which can be taken in at one view from almost any point, so as greatly to enhance the beauty of the scene.

In travelling from Dúzchah to Khandaḳ, in 1837, the river of Mílán was forded nearly to the W.N.W., but at the present season, the river being flooded, we approached the lake, crossed a deep cut, with water scarcely moving, to the N.W., and banks clad with a deep and rank vegetation, while a little further was the river of Mílán flowing with a quick current to the S.E.; we kept along its banks half an hour to the N., then turned E. to Dúzchah: it rained hard during all the latter part of the journey.

The number of columns, cornices, and fragments of Byzantine architecture in the burial-ground appear as evidences of some former splendour in this small station, identified by geographers with Duseprum, a site without a history: one capital of a column, forming a head to a well near the khán, was ornamented with well-sculptured doves encircled by wreaths. At present Dúzchah contains scarcely more than 20 houses.

---

* A foss or ditch, in Arabic.—F. S.

† Mílán ṣúï (Mílán water), Jihán-Numá, p. 650, takes its name from the Mílán, a rocky valley, where it rises (H. p. 647).—F. S.

‡ The Mílán river passes through the lake of Ifnánlú (J. N., p. 653), called Ifnán-lú-göl.

Sept. 26.—The town or village of Uskúb, vulgo Eskí Bágh, bore from Dúzchah N. 5° E. at the foot of the hills about 3 miles distant, but our route to it was, on account of the marshy character of the plain, very circuitous, and we had to ford the Mílán, flowing westwards. We found this place to have been once the site of a considerable town, part of which was contained within a strong circular wall round the hill, and still in tolerable preservation, while the remainder was *extra muros:* to the S. and upon the hill was also an aqueduct, but of a doubtful era. The modern village is for the most part within the old walls, and many of the streets are approached by narrow gateways, evidently belonging to an ancient style of building; the upper slab of one was 12 feet long and 8 by 3 in thickness.

We here copied several inscriptions: they appear to be only sepulchral. One inscription, however, was copied by Mr. Russell from a solid mass of stone, at a station where excavations had been carried on in some sort of outwork or temple: this appeared to have been the basis of a statue. Although no satisfactory evidence could be obtained from these inscriptions, still the position of the town near the banks of the Hypius river and at the foot of the Mons Hypius (Yaïlá Ṭágh), will probably satisfy geographers of the identity of this site with the ancient Prusa ad Hypium.

ΑΝΤΙΝΩ
ΘΑΛΛΩΛ
ΚΛΑΡΙΣΗ
ΑΝΕΣΤΗΣΕ

Rennell had already placed this city at a situation which he designates as Uskúb, from Eskúb, vulgarly called Bagh, as the district of Aḳchah Shehr is called *Uskúblí* Ḳazá-sí: this Uskúb is, however, on the banks of a lake at some distance from Duseprum.

On leaving Prusa ad Hypium we found further ruins in a very dilapidated state about a mile up the banks of a rivulet which flowed from the mountains higher up. This forest-clad chain we now approached, and began a long ascent, rendered the more difficult by the muddy state of the roads after the late rains. The forests consisted almost entirely of beech, with some oak and pine. The barometer at the summit level indicated a height of 1350 feet. The name of this chain, which extends to the E. almost uninterruptedly as far as to the Elḳás Ṭágh, has been variously written in the maps Tshila, Teheleh, Tshele, and Chila.* The word Yaïlá or Yáïláḳ signifies summer quarters. This phrase, which is applicable to any mountain pasture at a considerable elevation, is very common in Asia Minor, and has

* Chileh ṭághí, the mountains of Chilah. J. N., p. 653.—F. S.

been consequently mistaken by modern geographers for a generic term for any chain of mountains.

It is very interesting while traversing the forests of Bithynia to observe in practice at the present day the very same usages as were noticed by Xenophon centuries ago, trees being still, as then, fired at their base and then felled, while small waggons yoked with male buffaloes came from the shore to carry away the wood. There are no villages, and the driver sleeps in his cloak every night till his work is done; and the carts are so constructed that their slope becomes excessive without causing any danger of an overthrow: the wood is used for ship-building, partly at Aḳchah Shehr and partly at Constantinople.

It had been some time dark before we began to ford the Uskúblí Ḳazá-sí cháï, which we did two or three times previous to reaching the port, without a harbour, designated as "Baystairs," Chuvállí Iskeleh-sí, and consisting of a long range of wooden houses with a beach, upon which, in fair weather, the small coasting vessels are drawn up. A mile beyond this to the W. we came to Aḳchah Shehr, called in the maps Aḳ Hissar, where we were detained some time by bad weather.

Sept. 27.—Aḳchah Shehr,* "money town" (or whitish city), is now but a poor village of about 20 houses, where they were building one brig at the time of our visit, and one small Greek boat from Várnah, loaded with dried beef, was drawn up upon the shore. Great quantities of *Momordica elaterium*† grew upon the beach, and the inhabitants, which is scarcely credible, were ignorant of the squirting power of the seed vessel. It rained hard with much wind also all this and the following day. The journey from Uskúb to Aḳchah Shehr occupied us 9 hours, travelling 27 miles by a circuitous route.

Sept. 29.—We took advantage of a momentary cessation of rain, after midday, to start along the coast, passing the river of Uskúblí Ḳázá-sí, 10 yards wide by 1 deep, but swollen, and doubtless a mere rivulet in spring, ascended a woody hill by a narrow pathway, where our horses stumbled and fell in the clayey soil; from thence we reached the valley of the Aḳ sú or white water, 13 yards wide by 1 deep. We stopped for the night at a poor village of 7 houses called Akóyah Köï, and corrupted from Aḳ Ḳayá Ḳöï, "the white cliff village."

Sept. 30.—Started early in the morning, cloudy with rain and strong wind from the N.; ascent, as usual, up a forest-clad hill, with steep and slippery clayey road, and much obstructed by branches and climbing plants. Forded the river of Kójamán, about 13 yards wide by 1 deep, but swollen; crossed a small

* Or Shár. J. N., p. 654.

† Squirting cucumber.

stream and kept along the shore of the Black Sea. The long-continued northerly winds had caused much more sea than usual at the foot of the cliffs, which, combined with the heavy swell, rendered the road at times almost impassable; and after an ineffectual attempt, attended with no little risk, to proceed along shore and also to pass over the cliffs, we were compelled to return to Ak kayá köï, where we remained the following day till the weather moderated: latitude by meridian altitude of sun 41° 4′ N.

The rocks in this district consist of limestone shales, argillaceous limestone in thin beds, and of altered rocks in nearly vertical or in curved and contorted strata. The chief varieties are iron-shot, calcareous, and argillaceo-calcareous beds; blue and black carburetted limestone shale with Lydian stone, veins of calcareous spar, and some clay iron ore.

After passing the summit of the Yaïlá Tágh an evident change takes place in the character of the vegetation; the underwood of brambles, briars, and fern, is replaced by rhododendrons, oleander, myrtle, box, and, in flower even at this late season, *daphne*, *vaccinium*, and cistus; nearer to the sea heaths and fern became abundant: the number of species of forest trees was considerable, and among them the chesnut was now frequent.

The Mariandyni possessed in ancient times the coast between the Sangarius and Heraclea, where the sea forms a deep bay terminated by the promontories of Kalpe or Kirpeh on one side, and that of Posideum, now Cape Bábá,* on the other. But if the country possessed the same characters as it does at the present day, which from other circumstances (for from the time of Xenophon to that of Jaubert, 1806, travellers have uniformly taken to sea at Ereglí) appears likely to have been the case, the population and the productiveness of the soil could never have been great.

Oct. 2.—The weather being calm we started early. The Kójamán river had fallen half a foot since the preceding day; the next river, the Kókalá, offered no impediment, and early in the afternoon we arrived at 'Aláblí, a port and fishing village with about 40 houses and a large government house, where we were received for the night. The Elæus is here a fine river, 17 yards wide and 1 deep, crossed by a wooden bridge, and increasing in width immediately below to upwards of 50 yards. Although 'Aláblí has been identified with the ancient Elæum, we could find no remains here, save one fragment of a marble column at the governor's house. The hills throughout this day's journey were, as usual, covered with wood, in which variety and beauty were

* From the tomb of Hájí Bábá, a Turkish saint. J. N., p. 653.—F. S.

equally united; the rocks consisted of iron-shot trap and wacke, with altered rocks and limestone shales. On approaching 'Aláblí the country opened, and the prospect became more extensive; to the S. the high wooded mountains which form the continuation of the Yaïlá Ṭágh and bound the valley of Bólí to the N. terminate the view; to the E. a succession of hills and valleys rise up with the course of the Lycus as far as to the pine-clothed trachytic summits of the Ḳará Ṭágh; while to the N. the peninsula of the Posideum, with its lighthouse, and the walls and towers of Ereglí at the head of a calm bay, add to the variety and beauty of the scene.

Oct. 3.—As the Aghá could not provide us with horses, we took a boat to Ereglí, and starting early in the morning before the wind had got up, soon doubled the cape called Chingál Búrnú, composed of trap, limestone, and altered rocks; passed the mouth of the Ḳilij or sword river, the ancient Lycus, and arrived at Ereglí a little after 10, A.M.

It appears from our observations that there are no less than six rivers, besides numerous rivulets, between Akchah Shehr and Ereglí. The first, commencing from the W., is the Uskúblí Ṣúï, or river of Prusias ad Hypium, and therefore the Hypius; but there is here a difficulty. In the plan of Dúz-chah we found the river of Prusias ad Hypium, known as the Mílán, and it is a considerable river flowing into the sea 3 hours W. of Aḳchah Shehr, while the modern river of Uskúb is little better than a rivulet. In the Theodosian tables and in Arrian's Periplus, after the Sangarius comes the Hypius (Mílán), and Rennell identifies Lilium with Chuvállí Iskeleh-sí. Probably the river of Prusias was the same as the Mílán; while the river now deriving its name from Uskúb is called so on account of its coming from the neighbourhood of that town and passing through its ḳádílíḳ (Ḳáẓa).*

Oct. 4.—Heraclea, now Ereglí, as the Pontic Heraclea, "has filled the page of history by its grandeur and misfortunes; and its remains testify its former importance."† The celebrated botanist, Tournefort, in his coasting voyage from Constantinople to Trebisond, *passed a night* here, and, according to Gibbon, "His eye surveyed the present state, his reading collected the antiquities of the city." We have the same authority for the

* Mr. Ainsworth here examines with some detail the names of places as given in the Theodosian tables and Arrian's Periplus; but, as he well observes, it requires a far more careful examination of all the points in question than he had time for, and a reference to books which a traveller could not be expected to have with him, in order to arrive at any satisfactory conclusion on the subject. His discussion is therefore omitted, but the original MS. may be referred to in the library of the Society by those who feel interested in the subject.—ED.

† Rennell. Geo. of Western Asia, vol. ii. p. 115.

existence of a separate history of Heraclea in the fragments of Memnon preserved by Photius, and the authority of Rennell for a description of the same place by the Academician Beauchamp in the Mém. sur l'E'gypte, tom. ii.

Heraclea was the port where the fleet of the Goths awaited the return of the second expedition, that in the time of Galienus ravaged Bithynia and Mysia; but that that expedition should in its retreat have been attended by a long train of waggons laden with spoils, will appear doubtful to those who know the character of the country; and Gibbon has also been misled by Chardin when he asserts that to navigate the Euxine after the month of September is esteemed by the modern Turks the most unquestionable instance of folly. There were upwards of 30 craft in the harbour while we were there, and some going out or coming in almost daily.

Ereglí, or Bender Ereglí (the port of Ereglí), contracted into Bendereglí, as the Turks have named the ancient Heraclea, contains 250 houses of Mohammedans, and 50 of Greek Christians, who have one church. Having staid here 4 days, we had time to make a plan of the ancient town, and copied an inscription in the Acropolis. The walls are now in a ruinous condition, and constructed chiefly of the remains of a former rampart. In that part which fronts the sea, and where there are remains of an outer as well as an inner wall still existing, huge blocks of basalt and limestone are piled upon one another, and intermingled with columns and fragments of Byzantine cornices and tablets, with sculptured crosses and Christian inscriptions. The castle upon the height is in a very ruinous condition. Only part of the ancient town was contained within the wall; the outer portion, where we found mosaic pavements, extended in the form of a triangle to a valley with a rivulet, now called Gaúr Irmák, and formerly a harbour, defended by towers, the ruins of which still exist. Was this the Metroum?—Of the Acherusian peninsula we could find no traces; it might have been a point where is now the inefficient Pharos of the Turks, a spot where a few villages are exempted from taxes upon the responsibility of keeping a light burning before mirrors darkened with accumulated soot.

The latitude of Ereglí by mean of three meridian altitudes of the sun we found to be 41° 15′ 30″ N.; its long., by mean of several observations, 31° 30′ E.*; variation of compass, 9° westerly; we observed also for dip and magnetic intensity.

The formations around Ereglí consist of igneous and sedimentary rocks. The first exhibit themselves chiefly at Chísh Depeh and Cape Bábá, but also form the bases of the Heraclean rocks.

---

* Gaultier places the lighthouse in long. 31° 24′ 56″ E. of Greenwich.

They consist of basalts, dolerites, trap, and trap-conglomerate. The latter more particularly forms the foundation upon which the wall of Ereglí rests, and many of the huge stones which enter into its construction are formed of the same coarse material. To the S.W. of the town is a hill composed of argillaceous limestones of a pink yellow and yellowish-white colours. The pink variety is compact, fissile, and was much used as a building stone in the old city. To the north, and at the foot of the Acropolis, is a fountain of coarse gritty sandstone, very friable and passing, on the one hand, into a conglomerate, and on the other, into coarse limestone. It is shelly, and remarkable for containing littoral genera, such as patellum and astrea, and zoophytes of a similar character.

There is a general remark which might be ventured here, that it is impossible for the traveller to wander along the shores of the Black Sea without being struck with the rarity of shells common to the Mediterranean shores, as species of turbo, buccinum, purpura, solen, and mactra, replaced here by an occasional tellina, venus, or cardium, and the frequent occurrence of naïades belonging to the genus anodonta, while its waters abound in tunny and other sea-fish. In the gradual diminution in saltness which takes place in a Mediterranean sea like the Euxine, where the supply of fresh water is large and the waters of the sea itself as constantly going out, it would be curious in a geological point of view to ascertain whether the first forms of animal life affected by this new order of circumstances, be those which belong to the lower orders in the scale of creation? or whether their diminution in numbers be also indicative of a similar destruction of larger forms that haunt the deep waters?

Oct. 8.—We now turned our steps in an easterly direction up the course of the Lycus, and proceeded over low hills of sandstone and ironstone, by a road, part of which was paved with slabs from 2 ft. to 8 ft. in length, and from 1 ft. to 2 ft. in width. About 5½ miles from Èreglí we found on the roadside, and crowning an eminence, an old tomb of an oblong form built of large massive stones, and hollow within. It is called Ḳóchák Ṭásh, "the hero's stone." Three miles beyond this, we came to where the Lycus forced its way through rocks of sandstone in thick strata, dipping north, and rising with rounded but nearly vertical walls, over which fall numerous streamlets of water from the well-wooded hills above. Immediately beyond the pass, a large mass of rock 90 ft. high, now overgrown with wood, except on its most precipitous parts, has become an island, and in the centre of the stream it forms a singular and picturesque object. It began to rain in the evening, and we only reached the valley of Yálchílar (masons), about 12 miles from Ereglí, not far from which, in the forest to the N., are some cliffs apparently with hewn sepulchral caverns, now called Bál Kayá sí (honey-cliff), but which the bad weather prevented us from visiting.

Oct. 9.—It rained in torrents all night, and the Kilij rose nearly 4 ft., overflowing great part of the plain, and assuming the appearance of a small lake, covered with trunks of trees and wood of various kinds. The pass of Barakatlar (Blessings), so passable on the previous evening, was rendered totally impracticable. It is evidently from these rapid and very sudden risings that this river obtained the name of Lycus, from its resemblance to a wolf rushing upon the fold.

We started near midday during a momentary cessation of rain, but were soon turned out of our path by the swollen river, and obliged to ascend the hill. After an hour's journey through a picturesque country we arrived at a point where the Lycus was crossed by a bridge, and made a sudden bend from the N.E., receiving a considerable tributary from the S. The two rivers met below cliffs of trap and sandstone.

We soon arrived at the foot of hills consisting of trap and trap-conglomerate supporting coloured limestone, which a rivulet flowing from above had covered with a deep coating of travertino. After continuing our ascent 20 minutes we came to a village named Yáïlar, "summer-quarters," from whence we had a fine view of the Lycus flowing at first through a rocky country; and then through a fertile valley, from E. and W., and backed by the limestone hills of Ovah Tághí (plain mount), partly wooded and partly white rocky cliffs. To the S. the country consisted of alternate valleys and rounded hills, on one of which a spot was pointed out said to contain an old iron-mine. On this hill the barometer indicated an elevation of 840 ft. We continued along its crest for a short time, skirted round a hill, and came down upon the Lycus, on the banks of which we found a village of four houses, one of which was empty, and in it we found a refuge from the rain that poured down all night.

The country we had hitherto been travelling through consisted of more or less round and irregularly formed and scattered hills. It was almost impossible to reduce them to any system of arrangement: sometimes hard limestone or sandstone gave birth to cliffs, more or less picturesque, from their wooded knolls or steep bare acclivities; at other times, rounded summits commanded on all sides valleys containing villages and cultivated lands, or deep ravines with streams rolling swiftly below. The hills were sometimes, but very rarely, conical, and the serrated outline of the Kará Tágh generally backed the view: its summits were sharp and sometimes pointed, numerous and narrow, following one another in quick succession, nearly of the same height, and everywhere covered with dark forests of pine.

At this season of the year a large portion of the arable land is occupied by crops of flax which, now young and verdant, are allowed to be all winter protected by the snow. Maize and millet were yet in great part in the fields, cut, but not carried home. The garden-crops consisted of gourds and cabbages.

Oct. 10.—Our road to-day lay up the valley of the Lycus, which we first crossed upon a wooden bridge, above which are the remains of an older construction built of stone, and a little beyond it the river receives a tributary from the S.W., considerable when compared with the Lycus itself.

The path we followed was carried along the acclivities of wooded hills, of limestone, cretaceous marls, and sandstone, with trap-rocks and occasional dykes of basalt crossing from S.E. to N.W. The road wound round the acclivities and base of these hills from E. to N.E. At a distance of about 5 miles up the valley the Lycus receives another tributary, equal to almost half the body of its waters, from the S.E. Above the point of junction one of the Ḳará Ṭágh mountains rises like an isolated peak to a height of 900 ft. above the river. Six miles beyond this a river is met with flowing from the N., with a bridge and small khán, and receiving another tributary from the E.

Our ascent of the Ḳará Ṭágh began at this point, and lasted nearly an hour, when we attained its summit, near the village of Kará-bíńár* (Black Source.) The barometer indicated an elevation of 1500 feet, and the mountains around did not rise much more than 500 feet above this point. The view now spread out before us carried the eye down the Ḳará Dereh (Black Valley), over a hilly country to the basis of the Filiyás and Bártán rivers; and was bounded to the N.E. and E. by the lofty and bold rocky summits of the Ḳayá Dibbah (N. 55 E.), and the more tame and wooded outline of the I'ch-íl-ler Ṭághí. Ruins of a castle (?) are met with in the mountains to the N.

We descended the valley of Ḳará Dereh, and turned to the southward to Básh Burgház, a small village at the foot of Mount Ipsil,† a spur of the Ḳará Ṭágh, where we were to obtain a change of horses.

Oct. 11.—Although it had rained all the previous evening, we were enabled in the morning to obtain some lunar distances and altitudes of the sun before we started. Our road lay along the valley of the Ḳará Dereh, the waters of which flow eastward; the Ḳará Ṭágh forming the culminating ridge between the basin of the Lycus and that of the Billæus. Before mid-day we left the rivulet, at an assemblage of uninhabited buildings called Beg

---

* Pronounced Bunár; originally, no doubt, Bínár, and by the eastern Turks Buńár—ń in the French, ń in *en*, *mon*, &c.—F. S.

† Ipsilítágh, from the Greek Hypsile ?—F. S.

júm'ah-ṣí* (Bey's Friday market), and used as a market-place on Friday, the Mohammedan sabbath, for the two ḳáḍíliḳs or districts of Básh Burgház and Pershembah.† Hence we ascended N.E. to the chief place of the latter district, a rather showy place at a distance from its whitewashed mosque and large aghá's house, but scarcely containing 30 dwelling-houses. A meridian altitude of the sun gave the latitude of this place 41° 19′ N. Bar. 28·980 inch., att. ther. 62° Fahr. The country consisted of sandstone and limestone alternating in thin beds, and was covered with underwood, chiefly deciduous oaks and juniper. 5 miles beyond it we descended into a deep valley, with a rivulet flowing S., called Tursehgí Dereh-si. Our ascent was up an acclivity, amid trap-rocks and limestone; we then passed through a wood, and near two or three villages, when we came to the open valley of 'Abd-allah Páshá Dereh-si, so called from a mosque bearing the same name, placed in an insulated position on the S. side of the valley, at its termination above the Filiyás. This valley receives all the waters of Ḳará Dereh, and the Pershembah district, to pour them into the Filiyás, which is here a fine river, divided into five different streams separated by islands of pebbles, occasionally covered with plane, sycamore, tamarisk, and oleander, but sometimes stony and naked; the occasional floods of this river, to judge from its bed, upwards of a ¼ of a mile in width, must be very great, but it is soon confined in one channel; and at Tium, where it empties itself into the sea, though deep, is only about 100 yards wide.

The beautiful valley of the Billæus was crowded with villages, and the views on every side varied and extensive. The river is seen flowing N.N.E., in nearly a straight line, apparently from the very foot of the dark and frowning Yaïlá Ṭágh. To the E. is a hilly country, either cultivated or covered with wood, and interspersed with villages belonging to a second Pershembah district, where the plague had been raging but a short time before our arrival, while our road lay down the river N. 30 E. to *Chárshembah*,‡ the chief place of another Ḳáḍílik or jurisdiction, where we slept and changed horses.

Our attention was now directed towards the ancient sites on the banks of the Billæus, and every inquiry was made regarding the existence of ruins, &c.

Ptolemy has placed a city called Claudiopolis, and also Bithynium, on the river Elatas, which has been by some supposed to be either the river of Elæum or the Lycus, but Rennell and

---

* Beg jum'ah-ṣí bázárí.—F. S.

† For Penj-shenbeh, *i. e.* Thursday, put for Penj-shenbeh bázárí, Thursday's market.—F. S.

‡ Chár for Chehárshenbeh, Wednesday, *i. e.* Wednesday's market, an elliptical expression.—F. S.

D'Anville both agree in supposing this city to have stood on the Billæus, and Elæus or Elatas is substituted for it. From the exploration now made of the Lycus, which we had followed up to its source, we were convinced of the non-existence of any site of importance upon that river. We had crossed the Elæus (?) at 'Aláblí, and heard of no ruins up that river; and we now approached the Billæus at some distance from its embouchure, and to which point we directed our researches, having heard of no remains higher up the river.

There remained for us, besides Bithynium of Ptolemy and Strabo, Mantinium, and Tium or Tios, in the country of the Caucones, who succeeded the Mariandyni, occupying both shores of the lower course of the Billæus river. Of Tium Rennell says it is now represented by Filiyás, which is also the corrupt modern name of the Billæus, "but we hear of no remains of Tium."

Oct. 12.—After taking morning sights for me, we started, passing over a plain covered with sycamore and tamarisk, the river being enclosed between low ranges of wooded hills of limestone and marl, with every here and there a village peeping from among the trees. Hemp is much cultivated on the islands of the river, and on its banks. At a distance of about 3½ miles we forded a large rivulet, 7 yards wide by ½ a yard deep, called Dagermánós Dereh-ṣí,* with some small villages and one large one, Chamánlí Kóï, of about 40 houses, on the hill-side. A little beyond this, at a point where the river, after making a long bend, approaches the hills on its left bank, is a small village called Cháï Kóï, built in part upon a mound of ruins, where there are several large hewn stones, which make it not unlikely that this was the site of a guard-house or small military station. Beyond this, at Aḳ-bunár, we obtained a meridian altitude of the sun, which gave our latitude 41° 29′ N.

At Gölmekchí-ler (Pottersville), a village of 30 houses, on a low hill, by the river-side, and about 2½ miles beyond Aḳ-bunár (the white spring), we found many fragments of large hewn stones and marble columns, with distinct remains of an ancient causeway.

A short distance beyond this a mass of basaltic rocks, having a tendency to the prismatic structure, advances to the border of the river on the left, narrowing its bed considerably, and forming a kind of defile. Here are the remains of an old gateway, and a little beyond them a mound of ruins, overgrown with underwood, which may have belonged to a guard-house.

Beyond this was a noble plane-tree which measured 8 yards

---

* Degirmán Dereh-ṣí, Mill-valley ?—F.S.

round its trunk, at a height of 12 yards from the ground; but still more remarkable for its fine and symmetrical form than for its dimensions. Its branches spread nearly equal in all directions, as it towered to a height of upwards of 60 feet.

The river, winding round about 2 miles to the N.E., turns suddenly to the W. before reaching the sea, and sweeps round the foot of a hill, which bears a considerable structure of various ages, and which announces itself as the castle of Tïum.

There were several vessels of small burthen in the river, and a modern village designated as Saferjí O'ghlá. Crossing over the hill, we passed by an ancient gateway and entered upon the picturesque ruins of Tium, beyond which, after passing a fine village, Beglaruṅ Kőï, we came to a second, Ḥisár-Aṅlú, the village attached to the castle, the residence of an 'Ayyán, who found us a home for the night.

Oct. 13.—We walked the previous evening to the site of Tium, by an ancient causeway, hedged in on both sides by bays, probably sprung from olden roots, as the road from Antioch to Daphne is in part similarly ornamented; yet such plantations are rare among the Mohammedans. Passing over the walls, we found pillars and fragments of ruins rising here and there, but everything was covered with a dense and almost impenetrable shrubbery. The evening was occupied in measuring and sketching a beautiful ivy and shrub-clad ruin, that appeared to have been a church or a basilica. The next day we began our labours at a less picturesque edifice, probably a guard-house, with two stone platforms, descending into the town; beyond, and nearer to the centre, were a few arches belonging to an aqueduct; from this point we visited a mound where were some curious ruins, and numerous sarcophagi, the lids of which were of large dimensions, and cut in solid stone, but the coffins were made merely of fragments of pink slaty limestone.

The next object of interest was a small but very perfect amphitheatre, now buried amid trees and shrubbery; and from this we went to the castle, which we found to be the most altered and rebuilt, and consequently the least interesting of all the ruins. We met with no inscriptions, but transmit the details of our researches, which are only of a general nature.

About midday we were ferried over the Filiyás, the day being very warm, therm. in shade 72°, and the temperature of the water only 50°. Our road at first lay along a level plain of alluvium, formed by the river; and we were once turned back by some deep marshes, beyond which we entered a thick forest, and began to ascend along acclivities composed of trap-rocks and limestones. We were now in the district of Kól Bazár, but there were few villages; and in the evening, after travelling about 10½ miles, we arrived at Ḳisil Elmah, the residence of the 'Ayyán, in a pretty valley which extended northwards about 3 miles to the sea-shore. The 'Ayyán was captain of a merchant-vessel trading between Bárṭán and Constantinople, and being an intelligent man, I have ventured to insert on the map the details which he gave to us of the coast-line from Cape Bábá eastward to Amáṣerah.

Oct. 14.—We could not stop the Sunday in the crowded residence of the 'Ayyán, so continued our journey, which was a short one, over the same kind of country, low, hilly, and wooded, but with a few more villages, for about 9 miles, when we came to the crest of some chalk-hills, from which a picturesque view was obtained of the river and modern town of Bárṭán, like most Oriental cities, looking best at a distance.

In the calcareous limestones of these hills we found remains of marine algæ, but no shells. An abundant spring which issued from the same formations indicated a temperature of 57°, the air in the shade being 79°. We had still a marshy plain, impassable on foot, to ride over before we got into the town, where we were lodged in the Khán.

Oct. 15.—Bárṭán, a town little visited by Europeans, is situated in lat. 41° 36′,* as determined by a mer. alt. of the sun. It is built at the junction of two rivers, the Ḳójahnás, from a village of the same name, and near which it is said to have its sources, and the O′rdeïrí, which flows from the foot of the Durnah Yaïlásí, in the district of Za'farán Bolí. When the two rivers unite they are called the Ṣú Cháṭí. The Kójahnás Armak flows through a deep bed in alluvial soil, being liable to an occasional rise of from 8 to 10 feet. Its depth at the present moment averaged from 6 to 8 feet ; its width was 28 yards, and its rate 2½ miles an hour. The O′rdeïrí exceeds the Kójahnás in width, being about 30 yards across, but is neither so deep nor so rapid. There is one stone bridge over the Kójahnás, and a wooden one replaces another that existed formerly on the same river at the N.W. end of the town. The communication over the O′rdeïrí is kept up by means of a ferry, but there are also remains of a stone bridge. There were numerous vessels building at Bárṭán, some of which were of upwards of 100 tons burthen, but the port is 2 miles below the town, which is 4 miles from the sea by the river, and 3 by land.

The town of Bárṭán has 650 houses, out of which there are eight houses of Christians, who have no church. The Mohammedans have five mosques. The houses, on account of the marshy character of the surrounding country, are all built of two stories, the

* Gaultier says, 41° 33′ 52″ N., 33° 14′ 8″ E. of Greenwich.

upper one of which is alone inhabited. For the same reasons, the town is carefully paved with large limestone slabs, better so than any Turkish town we had yet seen; and some of us were doubtful if the pavement did not belong to a period anterior to the Mohammedans, but remains of antiquity are too scarce at Bárṭán to certify as to its being an ancient site.

The town is built upon two low hills of cretaceous limestone dipping S.E. at an angle of 20°. The houses also extend into the valley between these hills, which rise S.E. and N.W. of each other, and stretch to the banks of the Kójahnás on the one side, and to those of the O'rdeïrí on the other, rising up the hill-side to the N. or beyond the latter river. Strabo says the Parthenius rises in Paphlagonia, and derives its name from the cheerful meadows through which it flows. The Parthenius is generally considered the same as the Bárṭán river.

Oct. 16.—Leaving our baggage at Bárṭán, we hired horses to take us to Amáṣeráh (4 hours) and back again the same day. We crossed the O'rdeïrí, and soon afterwards turned up a narrow valley with a small tributary to the former, designated as the Ḳará Cháï, or black river. The valley soon narrowed, and was nearly blocked up by sandstone cliffs, which often presented a rude outline with fantastic forms, and in one place a rocking stone is curiously perched upon a pinnacle of the same rock. We now commenced the ascent of wooded hills, the road a bad one, and continued along these till we came within view of the sea, when we turned to the E., by a steep descent, with steps hewn out of the solid rock. It would have required little, in a country where so little public spirit in the way of internal improvement exists as in Asiatic Turkey, to have determined that this road had been executed by another nation of workmen, but it was not long before we came to a small niche in the rock, destined to hold apparently a figure, and beyond was a tablet containing an inscription in Latin, of which we could only make out a few words, as PROTAGE NORENTI CLAUDI GERMANICI * * *. A little beyond this is a tablet basement, supporting an arched frame-work, with the upright figure of a Roman in his toga, much mutilated and the head broken off, but the attitude is graceful and the detail good. Close by was a column and pedestal cut in solid rock, and supporting a colossal eagle, of which the head had also been struck off. There were also two tablets, of which the inscriptions were quite illegible. The column was 12 feet high, the statue of natural size. The base of the frame was 7 feet wide, the height 12. The base of the column was 3 feet wide, the height 12 feet, and the eagle was 4 feet 6 inches in height.

Further on, on the road side, was a semi-circular arch, formed of one ring of solid masonry, 14 feet wide by 7 high, and running back 15 feet; ½ a mile farther, upon an elevated site, probably visible at sea, were the remains of an oblong monument, apparently a mausoleum, near which was the lid of a large sarcophagus.

The road now led circuitously down a steep hill towards Amáṣerah, which, like most modern Turkish towns that occupy ancient sites, is picturesquely situated. The town, which consists of 145 houses, and has a population of about 800 persons, is built upon a rocky peninsula that has two necks, the first formed by a minor and greater bay of the sea; the second by a small inlet, over which a narrow causeway led to what was formerly the castellated or military portion of the town. Rude and nearly perpendicular rocks form the cape called Díwán Búrní, which rises above the little bay to the S.W., and the whole of which was formerly built in with large stone blocks, like a well-kept harbour. The bay to the E. is wide and capacious, and beyond the town there is a rocky mass, forming (apparently always) an untenanted island; and to the E. a lesser rock is connected with the mainland by a wall in a ruinous and dilapidated condition. The whole of that part of the ancient and modern town which occupies the peninsula was surrounded by a wall defended by towers, which appear to have been renewed at various times, but to have received their chief regeneration from the Genoese, whose Christian escutcheons are over every gateway, and whose ornamental taste in architecture has here and there interwoven Gothic tracery and Byzantine wreaths amid the solid blocks of Roman perpetuity; and even eagles, sculptured on white marble, are seen prostrate at the angles or corners of walls which they once adorned. The town overlooks the sea to the N., but its greatest extent fronts the interior or the S.; and there extends before it and in the same direction a well-wooded and picturesque valley, which is replete with ruins of various character. One of the most extensive of these is a large building of red tiles, supporting in an unscientific manner huge blocks of stone, and cut up by numerous irregularly-disposed and irregularly-formed arches. This place is called Badistán by the natives: it was approached by a handsome gateway with a semi-circular arch, and appears to have been a monastery. At the foot of the mountains to the W. is a fragment of wall with two tiers of arches, which perhaps belonged to an aqueduct. On the hill-side are other ruins, overgrown with shrubbery, amid which they were just discernible, while a more distinct arch stood prominent, high up on the hill-side, and assisted in filling up a picture such as lesser Asia is almost unrivalled in producing, and in which monuments of bygone times, belonging to such varied epochs and people, are gathered together in the same little centre of unchanging natural beauty.

We are now about to quit Bithynia, and, crossing the Parthenius, to enter upon the no less interesting districts of Paphlagonia, but before doing so we may be allowed one or two general remarks.

First, it is worthy of notice that, in all the contacts that were observed between the cretaceous or supracretaceous limestones and the igneous rocks in Bithynia, only two orders of modifications induced upon the original aspect of the rock were common,—the one into a compact limestone that was not granular, and the other, and by far the most common, was into a slaty pink-coloured rock, indicating a large evolution of gaseous matters at the period of the effusion of the igneous rocks, without any very considerable heat. This peculiar red colouration of portions or zones of stratified rocks has already been attributed by some geologists to an impregnation derived from igneous sources, as in the case of the red schists of some of the Swiss cantons, which have appeared to be derived from the prolongation of a metallic or plutonic vein (Boué, vol. i. p. 484). And the origin of the same bands of red limestone in the Alps has been placed in the same category by Studeo. In the Bithynian chalk and supracretaceous limestones it is only where that formation is in contact with or in proximity to igneous action (which is always made sufficiently evident by the flexuous and contorted or variously dipping strata) that the same phenomena are observed.

Secondly, an instructive comparison may be made between the sedimentary formations of Bithynia and those of Paphlagonia. It would appear that the limestones of the first country, from their uniform mineralogical characters when unaltered by contact with volcanic rocks, and the continued absence of organic remains, excepting a few marine algæ, had been originally formed in the deep sea; while the ostracite sandstones and highly fossiliferous limestones of Paphlagonia have evidently had a littoral origin.

This view of the subject would put the Nicomedian peninsula and a large portion of Bithynia at one period in submarine depths, which were limited to the S. by the Olympus, which at the same time bounded to the N. the central lacustrine deposit of Asia Minor described by Mr. W. I. Hamilton as bounded to the S. by the western prolongation of Taurus, and upon the south-eastern acclivities of which we have traced the same littoral formations of the supracretaceous epochs, succeeded in Northern Syria by deep sea and non-fossiliferous rocks, apparently of the same epoch as the Bithynian deposits.

Lastly, although the country we had traversed from Akchah Shehr to the Parthenius was everywhere hilly, and sometimes mountainous, still the country of littoral mountains could always be distinguished from the more southerly and lofty chain of

Olympus, which, although broken into different parts, and known by various names, is still always distinguishable from the lateral parallel and transverse chains which give origin to the numerous small streams that flow into the sea, or form tributaries to the greater rivers, as the Sangarius, the Lycus, the Billæus, and the Parthenius.

From this great E. and W. chain, others apparently start at nearly right angles, and approach the shore from S. to N., but this would be an incorrect way of expressing the fact, for these are distinct systems of mountains, having a different origin and structure, generally plutonic, and bearing up on their flanks the broken and tilted-up fragments of those sedimentary deposits which enter into the composition of the round and irregular hilly districts of Bithynia. Such, more particularly, is the case on the hilly chains of Ḳójamán and of Kará Ṭágh and Ipsil, with their trachytic cones and outlying basaltic dykes; and of similar nature are three distinct ranges of hills, which advance in as many different promontories into the sea, between the embouchure of the Filiyás and that of the Bárṭán river, formed of a nucleus of rocks of the felspatho-pyroxenic series, succeeded by trachytes near Amáṣerah, and having between them an irregular forest-clad and hilly country composed of rude sandstones, altered limestones, and limestone shales, the valleys of which, when filled with detritus or alluvium, are almost alone subject to cultivation.

Oct. 18.—We left Bártán and pursued our journey up the course of the O'rdeïrí in a south-easterly direction. A rugged and mountainous district, that of the Ḳayá Dibbah (hollow rock), lay to our left, as it had to our right in going to Amáṣerah. In this little alpine and picturesque district a total difference is observed from the generally tame outline of the Olympus, and of the transverse chains of Bithynia. The same craggy steeps extend by the Ḳará Ḳayá, or black rock, another lofty limestone precipice, as far as to the sources of the O'rdeïrí, in a south-easterly direction; but more to the S. they are united to the Paphlagonian Olympus by mountains which are less lofty, with a rounded outline and wooded acclivities, named the I'ch-íl-ler Ṭághí, and which attain by trigonometrical measurement an elevation of 1966 feet above the lower plain of the O'rdeïrí.

The O'rdeïrí forces its way through a pass in the I'ch-íl-ler Ṭághí, which we reached in 4½ hours by a winding route, after fording the river three times.

At the entrance of the pass two lofty mountains rose to the right and left, clad to the very summit with forest-trees of varied and beautiful verdure, while the river, now a mountain-torrent, rolled over a stony bed below. This pass opened into a pretty but uninhabited plain, and then again narrowed, the road being carried amid huge masses of sandstone and conglomerate, overshadowed by laurel, ivy, box, myrtle, oleander, and other evergreen and deciduous shrubs. After passing beyond this, through a forest of birch, we forded the river, and came amid plane and some pine, the seeds of which had been brought down by the torrents, to where the O'rdeïrí received a tributary from a portion of the Ḳayá Dibbah, to the N., while we followed the southeasterly branch, and at a short distance reached a poor hamlet called Ṣarnísh, in the 'Ayyánliḳ of Oluz.

Oct. 19.—There was no possibility of getting the requisite number of horses at Ṣarnísh, so we mounted the few wretched animals that could be obtained, and put the luggage into waggons drawn by buffaloes. At about an hour's travel we came to a mosque in the forest where the neighbouring villagers had collected for prayer, and were at the same time roasting two whole sheep, to feast upon after service. Crossing the river at this point, we commenced a long ascent through a forest on the hillside, the road being made of logs of wood laid transversely. The lofty precipice of limestone called the Black Rock opened upon us to the N.E., and a tributary of the O'rdeïrí came through a rocky pass near its base. The barometer indicated for the height of the crest we were passing over about 900 feet.

Descending again into the well-wooded valley of the headwaters of the Ordeïrí, we passed several good saw-mills, more especially on approaching Dúrsán-lí, a village in a valley to the S., and the residence of the 'Ayyán of Ováh or Ováh Ḳaẓa sí. Dúrsán-li is corrupted from Dórt Ḥasanli—the four of Ḥasan.

Oct. 20.—Our road still lay towards the sources of the O'rdeïrí, and after a short journey through woods of plane and cork, with underwood and coarse grasses, we passed the mosque and villages of Bágh Jeviz,* which extends far up a valley to the N.; while our route lay along a more expansive and wooded vale to the S.E. Everything was upon a large scale, and truly alpine: at the head of this valley was the mountain of Durnah Yáïlá-sí, with a forest of pine fringing its rude acclivities, but with a bald summit above all: to the S., wild crags and precipices, the home of the mountain antelope and the ibex, alternated with dark woody recesses, that appeared almost unattainable. There was here and there a village in the bottom of the valley, and a few houses (more indeed than might have been expected in so secluded a spot) were scattered upon the summit and acclivities of the hills to the N. These hamlets were tenanted by a race of a very dark and swarthy hue, with uncombed hair and a neglected appearance. The rocks in the neighbourhood consisted of sandstone and sandstone

* Walnut-Garden.

conglomerates, passing into millstone grit, and associated with limestones, limestone conglomerate, and limestone shales. It took us exactly 4 hours and 40 minutes from the time we left Dúr Sanlí to gain the crest of the watershed of the O'rdeïrí. The two barometers indicated for this point an elevation of 3200 feet; but, although we had been always ascending since we left Bárṭán, it was very different with the country now before us, which, forming the two Iflánís, and named by Rennell, after an Oriental authority, "the stony Iflání," extended to the E. in an elevated and continuous moorland.

The great features of the new country we had now entered upon consisted, first, in the high range of mountains which began at the Durnah Yáïlá-sí, and extended to the S. as far as the culminating point of the Ṣarkhún Yáïlá-sí, now covered with snow. This range is identical with the Mons Orminius of the ancients.

N.E. of this was the great upland of Iflání, or central Paphlagonia, which is formed of supracretaceous rocks. In advancing towards the S. this upland and the stratified rocks of which it is composed begin to be broken up by water-courses and ravines. When once the upper crust is broken, a softer and more friable material beneath is carried away with rapidity till it meets with another hard bed: at the head of the lateral valleys there is thus only one rock-terrace above the rivulet-beds; but, in descending (as is generally the case), the number of these rock-terraces increases, while, at the same time, the valleys widen, till, pretty nearly at the same point, namely, on arriving at the valley of the Ṣóghánlí Sú, the table-land, which has gradually diminished in elevation, terminates in rounded headlands between different rivulets. Beyond the same river, a vast pile of horizontally-stratified deposits, named Ḳází Yáchí (Goose's neck), towers above the river to a height of at least 1000 feet. This is no mountain mass, but a portion of the same upland, denuded by various causes, more particularly the action of torrents, and is found to be upon the same level as the plain of Iflani, although to a spectator in the valley of the Ṣóghánlí Sú it appears as a distinct mountain height.

It was here, and at an elevation of upwards of 3000 feet above the sea, that we first met with almost continuous beds of large oysters, and in the limestones below cones and spiral univalves, generally of a gigantic size. The whole of the rocks around Za'farán Bólí are redolent with fossil remains, and some beds are composed entirely of nummulites.

The same evening we arrived at Za'farán Bólí, and were glad to enjoy the repose which the sabbath offered to us in the quarter of the Greek Christians, which is called Ḳurán Koï.

Za'farán Bólí, a town almost unknown to Europeans, is built at the junction of two small streams. The one comes from the N., the other from the N.E., and the united waters flow under the lofty arch of a cliff-overhanging bridge, and down deep rocky dells, to the Ṣóghánlí Ṣú. To the S.E., the upland, terminating in abrupt but low cliffs over the town, is occupied by a new barrack and its attached mosque. In the valley between this and the central upland is part of the town and the Khán, while ½ a mile beyond is the large suburb called Ḳir Ḳullah; and at the entrance of the same valley, called Ḳayá O'ghlí, a detached mass of rock bears upon its summit the ruinous wall of a fort of no great antiquity.

The central upland terminates in a circular disposition of its low rocky cliffs, and is again continued in the centre of the town by a detached rocky terrace, which is also surrounded by ruinous fortifications, and is the present residence of the governor, and the prison of the place.

The next valley is divided into two minor ones, one of which is ornamented at a short distance by the suburb of Tóḳátlí, containing about 150 houses, embosomed in gardens, while at the top of a rock-terrace, and approached by a steep ascent, is the suburb called Kurán Köï, the residence of the Greeks. Beyond this again are two other large villages, Bághlah and Búláḳ, each containing from 150 to 200 houses.

The town itself is situated in the mouths of the different ravines, and contains about 3000 houses of Mohammedans, while the Christians have 250 houses and one church, that of St. Stephen. Thus, without the suburbs, Za'farán Bólí may be considered to possess a population of 15,000 persons. It contains a tolerable chárshí or market, four handsome mosques, besides several smaller ones, two large kháns, and four public baths. Its chief trade is in saffron, which is largely cultivated in the neighbourhood, and has rendered this place one of the most flourishing and populous towns of Anatolia.

The terminal name of this city indicates a Greek origin, whether Hellenic or Byzantine. The first is a later interpolation of the staple commodity of the place. The tradition preserved by the Greek residents is that of a church founded here by Theodora, the wife of Justinian, and consecrated by the gift of a limb of St. Stephen, from relics brought from Palestine. It is probable that this empress, the frail object of Gibbon's just but unsparing sarcasm, met with kindness, or dreamt her first visions of future greatness, in Paphlagonia, which it is recorded she last left with the pleasing assurance that she was destined to become the wife of a potent monarch. The memory of such an event may have led in after-periods of devotion to the foundation of a church at a spot for which the name of Theodoropolis is still claimed.

It is recorded by Gibbon, on the authority of John Malela, Theophanes, and the historian of Justinian's building exploits,—Procopius,—that in the journey to the Pythian baths, through Bithynia, she distributed liberal alms to the churches, the monasteries, and the hospitals.

In that part of the Peutingerian tables which contains a cross road by Otresa ('Osmánjík̇) to Amasia, there is a back road which Rennell has supposed to be the same as the coast-line from Tium to Sinope, but which contains two names on the road from Amasia to Sinope, viz., Stephane and Thomia. The first name might have belonged to that community where the gem-covered member of the martyr is still most reverentially preserved—the second to Kastamúní.

Oct. 22.—We made an excursion to Ḳará Bíñár (black spring), a spot on the stony upland, 7 miles N.E. by E. from Za'farán Bólí, where, in a small isolated grove of dark pines, and amid numerous Mohammedan tombs, were fragments and capitals of columns of a plain Byzantine order, while in their neighbourhood a slab of rude limestone bears a still more rude effigy of a female figure, of less than natural size, the breasts bare, the face mutilated by time, and the shoulders ornamented with what were probably a pair of wings. Tradition has preserved no memory of this spot, at once claimed by the Mohammedans and the Christians as belonging to their progenitors, but whether a temple, a monastery, or a mausoleum, might be a matter of discussion.

We started by the village of Tóḳátlí, and, leaving Ḳir Ḳullah on our left, crossed the easterly ravine of Za'farán Bólí, called Ḳaya O'ghlí, and a limestone plain having two farms upon its almost naked surface; beyond this, and 6 miles from Za'farán Bólí, we crossed the deep and rocky dell called Serb Dereh (rough valley), which we had also traversed at a point higher up, on coming from Dúr Sanlí. This deep ravine completely cuts Za'farán Bólí from the upland, and would form a strong natural line of defence.

We returned by another line, following the Serb Dereh, and keeping the rivulet on our left-hand side, as far as to the valley of the Ṣóghánlí Ṣú, which we found full of villages. We passed through one large one, Yasí Kóï, containing 300 houses of Mohammedans, and seventy-five houses of Greek Christians. It has three minarets, and the population is engaged in the cultivation of saffron, besides which the Christians manufacture wine and opium in small quantities.

Oct. 23.—Started early in the morning upon an excursion to see the junction of the Ṣóghánlí Ṣú with the river of Ḥamámlí. Our road lay across the Búlák̇ Dereh, a ravine picturesque as the others, and then along the banks of the Ṣóghánlí Ṣú, till about 7 miles from the town we arrived at the junction of the two rivers, and at which point a bridge was carried over them. The united stream was 42 yards wide, 2 feet in depth, and flowing at a rate of about 3 miles an hour.

We returned from the place, over well-cultivated fields, now in part clothed with flowering plantations of colchicum autumnale, to the village of Búlák̥, from whence, passing over a rocky ridge, we entered upon a beautiful mountain-enclosed vale, at the foot of Orminius, covered with vineyards, and diversified by small country-houses. We journeyed up this valley about a mile to where it turns N., and at the head of this, passing over igneous rocks which have disrupted and borne up vast cliffs of limestone, we found a fine stream of water flowing from beneath the limestone rocks. We passed over the rocky beds, through which the waters had forced themselves a passage, and found an old channel blocked up by huge masses, which had fallen down from above, and probably diverted the stream from its original course. The good people of Za'farán Bólí make picnic parties to this picturesque spot, which is also famous among the Christians as the site of two monasteries—one devoted to Theodorus or Theodora, the other to St. John. The temperature of the water was 45°; the air 50°. We returned by the suburb of Bághlar or Bógházlú, so that we had now encompassed the town on all sides.

But the important point determined by this day's excursion related to the distribution of the rivers of Paphlagonia, concerning which more errors have crept into the maps than even in the upper course of the Halys.

All the maps indeed agree in making the river of Bólí flow into that of Filiyás, but all of them equally agree in making the rivers of Cherkesh and of Bayándír tributaries to the river of Bártán. But while Kinneir, Leake, and Lapie make the river of Za'farán Bólí (Ṣóghánlí Ṣú) and that of Aráj flow into the Halys by Kastamúní and Ṭásh Köprí, Rennell has made the Aráj, flowing from Ḳastamúní, join with the Ḥamámlí or Bayándír river, and flow into the Bárṭán.

All these various and conflicting views, each incorrect in some one particular, are simplified by the knowledge that the great basin of the Filiyás or Billæus river receives the waters of the rivers of Bólí, of Bayándír, of Cherkesh, and of Aráj, which flows from the western side of the Ḳastamúní hills, and joins the Bayándír river 2 hours above the junction of the Ṣóghánlí Ṣú with the same river. The Cherkesh or Bayándír river, which we had seen in 1837 at both those places, is said to flow onwards, and 8 hours from Ḥamámlí to receive the waters of the Mílán, coming from a mountainous country, including the governments of

Sháh-butún and O'lák; the chief places of which are Ak Básh, 4 hours from Milán, and Pahlaván, 11 hours from the same place. Near Mílán are said to be some thermal springs, and opposite to the junction the village of Akchah Hisár.

The united rivers of Bayandír, Aráj, and Sóghánlí Sú flow through a pass in the Orminius at the foot of the Sarkhún Yáïlá-sí, and are said to receive the waters of the Bólí river 20 hours from this point, which must be very wide of the truth.

The basin of the Bártán river, or Parthenius, we have seen consists of two comparatively small streams, the Kójahnás and O'rdeïrí, both of which flow from the northern and western slopes of the Orminius and the Paphlagonian Olympus, and united are less in magnitude than the Filiyás before it enters the Orminius.

The latitude of Za'farán Bólí, by the mean of two mer. alts. of the sun, is 41° 13′ N.; its longitude, by chronometers, 32° 53′. We had much bad weather and snow while here. The mean height of the barometer was 28·450, indicating an elevation of about 1200 feet.

The rock formations in the Orminius consist of trap-rocks and trap-conglomerates, with limestones and sandstones which were non-fossiliferous; and of sandstones and superincumbent limestones in the plains, which abound in organic remains.

Retracing our steps to our former barometric station [3200 ft.] at Sabán Chilah, we found ourselves in the midst of snow and ice, and our road lay through forests, where the snow tumbled upon us from overladen branches of fir. At the village of 'Osmánzikí our attention was attracted by a dyke of compact quartz rock, rising like a wall 20 feet above the soil. The road was in other respects very uninteresting. The moorland being little diversified, when cultivated, there were a few villages,—when not, it was a continuous waste or forest-land. It was late in the evening before we arrived at an isolated house, where the 'Ayyán of that portion of Iflání which is under the jurisdiction of Za'farán Bólí resides. There were several robbers, in chains, walking about the house. In the valley of the 'Ayyánlik we counted fifteen small villages.

Oct. 26.—Crossing over some low sandstone hills, we came to a first valley of Bedil, with five villages, and then another, which expanded into a plain, cultivated in almost every part, and studded with villages, in the midst of which was the Bázár, or market-village, called Istánból Bázár. At Constantinople the term for the first day in the Mohammedan week is Bázár,* and for the second, Bázár Erteh-sí. In Anatolia the second day in the week is called Devrek, but, when applied to a market, they say Istánból Bázár, or Constantinople market (day).

---

* Market (day), gúní being understood. Bázár-erteh-sí, market's morrow, *i. e.*, day after market-day.—F. S.

At Chelebí Kői, N. 5 E. of the Bázár, an alt. of the sun gave the latitude of the place 41° 24′ N., bar. 27.00 inches. After a short journey of 4 hours, we came to a stony district, at the foot of which was the residence of the 'Ayyán of Iflání, under the jurisdiction of Kastamúní: hence the two Ifláńís are always distinguished as Iflání of Za'farán Bólí, and Iflání of Kastamúní.

Iflání of Kastamúní is stated to have twenty villages under its jurisdiction; but the difficulty of obtaining information upon this subject may be judged of by the fact that the 'Ayyán of Iflání of Za'farán Bólí asserted that there were only twenty-four villages in his jurisdiction, while the 'Ayyán of Iflání of Kastamúní asserted that the same government contained nearly eighty villages.

A number of villages often assemble together to hold a court, more particularly in reference to taxation: they call this one Díván, and, when several unite, they are numbered accordingly. This arrangement is frequent in this part of the country; and hence, in the map, four or five villages will sometimes be found marked with the same name.

The mean elevation of the great upland of Paphlagonia may be judged of by the height of these places, situate at a distance from one another, and upon what constituted pretty nearly the average between the level of the valley bottoms and the tops of the undulating territory: such are Iflání of Za'farán Bólí, 3000 feet, Chelebí Kői, 2780 feet, and Iflání of Kastamúní, 2840 feet. The plain containing the Bázár of Iflání of Za'farán Bólí is about 100 feet below Chelebí Kői, while the dominating land, as at Sabánchilah (yet not out of the upland), attained an elevation of 3200 feet. The district of Dádáhí may be looked upon almost as a portion of the same upland, only that, at a height of 2400 feet, it is surrounded by mountains, and is separated from Iflání by the U'zún Búrún chain, the summit level or lowest part of the crest of which has an elevation of about 3600 feet.

To the north the Paphlagonian upland is broken up by abrupt and sharp or rounded mountains, and intersected by deep, narrow valleys; which, with their several rivulets and rivers, soon find their way to the Black Sea, through various windings and picturesque glens and ravines.

On this upland the cultivation consists almost entirely of wheat and barley; indeed it may be considered among the most productive wheat-countries of Anatolia. Besides this they also cultivate a species of *Polygonum* in the fields, and a *Chenopodium* in their gardens, principally to feed fowls, the eggs of which form a large article in their diet; but these seeds are ground also, and used in making bread. The gardens also furnish a little maïze in sunny exposures, and plenty of cabbages and pumpkins. The climate and soil are well adapted for potatoes. The appearance of the fields, with their short stubble, the marshy spots covered with coarse sedges, and the green sward, with its long festucas, is very similar to that of many parts of Ireland. The land is both manured and regularly top-dressed. As a general average, 80 ákahs, or 220 lbs. of wheat, fetch 25 piastres, or *5s.*; the same measure of barley, *3s.*

Oct. 27.—Crossing the limestone rocks of Iflání, the road opened upon the cultivated valley of Ṣighir (ox,) beyond which was a small plain, with five villages under one court (Díván), named Tekiyyeh Kőï (convent ville). Beyond this, the character of the country completely changed, from a continuous upland, intersected by nearly circular plains and valleys, with gentle slopes, to more lofty mountains, chiefly with conical, although not actually sharp, summits; rapid, but not abrupt acclivities, and deep and narrow valleys, clothed to the base of their sides with forests of fir; which, on the mountain sides and summits, alternated with equally prolific, but now leafless, woods of birch.

One of these narrow valleys now opened before us, having a little cultivation, and corresponding groups of hamlets like *eyries* on its side, while a black forest spread out below. This district is called the Ḳará Agháj (black tree). It is in the 'Ayyánliḳ of Chílání, where we arrived after about half an hour's farther ride.

An hour's journey from Chílání brought us to the foot of the U'zún Búrún, and in another hour we reached the summit level or lower part of the crest, for which the barometer indicated an elevation of 3600 feet. This mountain chain extends nearly N.E. and S.W., and is formed of rounded mountains with gentle acclivities, covered with wood from the base to the summit.

The descent was more rapid than the ascent. We passed some fine specimens of pure quartzose rocks; but, unfortunately, night overtook us, and entering upon a plain studded with villages, and belonging to the 'Ayyánliḳ of Dádáhí, after a journey of 4 hours from the crest of the mountains, we arrived at the 'Ayyán's house (first passing the Bázár in the middle of the plain), and where we were, as usual, well received, and treated with a large fire.

The 'Ayyánliḳ of Dádáhí has under its jurisdiction about twenty-four villages: the cultivation is the same at an elevation of 2500 feet as that of the Iflání district; but maize, tobacco, and French beans are added to their productions: vines do not succeed.

The plain itself is formed by the union of several valleys, but more particularly two large ones, which extend to the S.S.E., and to the S.W. The first, Khónṣilar, contains eleven villages, of from seven to twenty cottages; the second, thirteen villages of a large size. The plain is surrounded on all sides by mountains, which are not continuous, but rather in groups. To the east alone are some limestone ridges above the isolated residence of the

'Ayyán; but to the S., the S.W., and the N.W. are groups of rounded schistose hills. The highest of these is called Gölgí Bel, now covered with snow at its summit.

Oct. 28.—Started at first in a N.W. and then in a N.E. direction, over the range of hills which bounded the plain of Dádáhí. Our road up the hills lay at first N., a little W., then N., and finally N. a little E. to the crest. The hills were covered with forests of oak and fir.

The descent was to the E. of N., by the village of 'Arabahchí-lar (Waggoner's-town), consisting of three parts on opposite sides of a narrow and steep valley. From this we turned E., through Gerish (three small villages), then over low, fir-clad schistose hills, down to a brook, where we first met with fine quarries of excellent roofing-slate, a fact that may not be uninteresting to the Turkish government, as the slates now used at Constantinople are brought, at some expense, from Europe, chiefly from England.

This valley soon opened and turned N. ½ E. to the vale of the Daurikán Irmáḳ. It was most remarkable for its abundance of cranberry-trees, now in full fruit. The trees were small, with a hardy but stunted trunk, somewhat similar to the olive. Jays, blackbirds, and fieldfares were feasting upon the ripe berries.

On entering the valley of the Daurikán Irmáḳ, here about 13 yards wide, by 1 foot in depth, our road turned due E., and, with some deviations, led us to Júrímarán, the residence of a mukhtár, or sub-governor, who was to provide us with horses for our further journey.

Oct. 29.—We followed the river's bank about E. by N. 3 miles. It then took a bend, crossing from N.E. from a different country, consisting of chalk cliffs and hills of flinty gravel, from which some neighbouring villages are employed in making gun-flints. Dereh Kői is a small village, prettily situated in a deep glen surrounded by precipices. To the S.W. is an extensive plain, cultivated and covered with villages, belonging to Kúreh Kaẓá-sí (Kúreh district). To the N. a hilly district, where is the bázár of Júrímarán, and a conical hill, on which are the vestiges of a castellated building: we obtained bearings of it from the next plain. Our muleteers professed not to know the road or the names of the villages, so we got a boy as a guide from the next we came to.

Continuing over the upland of Salmánlí, the chalk terminated in a long and abrupt precipice, forming a terrace of rock, which rests upon igneous formations, and extended from E. to W., fronting a broad, plain-like valley, which was cultivated, and abounds in villages belonging to the district called O'lunjeh.

Beyond this valley we passed through a pine grove on chalk and sandstone, at the head of which a rivulet flowed from a subterranean passage: beyond this we came upon carbonaceous and other schistose rocks, and among lofty wooded hills. Night again overtook us, and there was no possibility of stopping; for had we done so our muleteers, who had been rebellious all day, would have decamped during the night, and left us without horses, so we made a forced march, ascending for upwards of an hour the mountains of Bakir Kúreh-sí (copper district), and descending into the little mountain-basin, in which the town, attached to the copper-mines, is situated, arrived there at about 9 P.M.

Oct. 30.—The general effect of the town of Bakir Kúreh-sí (copper district) is very good. It is situated in a deep hollow, for which the mean of the barometer gave an elevation of 2800 feet, and is surrounded by mountains, of which the most remarkable is called Bakír Sultán. Its acclivities are covered with the red-looking refuse of former mining operations, and it terminates in a rocky pinnacle which rises 710 feet above the town. On this pinnacle is a tomb and a keeper's house. The view from the summit presented, in every direction, a continuous succession of mountains, rounded out with steep declivities to the E. and W.; broken up into bold, rocky, limestone cliffs to the N., and overtopped by the more distant snowy summits of the Alkás Ṭághí*, to the S.E.; while clouds lay over the Black Sea like a white shroud spread at some distance beneath our feet. The effect of these mountains upon the climate may be judged of by the fact that the previous evening we were almost wet through by a drizzling rain, while the sky above was nearly cloudless, and the moon shone bright.

To the S.W. of the town is a conical mountain, named Ḳizil Ḳárá Ṭághí; to the S.E., Ḳirnáḳ Ṭághí, which we crossed on our arrival and at our departure; to the N.E., Ḳáz† Yúsof Ṭághí; and N. by E., the limestone cliff called 'A'r-Sizler Ḳayá. Such are the names of the principal mountains which are grouped around the copper-mines, comprehended by the Turks under the name of the Bakir Kúreh-sí, *i. e.* copper district.

There is every probability, from the character of the scoriæ, that the veins or beds from which the ore was formerly obtained consisted of compact copper pyrites; but although we descended some distance into a shaft, we were unable to obtain a specimen, nor were any in existence in the town itself. It is now many years since the mines were wrought. In some cases the walls and roof of the galleries fell in; in others they were filled with water, although there still exists some doubt whether the deepest part of the shaft goes below the level of the valley, and if not, the mines might be recovered by carrying horizontal or slightly-inclined galleries directly to where the waters are accumulated: at all events

* J. N., p. 648.

† Ḳází?—F. S.

a little enterprise might, with the aid of machinery, recover these long-lost mines.

That they were formerly very productive may be deduced from a statement made in Gibbon, chap. lxix., that, in the time of Mohammed II., Ismaïl Beg, prince of Sinope, yielded to the conqueror of Constantinople, on his summons, a city and a revenue of 200,000 ducats, derived, it is said, chiefly from the copper-mines, an amount which, Gibbon says, appears enormous. The Turkish geographer of Anatolia has said after Strabo (p. 562), and the circumstance has been repeated in modern geographical works, that the people employed in these mines emit a horrible stench from their bodies when they come to the surface. This, it is evident, could only apply to former times; but when in the old galleries, we did not perceive any either uncommon or unpleasant odour. The Sandaracurgium * of Strabo, which, according to that geographer, was rendered hollow even in his time by the continual mining operations carried on in its interior, was a branch of the Olgasys, Alḳás Ṭagh. Now as the Baḳir Kúreh-sí are situated in the Yerálàh Göz, a rocky region, only, as it were, detached from the loftier range, from which it also borrows its name, it is not impossible that the places may be identical.

In the present day a few thieves and convicts only, besides some more respectable persons, are engaged in sifting and collecting from the refuse of former times stones that may again be sent to the furnace with some chance of profit. It is curious, in examining the vast piles of refuse among which they are carrying on their labours, to observe how regularly the stones have become stratified, and how frequently the formation of carbonate of copper †, even in a botryoidal form has taken place. We also found a specimen of bone passing into a substance resembling turquoise, from impregnation with copper.

The furnaces were formerly upon the brow of the hill, where the former refuse exists, but they are now 16 in number in the valley, close to the water, by the aid of which bellows, of small size, are made to work.

The town itself has a handsome mosque, and upwards of 200 houses, of which one-half only are substantially built, the remainder are the dwellings of poor miners, and often ruinous and untenanted. With the loss of its resources, the town has fallen into great poverty. Its latitude by the sun's mer. alt. 41° 47′; its long. by chronometers 33° 50′ E.

Nov. 1.—We left Baḳir Kúreh-sí by the gap in Ḳirnáḳ Ṭághí,

* *Sandaracurgium* signifies "arsenic works," and is therefore a mine; consequently hollow as far as it has been worked.—F. S.

† Bicarbonate of copper and hydrate of copper.

and descending into the valley, followed the upward course of a rivulet flowing N.E., and round a mountain called I'kinjíler, composed of coarse clay-slates and ironshot schist, and covered with forests of pine and birch. We continued to ascend for a period of 3 hours, through forest and bad roads, made worse by increasing rain. The country then opened, and became less wooded; and after we had passed a high limestone cliff, we found the same rock beginning to abound, and forming a new feature in the district. A few miles farther, ostracite sandstone made its appearance, and we entered upon the plain of Daurikán, or Kúreh Kaẓásí (the Ḳáḍílik of the mining district), full of villages and well cultivated, and bounded by hills of pink and white cretaceous rocks.

Nov. 2.—The valley of Daurikán contains the river of the same name, which we had already seen at Júrimarán in a rocky district. It extended from N.E. to S.W., and contained fourteen villages, most of them large. We left the village of Daurikán the next day, and our route lay over hills of trap, black schist, and limestone, without wood, but pasturing cattle. Crossing two valleys we came to a low crest of wooded chalk hills, where is a guard-house; a little beyond these the valley of the Gök Irmáḳ was spread out at our feet, studded with villages and plantations, and backed by the city of Ḳastamúní, above which towered an old castle placed upon a rock.

From Za'farán Bólí to this place, except in the low valley of the Júrimarán, our elevation had probably never been less than 2000 feet above the sea. At Bakir Kúreh-sí 2800 feet, and at Daurikán, on the high upland of Yerálah Göz, and near the sources of the river of the same name, an approximate observation gave 3240 feet. This fact, which had hitherto been rendered most apparent to us by the climate and vegetation of the country we were travelling through, was now made actually visible, for, without having made any particular ascent from the general level of our road, the head of the Gök Irmáḳ, which river had still to flow some distance before it joined the Ḳízíl Irmáḳ (Halys), and must then still have a descent towards the sea, was at a level of several hundred feet below us.

On our road to Ḳastamúní, where we arrived early in the afternoon, we passed the river of Dádáhí flowing into the Gök Irmáḳ, which it exceeded in size, being about 10 yards wide by 1 in depth, while the river of Ḳastamúní is only from 7 to 8 yards wide where 1 foot deep, and becomes very shallow when spread over a large surface.

Nov. 3.—Ḳastamúní is a large Turkish town, situated in a valley from ½ to ¾ of a mile wide, which it completely fills up: a break in the hills which bound the city on the west formed another valley, which is filled up with suburb called Ḥisár Ardi,* while upon the rocky cliff above stand the ruins of an ancient castle.

The total number of houses is said to amount to 12,000, giving a population of 48,000 persons. From our inquiries the Greeks have only 110 houses, and the Armenians 20. The former have a small church dedicated to St. John the Baptist. The Armenians meet for prayer in a khán. In the Mohammedan city we counted thirty-six minarets, and there are twenty-four public baths.

The principal trade of Ḳastamúní is in wool, of which the the neighbourhood is said to produce nearly as good as that of Angora. The men also work largely in copper, and the women in cotton brought from 'Adanah in Cilicia, and of which sails for shipping are made and sent to Constantinople. They also print cottons and tan leather, but in the latter article Ṭásh Köprí excels them. There are said to be thirty-two printing-houses, having from four to eight presses each, also twenty-two dyeing-houses, of which six are for red and sixteen for blue dyes. There are only two tanneries. The country produces no grapes: wine is brought from Ṭósiyyah, rice from Ṭósiyyah and Bói-ábád, a little silk also from the latter place, and water-melons from Ṭásh Köprí and the gardens along the Gök Irmáḳ.

There are in the city four monasteries (Tekiyyeh) of stationary, and two of itinerant Dervíshes. The castle is a very rude structure, built of the same coarse sandstone as the rock on which it stands. The mortar is a mixture of lime and pebbles. Some of the towers, three of which are round, are nearly 50 ft. high; another is partly built of tiles, and some square ones, more especially such as flank the outer wall, are of better construction, and formed of larger stones, probably belonging to a more remote era than the rest. A plan was made by Mr. Russell, whence it appears that the castle is of an oblong form, 414 ft. long by 60 wide.

Ḳastamúní has always been under the Turkish dominion, the capital of a province or sanjáḳ, and was a long time the residence of a páshá, but, under the economical reforms of the present sultán, it has been made the seat of a mutesellim under the Páshá of Angora. Little more than a century ago the Christian inhabitants were expelled from this city and forced to take up their residence in a village on the Gök Irmá , still designated as Gaúr Köï. When re-admitted to live and trade within the precincts of the town they had no church, and only their old burial-ground, till under the present sultán, a fermán was granted, allowing them

---

* J. N., p. 648. Vulgo Ardhí.

to build a church and bury their dead near the abode of their forefathers.

The population and extent of Kastámúní claim some attention, though its situation is not among the most picturesque in Asia Minor, and notwithstanding its commerce is very inconsiderable. Some of the mosques and the new barracks rise above the surrounding houses, and are rather superior to common buildings; but the houses, although of two stories, are in general ill built; the streets are narrow and dirty, and the centre of the town is washed by a deep kennel, into which the filth of the whole place is collected. There are no open quays to enliven the scene, and only here and there a covered wooden bridge, across which the Mohammedan has to pick his way, lest he should wake the sleeping dogs, and be defiled by touching them.

Kastamúní is not unfrequently visited by the plague, and is always liable to bad fevers, more particularly malaria, which is said often to assume a very fatal type. At an altitude of 2350 ft. above the sea, the snow is said to lie two months upon the ground, and the summer to be very hot.

It is, however, in its past history that Kastamúní, with its commanding though ruined castle, presents itself to us with features of peculiar interest. The ancient title of Constambol, the city of Constantine has been claimed for it (Bell, vol. iv. p. 107). It is still called Constambel in a modern Greek work on geography published at Smyrna; and Kastambol is the most frequent name by which it is known out of the precincts of the town. Rennell even claims for it the still more ancient name of Germanicopolis, but it is better known in the pages of modern history under the name of Castamona (Castamon, Leake*), as the patrimonial estate, if not the capital of an independent kingdom under the Comneni family, before that family attained the power and the eminence which upheld for a while the fate of a sinking empire (Gibbon, chap. xlviii.). And the same fertile district, secluded by mountains on every side, was also the stronghold of an independent prince expelled by Ílderím Báyázíd, re-instated by Tímúr, and again expelled by Muḥammed I.

We have already had occasion to comment upon the comparative geography of the river Kastamúní. This river, designated as the Karáṣú on the maps, is known by the name of Gök Irmáḳ from its sources, 8 hours above Kastamúní, to its junction with the Ḳizil Irmáḳ or Halys. At Ḳastamúní it is a mere rivulet 7 yards wide by 1 ft. deep, but 6 miles below it receives the river of Dádáhí, 10 yards wide by 1 in depth. The united streams flow onwards by Ṭásh Köprí and Bói-ábád, to the Ḳizil Irmaḳ. The

* A probable abbreviation of Castra-Comneni.

river of Sarpunja is placed by Kinneir as a tributary to the Kastamúní river; but from all the information we could obtain, it appeared to flow into the Aráj. Kinneir first showed the connexion of the river of Kastamúní with the Halys. Rennell by mistake introduces two rivers, one of Kastamúní, the other of Tásh Köprí, and both flowing into the sea at different places.

The weather was very cloudy and uncertain during our stay of 3 days at Kastamúní. The lat. by the sun's mer. alt. was 41° 21′ N.; long. by two chronometers 33° 56′ E.; bar. 27.284 inches, indicating an elevation of 2400 feet: we also made observations for dip and magnetic intensity.

The rocks around Kastamúní consist, below, of primary schists; above these are cretaceous rocks, white, red, or brown, with few organic remains. The upper beds of chalk alternate with sandstone conglomerate. This rock crowns the castle-hill, and passes, in the hills E. and W. of the town, into sandstone. To the W. of the castle-hill the sandstone dips 7° E., but beyond this, and over the adjacent valley, the same beds dip W. at an angle of 9°.

The contrast of the dark-coloured schists with the red and white cretaceous beds furnishes some of the most remarkable features in the neighbourhood, more particularly in the glen at the N.W. extremity of the town.

Nov. 6.—We continued our journey, proceeding along the valley of the Gök Irmák, which flowed with few windings to the N.E. about 5½ miles, at which point low sandstone hills advancing from the S. from the neighbourhood of Kastamúní come down to the edge of the river, which, curving round the hills, afterwards take a course of E.N.E. We stopped at this point to obtain the sun's mer. alt., which gave the lat. 41° 26′. A little farther on the Gök Irmák receives a small tributary from the S. The temperature of the water was 13° cent.; of the air 13½° cent. 1½ mile farther we crossed the same river by a covered wooden bridge, beyond which the river stretched to the eastward along a pleasant valley, cultivated, full of villages, plantations, and gardens. 7 miles beyond the bridge the river enters into a rocky ravine, and is lost sight of for about a mile, when turning to the N. of E. it enters upon the valley of Tásh Köprí, surrounded by rocks and mountains, but wooded and cultivated in its centre, and abounding in villages.

Nov. 7.—Tásh Köprí (stone bridge) is on the right bank of the Gök Irmák, and is approached by a bridge 75 yards long, which formerly consisted of four arches; two of these now remain, and two others, which were carried away, are replaced by three low and badly-constructed modern arches: the river, which here flowed from W. to E., divided itself into four shallow streams.

The town is said to contain 1500 houses: we counted 10 minarets, 2 kháns, and 2 baths: tanners and blacksmiths form a large part of the population.

The number of architectural remains in Ṭásh Köprí attest an ancient site: old columns and hewn stones, cornices, &c., abound in the cemeteries. We visited a small building yclept a madreseh or college—a collection of hewn stones and remains of antiquity, put together in the form of a parallelogram, with an open space in the centre, and two rows of ancient columns, not two of which had capitals of the same order.

Near the building was a beautiful sarcophagus of white marble, 7 feet 9 inches long, 4 feet wide, and 3 feet 6 inches high, and ornamented laterally with wreaths encircling a mutilated human face: bulls' heads were sculptured on the sides and rams' heads at the corners, with bunches of grapes beneath. On the left bank of the river is a low hill, which is pointed out as having once borne a castle, of which there are no ruins at present. On a stone near the gateway is a mutilated inscription, of which only a few letters are legible. Mr. Russell also copied other inscriptions from the walls of the college, from a modern fountain, and one from the interior of a tanner's house. The following, which confirms the supposed identity of Ṭásh Köprí with the ancient Pompeiopolis, is inscribed on a large slab of stone inserted laterally into the wall of the building above alluded to as constructed of materials of former edifices and called the college. As an excuse for any inaccuracies, I should state that both this and the following were copied by Mr. Russell during a heavy shower of rain:—

ΑΓΑΘΗΙΤΥΧΗΙ
ΚΑΙΟΝ ΚΛΑΥΔΙΟΝ
ΓΑΛΙΤΤΙΑΝΟΝ
ΤΟΝ ΠΑΛΛΙΚΟΥ ΑΓΡΟ
ΝΟΜΕΑΝΤΑ ΦΙΛΤΑΤ
ΩΣ ΗΒΟΥΛΗ ΚΑΙ
ΟΔΗΜΟΣ ΤΗΣ ΜΗ
ΤΓΟΠΟΛΕ ΩΣ ΠΟΜ
ΠΗΙΟΠΟΛΕΩΣ...Ω
ΑΝΕΘΕΝΤΟ ΑΝΔ'ΡΙΑΣ
ΕΝΕΚΕΝ.*

The following was in the front of a fountain in the N.W. quarter of the town:—

* To Good Fortune. Caius Claudius Gallitianus, the son of Pallicus, the kind administrator of the country, the senate and people of Pompeiopolis, the metropolis, have dedicated on account of his courage.

L. FLAVIVS ASCLEPIADES
PETRONIANVS HIC POSITVS EST.*
ΛΟΥΚΙΟΣ ΦΛΑΥΙΟΣ ΑΣΚΛΗ
ΠΙΑΔΗΣ ΠΕΤΡΩΝΙΑΝΟΣ
ΕΝΘΑΔΕ ΚΑΤΟΙΧΕΤΑΙ.

It is well known that Kinneir, and after him Colonel Leake in his Map, identified Ṭásh Köprí with Pompeiopolis of the tables, and which site Rennell has sought for at Ṭósiyyah. The information given by the tables is so inaccurate, that, without an inscription, a coin, or monument, to satisfy the mind, the rest is all conjecture. If we take the distance given by the tables of Pompeiopolis from Gangra, the former comes nearer to Ṭósiyyah; but will the 27 m.p. given from Pompeiopolis to Sinope agree with the same place? Again, if Ṭásh Köprí (of which there now seems no doubt) be Pompeiopolis, it is much more than 35 m. p. from Gangra.

Having so far examined the antiquities of Ṭásh Köprí, we started, quitting the river flowing N.E., and began an ascent of the Ilik Tágh, a spur or rather portion of the Alḳás Ṭágh, which here crosses the valley of the Gök Irmáḳ, and unites with the Yerálah Göz, beyond which again, and nearer to Sinope, were some high and lofty pinnacles of limestone reposing upon primary schists.

Our general direction was E. by N., and we stopped upon an adjacent plain to take the sun's mer. alt., which gave for this place N. lat. 41° 29′. 400 yards to the W.S.W. was a small village of 7 houses called Khazíneh-dár Köï (Treasurer-ville), and about ½ a mile N. 52 E., but separated by a deep ravine, the dismantled walls and crumbling fragments of a castle occupied the summit of a nearly insulated rock. This castle was named Ḳiz Ḳal'ah-sí (Virgin-castle); not an uncommon name in the East, as, for instance, near Köpri, and probably referring to the castle's never having been taken. It appears to be the same as the Ḥiṣár I'má of Rennell, from which he named the chain of mountains which are now particularly known for furnishing the best masts for the Turkish navy, and consequently agreeing in this point with the Ilik Tágh.

Leaving the castle on our left we descended into a ravine with rivulets flowing to the Gök Irmáḳ, where we found fragments, apparently of old arches, called Cheḳmáḳ Ḳápú-sí (Tinder-box bridge); but whether a bridge or a defence to the defile it is difficult to determine.

Our ascent of the Ilik Ṭágh commenced from this point, amid rocks of mica slate, &c. We soon entered upon the pine forests, and attained the highest level of the range, where the barometer

* Lucius Flavius Asclepiades Petronianus is placed here.

indicated an elevation of 4000 feet. The only species of pine was the *Pinus pinea.* Some trees which we measured were upwards of 100 feet high, and 3 feet 4 inches in diameter, cutting into timber of 1 foot 9 inches square. Our ride through the forest was dreary and monotonous, and we only arrived after dark, and amidst the din of jackals, at Kavváshah Tekiyyeh-sí, as its name shows, a convent of dervishes, of whom there were four, out of the inhabitants of about 15 cottages.

Nov. 8.—Started early up hills of crystalline schists, the country around being now more bare of trees and in part cultivated, but at an elevation of about 3200 feet. After travelling 2 hours through this tract we opened upon the valley of the Gök Irmák, which immediately upon leaving the mountains was as crowded with villages as when we last saw it. Limestone had for some time past succeeded to the pine-clad schistose rocks, and they were themselves now succeeded by trap rocks and altered formations; but the valley of the river derived a new feature from the predominance of vast deposits of detritus; the acclivities, offering little resistance to the action of rain and springs, were deeply furrowed by ravines without any vegetable covering, and presenting a curious succession of different coloured indentations.

At an elevation of about 1770 feet above the sea, and about 700 above the level of the river, we stopped to get the sun's mer. alt., which gave the lat. 41° 28′ N. The hills around were clad with juniper (*J. Phœnicea* and *J. macrocarpa*) and with dwarf and prickly evergreen oak. A little more than an hour's journey brought us to Bóï-ábád, a small town beautifully situated upon a tributary to the Gök Irmák.

Nov. 9.—Bóï-ábád contains about 300 houses, in which the population is said to consist of 1000 females and 800 males. The town is divided into 11 mahallahs or divisions, each said to have its mosque: this is an exaggeration. There are 3 kháns and 2 ballis. The town is rather scattered, which adds to its beauty, and occupies a valley to the E. of the rivulet. The bed of the latter is filled with luxuriant gardens, full of fine fruit-trees overrun by vines. The rivulet of Boï-ábád is called Káz Dereh-sí (Goose-valley), and to the N. flows, in an open and cultivated vale about a $\frac{1}{4}$ of a mile in width, to its junction with the Gök Irmák (Amnias); but to the S. it issues from a narrow and secluded vale, containing a large village bearing the same name. This upper vale of Káz Dereh-sí is separated from the valley of Bóï-ábád by the limestone rocks which bear the remains of the castle or hill-fort, and the rivulet passes from the one valley into the other through a ravine with perpendicular sides at least 300 feet in height.

Boï-ábád appears to be partly a Persian name, in which language A'bád means a city. Bői has scarcely a Turkish meaning; in Wallachia and Moldavia Boï or Boïavar is a name applied to landed proprietors or lords of the soil. The best-informed natives had no idea of the meaning, but the Turks corrupt it sometimes to Boïvad or Boïavat, as it appears on some maps.*

The castle, which is attributed by the natives to the Genoese, but bears no inscriptions, was long the seat of an independent chieftain. It still contains within its walls about thirty dwelling-houses, which are said to have been deserted only about 8 years ago, when its lord was a certain Ḥuseïn, the last of the Chipawis† or Sipáhís (Gibbon, c. lxv.), a term used by the people of the country, and corresponding to the original of our Anglo-Indian term "sepoy." The last chieftain is said still to be alive in Constantinople; but his property was confiscated, and his followers have turned to the humble occupation of gardeners. When we visited this hill-fort, of which a plan was made by Mr. Russell, there was not a person within its walls, and the houses, still new although built of wood, were deserted as if yesterday: the pathway being overgrown with viper's bugloss (*Echium Creticum*), spurge, and other weeds, gave an almost painful idea of sudden desolation; but the inhabitants below spoke of the thing in a manner highly characteristic of the feeling at present predominant in Anatolia. "Of what use is it to live secluded on yonder mountain? Is it not better to dwell among gardens and corn-fields?"

By three sets of lunar distances we made the long. of Bői-ábád 34° 51′ E.; lat. by sun's mer. alt. 41° 27′ N.; variation 9° W.; barometer 28° 8, indicating an elevation of about 1000 feet above the sea.

Nov. 10.—Continued our journey along the valley of the Gök Irmáḳ, passing villages at almost every mile and a half, till we had travelled 4 hours, when we turned to the right, about a mile along the banks of a small rivulet, to the village of 'Alí Páshá-Shálí, where we stopped for the night, in a room similar to what we had at Bői-ábád, without a window, and dark as a dungeon.

The valley of the Gök Irmáḳ averaged to-day a width of from ½ mile to 1 mile; and from its numerous windings, wooded hills, and rocks, its general luxuriance of vegetation, and its villages and corn-fields, furnished a continued succession of fine and varied landscape.

The recent breccia and sandstone here rose in hills to an elevation sometimes of upwards of 800 feet; and although at first a mere pebbly detritus, as we proceeded down the valley, the same

---

* It is properly Bőyá-ábád, dye-town. Bőyá, colour or dye, is Turkish; ábád (abode), Persian. Such compounds are not uncommon.—F. S.

† Probably of the Chapwán family. Sipáhí is a Persian word, signifying "soldier;" misspelt *Spahi*, and *Sepoy*. It could hardly be changed into *Chipawi*.—F. S.

deposit assumed the character of true sandstone and sandstone conglomerate, rising often in lofty and nearly vertical precipices, in which the character of the beds varied much, from coarse to fine breccia, and from rude conglomerate rock to siliceous and calcareous freestone. From various circumstances there seems reason to believe that this valley had its existence, at least in part, anterior to the last igneous agency which disturbed the surface of Eastern Paphlagonia.

Nov. 11.—Quitting the valley of 'Alí Páshá-Shalí, we again entered upon that of the Gök Irmáḳ, and passing Ṭahirán, now a village, but formerly a small Moḥammedan town, found the valley beginning to lose its dimensions, and to be shut up by diallage and euphotide rocks, supporting compact limestone, and advancing in wooded precipices or rocky promontories (Ḳará Deñín) upon the bed of the river. A little farther we came to the junction of the Gök Irmáḳ and the Ḳizil Irmáḳ, which occurs in an open space; after this the latter having passed through high and precipitous rocks to the S.S.W. descends to the N. by E., and then suddenly sweeping round to the S., forces its way through a defile formed by perpendicular cliffs of limestone, called Ḳará Depeh (Black vale), while the rocks which it thus encloses to the S. are named the Adá Ṭagh (Island-mountain). The latitude of the point of junction of the two rivers by the sun's mer. alt. was 41° 22′ N.

It was our intention to have proceeded up the banks of the Ḳizil Irmáḳ to Ḥájí Ḥamzah, and from thence to 'Osmánjiḳ; and notwithstanding the many representations made to us of the impracticability of this route, we were determined to attempt it. Proceeding, however, up the Kizil Irmáḳ, about a mile from where it receives the waters of the Gök Irmáḳ, we came to the village of Beg Kö̈i, beyond which the former river comes through perpendicular rocks of schist and limestone, which afforded neither passage below nor above on either of its banks. Under these circumstances we could not do better than proceed down the river to the ferry of Vezír Köprí, and, from that place, cut off the westerly bend of the stream towards Ḥájí Ḥamzah, meeting the river again at 'Osmánjík.

Accordingly we forded the Gök Irmáḳ near its junction, and proceeded through the pass of the Ḳará Depeh. This pass, unless Paphlagonia extended farther east in the time of Xenophon, would appear to apply itself peculiarly to the description given by Hecatonymus, one of the ambassadors from Sinope to the Greeks at Cotyora, who said that Paphlagonia must of necessity be entered by but one pass, and that lay between two points of a rock exceeding high. The river runs through a gorge in a limestone range, which extends from S.W. to N.E., forming on the

S. bank a conical rock about 250 feet high, and then rises in a wide rampart of rock to the N.E. till it forms cliffs, at the distance of scarcely 2 miles, nearly 1000 feet in height. To the S.E. numerous rudely-shaped pinnacles of limestone crown the mountain summits, upon one of which there are some ruins, apparently of a monastic character, concerning which we could not obtain any information.

This defile has, in modern times, obtained notoriety from the frequency of the robberies committed in a neighbourhood so well adapted for such exploits; and this circumstance led the Turkish government to build a guard-house in the pass, which was tenanted by two useless veterans, who, however, gave us a hospitable shelter for the night. Long. by chron. 35° 14′ E., lat. 41° 21′ N.

From the upland plains of Iflání, at a mean elevation of 2500 feet above the level of the sea, we had descended to Ḳastamúní, at the head of the valley of the Gök Irmáḳ, about 150 feet. In following the deep declivity of that hollow, in part filled with detrital deposits, already at Ṭásh Ḳöprí, a distance of about 24 miles, the level had descended 550 feet more; at Bőï-ábád, by 800 feet more, and at the junction of the same river with the Ḳizil Irmáḳ, by 600 feet more; the level at the confluence, by barometer, being about 450 feet above the level of the sea: but through the rapids of Ḳará Depeh alone, the Ḳizil Irmáḳ descends at least 100 feet; the elevation at the Guard-house, below the pass, being only 350 feet, or 100 feet less than at a distance (by windings) of about 8 or 10 miles; and consequently having a fall of more than 10 feet to a mile. The river was flowing at a rate, by our measurements, of upwards of 5 miles an hour.

The influence of so rapid a decrease of elevation, combined, more especially, with the increased temperature of a valley where vegetation is protected by hills, the radiation of the sun from their sides, and the evaporation from numerous rivulets, is equally marked by the successive changes of the vegetable productions.

On the plains of Iflání, as previously seen, barley and wheat form the chief, if not, besides some culinary vegetables, the sole produce; but even at Ḳastamúní, in the valley of the Gök Irmáḳ, maize, tobacco, gourds, and a few melons, are produced; grapes, however, do not ripen well. The same fertile valley, down to Ṭásh Köprí, is for the most part cultivated, as far as to the base of the hills; while the water-side is occupied almost everywhere by villages and gardens, in which grapes begin to abound, and melons of various kinds are raised. Walnut-trees, deciduous oak, willows, and tall poplars are frequent.

The great forest of the Ilik Ṭágh, stretching from the water-side up to an elevation of 4000 feet, consists almost entirely of pine-trees *(Pinus pinea)*, which attain a height of upwards of 100 feet. These forests, therefore, are still, as they have long been, the most valuable of any in the Sultan's territories for timber of that description.

In descending from the Ilik Ṭágh to Bói-ábád, some rocky tracts afford a flourishing vegetation of shrubs of cypress, two species of juniper and evergreen oak. The castle-hill is covered with spurge. The fruitful valley below affords food for the few silkworms kept in this neighbourhood; vines, without care or cultivation, climb over the tops of the highest trees; and the gardens of the Ḳáz Dereh-sí produce the most delightful fruits and vegetables in Anatolia.

In the valley of the Gök Irmáḳ and the lower part of 'Ali Páshá Sháli, much rice is cultivated, as usual, in small fields, for purposes of irrigation: this, with that from Ṭósiyyah, supplies all the markets of this part of the country, and leaves a surplus to be shipped at Sínúb (Sinope), for the market of the capital. In this part of the valley the Acacia spinosa begins to make its appearance, and soon afterwards, with the evergreen oak, covers extensive slopes, and forms, together with it, the chief vegetation of the district. The larch is now not unfrequent in the ravines, and on the sides of hills a species of privet also becomes common—rose-trees abound; Arbor Judæ (*Cercis siliquastrum,*) is more rare. The Astragalus tragacanthus and Acacia spinosa render the thickets almost impassable. Broom and box are interspersed here and there; and that useful plant, *Rubia tinctoria,* flourishes under the hedges of madder, bramble, and privet.

At length on the banks of the Ḳizil Irmáḳ, coarse grasses are replaced by the three-fingered grass (*Dactylaria dactylon*). The cypress becomes a tree of some size. The pine is succeeded by the Pinus pinaster, a stunted tree which climbs up the acclivities of the Ḳará Depeh, till stopped by vertical precipices of limestone, over which, again, it often shows itself, but in separated patches or solitary trees, with roots issuing from crevices in the rocks, at the foot of which, even at this advanced season of the year, a scentless but pretty-coloured pink was in full flower. The banks of the river were clad with Syringa argentea, and the Tamarix. There are no Oleanders, but wild vines, like the climbing plants of South America, convert the trees on the banks of the river into shadowy bowers. There is also some underwood, chiefly tamarisks and blackberries, with sedge and tall grass, which shelters many wild boars. The evergreen oak still possesses itself of pebbly promontories and declivities of detritus. These features in the distribution of the vegetation disappeared when we left the river to approach Vezír Köprí, and an undulating and low hilly country of schist was uniformly covered with a shrubbery of deciduous oak and juniper, as well as some privet, till the same soil was reclaimed by cultivation, through the industry and wants of the inhabitants of a town and numerous villages.

Nov. 13.—We travelled along the banks of the Ḳizil Irmáḳ, 9 miles from the guard-house to the ferry, which we crossed in a rude boat. There were few villages in this part of the valley. We passed a ruinous khán, and a bridge which had been commenced, but never completed.

The river at this point was 80 yards wide, had a current of about 4 miles, and was from 3 to 4 feet in depth.

Following the river a short distance, we came to a steep cliff advancing upon the river, which appeared, at the distance of a mile or two, to pass through lofty cliffs of limestone, somewhat similar to those at Kará Depeh. There is no road along the river's banks to Báfirah. Turning at this point to the S.S.E., we passed the village of Cheltijak, containing about twenty houses and a mosque, and entered upon the country of low shrubs, previously described, reaching Vezír Köprí just as night set in, about 9 miles nearly S.E. of where we left the river's side.

Nov. 13.—The town of Vezír Köprí is situated in a hollow, containing a mere rivulet, which soon flows into a country of rude limestone rocks, while the more fertile district is backed to the south by the Ṭavshán Ṭághí. It contains about 1000 Mohammedan families, fifty Armenian, and twenty Greek, each of which sects has a church. The town is divided into quarters, sometimes separated by party walls; and the market is divided in the same way, each portion having regular gates for its protection. There is a bezestán, or covered market, for silks and fine goods, which is a tolerably well-looking edifice, with four domes built of tiles. At each of the gateways of this building a tombstone is dovetailed into the wall. One of them contains a mutilated inscription, and the sculptured insignia of a Greek priest. The other also bears an inscription, which, although turned upside down, was more legible:—

ΣΕΜΝΩΣ
ΚΑΙΖΗΣΑΝ
ΤΑΚΟΣΜΙ
ΩΣΕΤΗΜ
ΚΥΡΙΛΛΗ
ΣΥΤΟΥΤΕ
ΚΝΩΣΑΣΑ
ΜΕΤΑΥΤΟΥ
ΜΝΗΜΗΣΧΑ
ΡΙΝΕΝΤΩ
ΡϘΒΕΤΕΙ.*

* "Honourably, and having lived respectably forty years, Cyrilla, daughter of Sytus, who bore children to him. In memory [of him erected this monument] in the year 172."

There are no ancient buildings in the town, but some of the hewn stones alternating with tiles, which have been used in the construction of the baths, appear to be of better quality than are generally found in mere Mohammedan buildings. Fragments of columns are also not unfrequent. The four kháns are poor places; and only one of the four baths is good. The lat. by sun's mer. alt. is 41° 7′ N.; the long. by chron. 35° 35′ E. The mean of the barometer was 29·100, and at this low elevation, of about 800 feet, cotton and mulberry trees are cultivated; and storks' nests make their appearance on the chimney and house tops.

Vezír Köprí was formerly called Kedí Ḳal'ah by the Turks; and 4 hours to the south there is a castle situated upon one of the highest and most conical peaks of the Ṭavshán Ṭághí, called Ṭavshán Ḳal'ah-sí (Hare-castle). We would have visited this castle from Vezír Köprí, but did not know of its existence till we observed it far away to our left, on our route to 'Osmánjiḳ.

Nov. 15.—Our route lay up the valley of the river in a direction W. by S., towards the foot of the Ṭavshán Ṭághí (Hare-mount), with few villages and a scanty cultivation, also blocked up at a distance of 6 miles by rocky knolls of limestone. After travelling 3½ hours we came to the foot of Ṭavshán Ṭághí, and commenced our ascent amid forests of deciduous oak gradually becoming trees, interspersed with pine and beech. The ascent lasted 2½ hours, the rocks consisting of sandstone and limestone. The sandstone continues to the summit of the range, when it becomes laminar and schistose, dipping at a high angle to the S.W. At the crest the vegetation consists chiefly of small birch trees, covered with lichen, and in a state of decay, with shrubby vaccinium at the foot. At the summit, the barometer indicated an elevation of 3690 feet above the sea. After continuing along the crest for upwards of an hour, we descended the opposite declivity, 680 feet, to Ḳósájak, a village situate in a ravine amid cliffs of schistose and limestone, and inhabited by charcoal-burners attached to the silver-mines of Gúmish Khán. A curious phenomenon exhibited itself here in a huge mass of limestone, about 300 feet long by 100 high, which lay detached upon the declivity of a hill of mica slate. It certainly was not a boulder-stone, and was more probably tilted up into its present position.

Nov. 16.—Continuing along the valley of Ḳósjak in a W. ½ S. direction, we descended gradually towards the valley of the Ḳizil Irmáḳ. At a distance of about 9 miles we entered upon a plain bounded to the N. and E. by limestone cliffs; to the S.E. was the defile through which the great Constantinople road is carried from Merziván to 'Osmánjiḳ, which we soon afterwards joined at two or three straggling cottages called Ḥájí Ḥasan. The sun's merid. alt. gave the lat. of the centre of this plain 40° 58′; bar. 28·230.

From this point to the gap in the mountains above the valley of the Karchaḳ Chái, about 5 miles in distance, the road lay to the S. of W. Hence the Constantinople road is carried by the pass of Menzil 'A′shiḳí to the N. of W., a point of importance in the construction of this part of the country.

'Oṡmánjiḳ, with its trachytic pinnacles and castellated ruins, has been so frequently described, that we need not enter into any detail respecting it. The Ḳizil Irmáḳ at this place flows from S.E. to W.N.W. Bajazet's Bridge (Báyazíd's) consists of thirteen arches, and is 283 yards long, 8 wide, with water at this season of the year in only half the channel. The town has 300 houses, five mosques, and three kháns. The lat. by altitude of α Polaris 40° 57′ N.

Nov. 17.—We pursued our journey southwards, following the banks of the stream for 10 miles, when we left it to the right, the river crossing through a gorge in the mountains formed by the junction of the Ḳirḳ Delím mountains and those of Iskilúb. The Ḳizil Irmáḳ is in every respect a fine river, both below and above this pass, where we ferried across it on our way to Iskilúb; but it does not equal what would *à priori* be expected from a river of such a long course, and fed by so many large tributaries. In the valley of 'Oṡmánjiḳ, and in that of Kánḳarí, it is no more navigable than it is in the districts of the Ḳirḳ Delím, or the Adá Ṭágh. An exaggerated idea of the magnitude of this river has obtained credit in Europe from the vague reports of travellers, who have not adopted the test of actual measurement. Tavernier calls it a large river at Kesrá Köprí; Tournefort compares it to the Seine near Angora; Mr. Charles Vaughan estimated it pretty accurately at 160 yards at 'Oṡmánjiḳ; and at a different season of the year it had appeared beyond Suaz, to some of our party, almost as large as it now was in the vale of Kánḳarí. Several writers have called it Aitoe-su and Atoe-su;* but below Angora we found it generally recognised as the Red River, or Ḳizil Irmáḳ. The fact mentioned by Tournefort, and noticed by Rennell, on the authority of Ḥají Khalífah, of its passing below the parallel of Angora, through formations of red saliferous sandstone, is quite accurate.

The sun's mer. alt. to-day gave the lat. 40° 50′ N. The Ḳizil Irmáḳ W. 700 yards distant; the pass in the mountains S. 50° W.; and village of Ḳíjanláḳ N. 12° W. ½ mile. On quitting the river's side we crossed some hills of trachytic rocks, and

* A′tó-sú, or Aïtó-sú (A′tó, water), is some local name. The common Turks name their rivers from the neighbouring town, and consequently the same river has frequently several different names.—F. S.

descended into the valley of Ḥamám Gözí (Warm-bath's eye), so called from warm springs situated at the head of the valley. This valley contained two villages of Turkománs, the first of that nation we had met with in proceeding southwards. At one of these (Mujtelí) we stopped for the night. It is remarkable as containing some fragments of modern Greek buildings, columns, hewn stones, &c.; and in the side of a fountain there is a tombstone with scollop shell, a pilgrim's crook, and a deacon's badge, such as are used by the modern Greeks. Half an hour farther on was a village of Greeks, called Rúm Köi.

Nov. 18.—We approached Ḳirḳ Delím by a narrow pass in limestone rocks, in which were several caves; and on the eastern side was a large tomb hewn out of the solid rock, like the tombs of the kings of Pontus at Amasia. This huge relic of human labour was at a height of about 100 feet above the valley, and cut into the side of a precipice which fronted the N. It consisted of a hollow stone-coffin, hewn out of the solid rock, with which it was still connected at the top, but separated at the sides by a passage 4 feet 9 inches in width, and 31 feet in depth, but crossways, at the base, was 44 feet 9 inches. The tomb was ornamented with two lateral pillars, in low relief, and could only be entered by a small aperture about 4 feet high, and 15 feet from the ground; the total height being about 30 feet, and the width 44 feet 3 inches. Above the aperture was inscribed in colossal letters—

ΙΚΕΣΙΟΥ*

Some of the letters were painted red, and it is probably to the modern Greeks of Rúm Kőï that this monument is indebted for the inscription, as well as for the addition of a red cross and some rudely-painted letters.

Our ascent up Ḳirḳ Delím took us exactly one hour, when we attained an elevation of 3090 feet, and from which point the country soon extended in a high alpine plain, with a small lake, and in good part cultivated. The land gradually lowers beyond this to the great plain of Chúrúm, which, in its lowest part, is only 730 feet below the crest of the Ḳirḳ Delím. The intervening country is destitute of wood, thinly peopled, and dreary in aspect, but in parts cultivated by different crops in rotation. The sun's mer. alt. gave the lat. 40° 37′ N.; the Castle of Chúrúm bearing S. 30° E.

The Ḳirḳ Delím mountains have a nucleus of mica slate, but are composed principally of trachyte, earthy, compact, and pyroxenic, and of superimposed and tilted-up limestones. The chain is well wooded in parts with oak and pine. It extends from

* "Of Icesius."

the Ḳizil Irmáḳ, in the W., to the hilly district containing the silver-mines W. of Merziván, which by its southern branches separated Pontus from the Trochmian Galatia (Strabo, p. 561), and is considered by Rennell as a spur of Paryadres. But although from the plain of Merziván it has somewhat the appearance of a distinct chain, from so elevated a plain as that of Chúrúm, it loses its mountainous character, a circumstance which also applies itself to the Köseh Tágh, which, like low hills when seen from Chúrúm, descend, on the western side, down to the valley of the Ḳizil Irmáḳ, upwards of 1500 feet in a few miles. As Chúrúm has been visited lately by Col. Chesney and Mr. W. I. Hamilton, it need not therefore detain us long. The number of houses is said to amount to 1800, but they are generally of one story, and the population can scarcely be estimated at more than 7600. There are four kháns and as many baths, and we counted sixteen minarets from the castle. There are but few Christian families, and these are chiefly tilemakers and potters; there are also tanners at the place, and much wheat is sold in the market.

The castle, of which we made a plan, is nearly square, walled round, with towers at the angles, and two square towers between these on each side. The interior is occupied by dwelling-houses; the walls are of various dates, and have often been repaired or rebuilt, the original plan of the building having apparently been preserved. The principal stone is a fine compact white limestone, but an easily disentegrating sandstone has also been used. A great number of white marble columns has also been been worked into the wall, besides many Greek tombstones, with crosses and sculptures, and various inscriptions, most of which are illegible. We copied one of the most legible, near the gate. The lat. of Chúrúm, by the sun's mer. alt., was 40° 31′ 47″; and the long. by chron. was 34° 51′ E.; the mean of the bar. was 27·374, indicating an elevation of about 2360 feet.

Chúrúm, after the subjection of Chapván O'ghlí (the well-known Turkomán chieftain of Yúz-ḳat) was made the seat of a pasha, but soon sunk, from its want of resources, to a mere 'Ayyánlíḳ, under which are Ṣáz, 'Osmánjiḳ, Ḥájí Ḥamzah, Iskelíb, and Ḳaṣár: the two latter are towns unknown to Europeans.

Crossing the plains of Churúm iṇ a north-westerly direction, we approached the foot of the Köseh Ṭágh, which we entered by a valley planted with gardens and vineyards, and then ascended till we reached an alpine plain covered with pine, and having a village upon it. Beyond this we passed over the crest of the ridge at an altitude of 3330 feet, and then began to descend, passing a guard-house in our way. The sun's mer. alt. gave our lat. 40° 33′; and immediately beyond this we first entered upon beds of gypsum, which gave origin to a low, undulating country, with wide, grassy plains or nearly level uplands, occupied by migratory Turkomán tribes, and filling up the valley of the Ḳizil Irmáḳ from the pass in the Ḳirḳ Delím to the igneous rocks of Ḳal'ah-jiḳ.

The formations of the Köseh Ṭágh consisted of carbonaceous rock, &c., with altered limestones, and on the west declivity gypsum. It would be out of place here to enter into the detail of the mineralogical peculiarities, which we observed in these mountains: suffice it to remark that the various associations of the rocks are highly interesting. A short ride from the place where we entered upon the gypsum pasture lands, led us to the banks of the Ḳizil Irmáḳ, which we crossed without delay in a ferry-boat, and turned to the N. by W., the table mountain of Bayád bearing N. 65° W., and the pass of the Ḳizil Irmáḳ, through the Ḳirḳ Delím, N. 65° E.: we crossed similar marshy and grassy lands of gypsum till we arrived at the foot of the hills of Iskelíb. There was a great variety in the vegetation of these tracts, more especially with respect to that interesting point the social propagation of the same tribes; and it was evident that, as in the plains of Syria and Mesopotamia, a predominance of *Compositæ* and *Umbelliferæ* characterized the autumnal vegetation, and that one or two species had generally a large local development to the exclusion of others; and that these predominant species differ from the slightest variation in the soil and position. The season of the year, however, precluded any very accurate researches of the kind, except to an experienced botanist.

Leaving the plains, we entered into a small valley enclosed by rocks, chiefly cultivated with vineyards: volcanic rocks began now to protrude through the gypsum, and were succeeded by volcanic tufa, and conglomerates, dipping N.E. The castle of the town of Iskelíb, most singularly placed, then broke upon us, occupying the summit of a conical hill of white limestone embosomed in the centre of naked volcanic rocks and conglomerates, the barren and stony declivities of which gave little promise of ever becoming the site of a town. Minaret after minaret, and houses crowding from the deep valley up the rocky side of the hills, gradually opened upon us till they were seen sweeping circularly round the castle, almost as far as the eye in the dusky eve could reach. It presented the very remarkable picture of a town perched upon a naked rock; but the gardens in the bottom of the valley were productive of all sorts of fruits, although wood was rare.

Nov. 21.—Iskelíb, commonly pronounced Eskilúp, contains

1500 houses, chiefly of two stories in height, with tiled roofs: of these about thirty are within the castle. The remainder of the town is much scattered through the valleys round the castle, each of which is watered by its own rivulet, and when not occupied by buildings converted into fruitful gardens. The number of houses was obtained from the Shehr Kayásí* by Mr. Rasám, and the population may be estimated at 9000.

The castle or hill-fort is an old structure of irregular form, and very much dilapidated. The remains of towers at the opposite angles are still evident. The gateway and most of the walls have been repaired in modern times. At the southern foot of the rock on which the castle stands, and fronting the centre of the town, are several sepulchral caverns, two of which have some claims to elegance, being ornamented with sculptures: one of them is much mutilated, but upon the other two winged angels are clearly discernible, the one bearing a round cup, the other a branch. These two grottoes have also each of them two columns of incongruous orders, and, with the sculptures, are indifferent in execution, and evidently of the lowest Empire.

This town, like Kánkarí and Kal'ahjik, notwithstanding its size, its castle and antique remains, is in so secluded a spot in the centre of a mountain district, that it did not appear to have been visited by Europeans; and some jealousy was shown on our reaching a place where no Christian subjects of the Porte are allowed to take up a permanent abode. The few Rayás† that pursued their avocations here lived in the kháns, for by a law peculiar to the place, founded upon some superstitious notion, no Christian is suffered to live there for more than nine months in the year. Their wives are consequently left at Kánkari.

The weather was cloudy, with rain, during our stay at Iskelíb, and did not allow of any celestial observations. The mean of the barometer was 27° 348; mean temp. 51½°. Elevation about 2300 feet.

We quitted Iskelíb in a dense fog, and after about an hour's ride we left the igneous rocks, and came upon upland gypsum, bounded to the W. by a hilly range, the central portion of which, with conical summits, was covered with a thin vegetation of pine-trees. At 6 miles from Iskelíb we passed a large Kayá Aghzí (Pass's mouth), the weather had cleared up a little, and 2 miles farther, we came to the valley of Cháï Kóï (River-town), which brought us nearer to the foot of the hills where large beds of pitchstone and pitchstone porphyry hills appeared among trachytic rocks. Ascending the next hill, a singular,

* Shehr ketkhudá-sí, pronounced Shehir kyayá-sí, *i. e.* Deputy Governor of the town.—F.S.

† Properly Ri'ayah, or Ri'áyat, spelt Ryot by the Anglo-Indians.—F.S.

naked rock presented itself to our view, beyond and on the southern side of Ḳizil Irmáḳ, isolated upon a plateau of gypsum, and bearing E. 65 S. We could not divest our minds of the idea that this was a castle, occupying the site of Stabium, which we must have left, although sought after diligently, very little to our left, on our road from Chúrúm to Iskelíb: the natives, however, knew of no castle in that direction, but the intervention of the river renders the intercourse very small. After a journey of about 15 miles, we came to the valley of Bayád, in which are four villages; on the side of the hill, Depeh; at the head of the valley where we passed the night, Naháden; 600 yards below, Yátiḳchí; and high up on the hill side to the W., Urulán, a small village. The four villages are united under the name of Bayád; and that of Naháden is enclosed by a dyke of pitchstone porphyry, running E. and W. in a kind of recess, while a small river coming from the hilly range flowed nearly from N. to S.; over it was a bridge with some appearance of former solidity, if not antiquity; and near the mosque of Naháden was a mass of pitchstone porphyry hewn into a singular form, similar to that of a baptismal font. On the summit of a hill close to Naháden, and separating the recess in which the village stands, from the upper valley of the river, were the rude walls of an old hill-fort, which must in its best day have been but a very rude structure, not improbably a stronghold of Gallo-Grecians. It is also said that there is a castle one day's journey N.W. from this place.

Nov. 23.—Our route lay over a country nearly similar to that of the previous day. Alternating hills and valleys in the gypsum formation bounded to the W. by hills of sandstone. At a distance of about 6 miles from Bayád we approached these hills, in the narrow valleys opening from which were several Turkomán villages. The sun's mer. alt. gave our lat. 40° 34′ N. Beyond this we came to a deep ravine with a guard-house, succeeded by a dreary upland of unproductive gypsum, terminating in a valley of sandstone, and the village of O'lájik, where we stopped for the night. Here a change of soil is accompanied by fertility.

Nov. 24.—We started in a W. by N. direction, over cultivated lands, followed by plains of gypsum, varying our direction till we arrived at a spring at the head of the valley. Here we sent on our driver with the baggage-horses to Ḳángharí,* while we ourselves turned to the S.W., by the village of Belí Bágh, descending into a deep valley composed of alternating beds of red sand sandstone and gypsum, among which there was a mine of salt which we had some difficulty in finding.

* Or Ḳánḳarí. Pronounced vulgarly Chángrí; this substitution of the sound of *ch* for *k* is common among the Arabs S. of Baṣrah. See Niebuhr's Travels, Vol. ii. —F.S.

This mine occurs in clay beds in the gypsum, at an elevation of about 2500 feet above the sea, dipping about 70° to the N. The entrance fronts the S.: the slope varies, but mules go to the bottom of the mine with facility, and return loaded. The width varies from 7 to 28 yards, and the height from 4 to 6 yards, the works being carried on without order or regularity. The inclination at the bottom averaged 1 yard in 17. The total length was 400 yards; the direction very winding. There is a superintendent who receives from 1 to 1½ piastres per load: the people who come for salt have to dig for themselves. This salt, like that of Persia, which is found in formations of the same character and age, is remarkably pure, and in homogeneous masses, more or less transparent.

Salt is procured from other mines in this neighbourhood, but this is by far the most extensive bed. It is also procured in summer in the neighbourhood of Káṇḳarí from a small lake, round which the salt crystallizes when the water evaporates. The therm. indicated at the bottom of the mine a temp. of 23° cent., (73½°, Fahr.), the temp. of the external air being 13° (55½° Fahr.). The miners were working almost in a state of nudity. Their hammers were attached to very thin and flexible bars of wood, upwards of 3 feet in length.

The gypsum of Galatia resembles in its characters the gypsum of Paris in part, and still more closely that of Mesopotamia and Arabia. But laying aside the question of the division of that great formation into two distinct portions, there is little doubt but that the gypsum of Galatia or of central Anatolia belongs to the supracretaceous epoch.

The saliferous red sandstones, which are beneath the gypsum, and which constitute the great range of the Káṇḳarí hills, are, with the gypsum, broken up and altered by the eruption of trachytic and volcanic rocks, as in the neighbourhood of Iskelíb; and the lower beds of the same formation are composed of a rudely-aggregated mass of trachytic and volcanic rocks, with chlorite and jaspers; so that on first investigation it appeared like a mere local breccia. It was only after examining the development which the same formations subsequently took in the hilly districts of Galatia, where the red sandstone conglomerate passes into red and brown slaty sandstones, together with red sand and saliferous clay, and occupies large tracts of country, that the real importance of the formation declared itself, and it appeared as a part of the same saliferous sandstone which Mr. W. I. Hamilton met with between Yúz-ḳát and Aḳ-serái, and to which he called our attention, as the sandstone through which the Ḳizil Irmáḳ flows between Angora and Ḳaïṣariyyeh.

Now, it is important to remark that although this sandstone, like most saliferous deposits, contains few fossils, and that its age is difficult to determine, that still in their upper beds they alternate with the gypsum and its marls and clays; and this alternation in the low country near the Ḳizil Irmáḳ, between Kánḳarí and Iskelíb, is repeated as often as twenty or thirty times, establishing the contemporaneous origin of at least a part of these formations, as well as their almost immediate succession and close relation to each other: if, therefore, the gypsum can be demonstrated to belong to the tertiary epoch, so also must the accompanying sandstone and trachytic conglomerates; a result which will establish a further relation between the formations of the uplands of Asia Minor with the plains of Mesopotamia, Arabia, and parts of Persia and Syria. The non-occurrence of cycladic limestone, or lacustrine marl associated with the gypsum in south-eastern Galatia, cannot be properly considered till we have a more perfect acquaintance with those formations.

A couple of hours' ride over upland gypsum led us to a broad valley, crossed by a rivulet coming from the N. a little E., and met by another valley of less extent, having its rivulet coming from the N.W. The town of Kánḳarí, of considerable extent, occupies the base of a promontory of breccia, which advances between the two valleys, and is crowned by a ruined castle, the interior of which is filled with dwelling-houses.

Nov. 26.—Ḳánḳarí is, as just mentioned, a large town, containing about 3000 houses and a population of 18,000 persons. Out of these there are 30 Greek families and 16 Armenian. The Greeks have a church dedicated to St. Obadias: the Armenians have no place of worship. The Mohammedans have 8 large and some smaller mosques; there are 6 kháns and 4 public baths; the chief trade is in salt and wool; red berries* are also an article of trade. The town is built like Iskelíb at the junction of two valleys, but is all grouped, at the southern foot of the hill, on which the castle stands, in one large and irregular mass, only diversified by a few buildings placed apart, as the new barracks are which occupy a low and unhealthy position in the bottom of the valley, whilst a fine natural plateau, out of reach of malaria, exists close to it. Another is a singular Mohammedan building, of which we made a plan. It is called the Mejíd Ṭásh or "glorious stone,"† and appears to have been a convent of dervishes and a burial-place for holy men. It was said by a resident dervish to have been built in the time of Hárún el Ṛashíd, but an inscription on it shows that it dates from the reign of the Eyyubíte sovereigns of Syria, and John Lascaris at Constanti-

* Probably *yellow* berries (*rhamnus tinctorius*).—F. S.

† There is probably some mistake here, as such a compound is very unusual.—F. S.

nople, not long before the overthrow of the Khalifate by the Moghuls.

At Yapráḳli, 6 hours from Ḳánḳari, 1 from Túelit, and 10 hours from Ṭósiyyah, an annual fair is held, which begins on the 17th September and lasts 7 days: it is celebrated over all Lesser Asia. This place is also noted among Christians as being the site of the tomb of the prophet Elias, a rather dubious point, but there are said to be no stationary houses at the place, which is described as being upon a mountain height.

Our stay at Ḳánḳarí was prolonged a few days in order to get some astronomical observations, as we had been disappointed at Iskelíb. The weather was, however, very unfavourable, generally cloudy and much snow falling. We succeeded in getting lunar distances from Pollux; two sets of sun's alt. for time at this place; and sun's mer. alt.: these gave its long. 33° 49′ E., and 40° 35′ 50″ N. lat.: the mean of the barometer was 27.340; approx. elevation, 2350 feet.

Dec. 3.—We left Kánḳarí in a dense white fog and hoar frost below: our road lay along the valley of the united rivulets, the north-western of which was 12 yards wide at Kánḳarí. Soon after quitting the environs there is no cultivation, except in one small valley, 2 miles below, and then little else but vines and stunted tamarisks. The hills on both sides consist of gypsum, much waved and twisted in its stratification, with occasionally red sand and sandstone. After following the river about 6 miles we turned to the S.W. over a hill of gypsum; descended into the valley of Beg Dúzí; re-ascended, and crossing another bed of upland gypsum, again descended into the valley of Aḳghorán, which we followed in a S.E. direction. We were much hurt here at seeing some travelling Rayáhs scoffed at and assailed with the most contemptuous language by a parcel of boys and children. One more bed of upland gypsum brought us to the valley of the Túnáï river, which there is reason to believe comes from the mountains about Yapráḳlí. It is a small stream from 12 to 15 yards wide, which empties itself at a short distance from that point into the Ḳizil Irmáḳ. The valley was about ½ a mile wide in some parts, less in others, and there were two ḳishlás or winter villages, one belonging to the people of Aḳghorán, the other to the villagers of Inánduk, whose actual residence we soon observed at a short distance up a small valley to our right. Travelling up the valley of the Yapráḳlí river in a direction of S. 70° W., we arrived in little more than an hour's time at Túnáï, a small and dirty village, inhabited by guards, who are required for the protection of travellers at the time of the Yapráklí fair.

The fog had continued all day almost without interruption; at times we rose above it, on the high uplands, where the sun was seen through the mist; towards evening the effect of the colour of the sky, by the decomposition of the least refrangible rays, gave to the cliffs above Túnáï a beautiful and remarkable appearance. At daybreak next morning the sky was clear, the ground covered with hoar frost, and the summits of the neighbouring hills tinged with the first rays of the rising sun; but a dense bank of mist lay along the valley: in a few minutes the diffusion of vapour became general, and everything was wrapt in a fog, which continued more or less all day, only clearing up at intervals, but never leaving the sky cloudless.

Dec. 4.—The air at an early hour was cold, and our progress cheerless in a fog which scarcely permitted us to see the objects around; but had it been clear, there would have been little variety, for after leaving the valley we came upon a level upland of gypsum, on which for 4 miles there was only one stone, which birds had perched upon and whitened with their dung.

This character of country was, however, soon interrupted by the occurrence of igneous rocks which had broken the upland into fragments, leaving solitary round or conical hills with narrow valleys and passes between them, which became more extensive as we re-approached the gypsum: in these valleys there is either pasturage or cultivation. To the left was the large village of Chándur, before us a narrow rocky pass, beyond which a herd of camels were browsing or reposing with their heads towards the place of the sun, which was not visible, while to the extreme right was the village of Ḥájí Köi, backed by low hills, and to the left igneous rocks and gypsum stretched down to the banks of the Ḳizil Irmaḳ.

Crossing a more level country about 4 or 5 miles, with the remarkable conical hill, on which the castle of Ḳal'ahjiḳ * is placed, bearing S. 32° W., we arrived at that town, and were kindly received in the Christian quarter.

Ḳal'ah-jiḳ is remarkably situated at the base of an isolated and nearly conical hill, upon the summit of which is a fine old castle filled with dwelling-houses, which in some places assist in forming part of the walls: upon closer examination the castle is found to be in a very dilapidated state: still, from the precipitous nature of the rocks on most sides, and its own strength, it must in ancient warfare have been a very strong place. The remainder of the houses are congregated round the base of the hill, which the town completely encircles, and, except on the southern side, it scarcely extends beyond the base, nor

* Ḳal'ahjiḳ (little castle).

above a certain height up its sides. This place, which was sacked at the time that Ibráhím Páshá came to Angora, is now in a state of great poverty and partial ruin. It is said to contain 800 houses of Mohammedans and 60 houses of Armenian Christians, 5 of which contain 2 families. The town is divided into 14 maḥallahs or districts, each of which contains a mosque: there is but one khán and one public bath. The Armenians have a tolerable church, dedicated to St. John the Baptist, and in the neighbourhood are the ruins of an old monastery.

The circumstances which led to the partial destruction of this town were a general rebellion against their governor, Ḥájí Aḥmed Bey, on the occasion of his levying an oppressive tax. The governor was besieged in his own house: a swivel was brought down from the castle, but having been found useless, the house was set on fire, and the governor (whose brother was at the same time governor of Ḳánḳarí) was ultimately killed by the populace. They then placed themselves under the protection of Ibráhím Páshá, who sent hither 4000 soldiers under Ḥasan Yárdahchí. The Turks, however, were enabled to send 10,000 troops from the ḳádílíḳ of Ḳurán, and the result was that the Egyptians, having a smaller force, were obliged to retreat, and the unfortunate town was exposed to a most severe retribution.

Dec. 5.—At 2 h. 26m., A.M., there was a shock of an earthquake, which threatened to tumble about our ears the house in which we were lodged. The movement was in undulations and not irregular, and the house rolled for a second or two like a ship at sea: the sensation was equally unpleasant, while the mortar falling from the rafters and dirt coming down the chimney, with the strain which the building underwent, gave a momentary alarm, but nothing fell near us, and there were only two houses thrown down in the whole place. The dogs began to bark in every direction, and people were running about half naked with lights in their hands. This first shock was followed shortly afterwards by a second, but so feeble as to be scarcely felt. The barometer, a few minutes after the first shock, stood at 27.588 in.; at 6, A.M., it indicated a fall of .056: the ensuing day was as cloudy and misty as ever.

At 2 h. 35 m., P.M., we had another slight shock, rather circularly undulatory; bar. 27.456, ther. 51°; no wind, but cloudy. 8 h. 46 m. a fourth slight shock: bar. the same.

Dec. 6.—At 8 h. 17 m. and 10 h. 26 m., A.M., an irregular rumbling shock: bar. stationary. At midday the weather cleared up a little, and we could distinguish the castle for the first time since we had been here.

Dec. 7.—2 h. 35 m., A.M., a rather rude shock: bar. 27.550; ther. 49°: mist as usual. There was no wind, and the weather was calm both before and after the shocks: there was only upon one occasion a distant rumbling noise that accompanied the shock: there was not the slightest odorous exhalation perceptible, at least where we were. The effect upon the soil was imperceptible, nor could we hear of its affecting any of the neighbouring springs. The electrical condition of the atmosphere must have been, from the previously described state of the weather, subjected to great tension and great extremes, but we had no portable instruments to determine the extent of this phenomenon. On the night of the 3rd of December Mr. Russell had noticed a large meteor shooting from N.E. to S.W. We endeavoured to observe the direction of the oscillating wave by means of a basin filled with water, but the indications were unsatisfactory: to the feelings of all of us, however, the direction appeared to be from N.E. to S.W., which coincides with the direction in which the igneous rocks of the country have extended the line of their upheaving force, and is the same as the direction, not of the dip, but of the beds of the sedimentary formations.

The rock on which the castle and town are built is a trachyte. The only observations we got at Ḳal'ahjiḳ was an altitude of α Polaris, which gave its lat. 40° 9′, and one set of evening sights, which gave its long. by chronometer 33° 35′ E.

However interesting it might be to watch the phenomena of an earthquake, we were not at all sorry at leaving this unstable place, on which the fog lay as thick as ever, but when we ascended the hills to the S.W. and got above it, we found that it occupied chiefly the valley of the Ḳizil Irmáḳ and some of the adjacent cold uplands of gypsum. At about 4 miles from the town we met, among the various igneous and altered rocks which we crossed, a remarkable and interesting formation of vitrified schist.

On our right we had a mountain now partially covered with snow; and passing by a guard-house, we stopped a little beyond it, to take the sun's mer. alt., which placed us in lat. 40° 1′ N.

As this place is said to be infested by robbers, we were somewhat alarmed by the appearance of two horsemen armed with pistols, who came suddenly upon us; but when they saw that we were prepared to give them a warm reception, they left us unmolested.

A few miles beyond the guard-house we found the remains of a poor wretch who had been impaled there eight months before. Three other individuals had been impaled, at the same time as this miserable being, at Angora.

We descended from this point among rocks abounding in partridges, into a picturesque glen, at the head of which was a village where the robbers who rendered this district so insecure in part took up their abode. The country after this began to improve, some cultivation showed itself, and we arrived after a journey of 18 miles, at the large village of Ḥasan O'ghlán, inhabited by Turkománs, who are pastoral, and whose women make carpets.

Between the valley called Dereh Sheïkh, beyond the impaled man and the village of Ḥasan O'ghlán, we had crossed the line of division between the Ḳizil Irmáḳ and the Saḳáriyyeh. At the last-mentioned village there is an abundant spring, the water of which flows by Angora into the river of Aḳ Köprí. A hilly district, called Ḥasan Ṭághí, bounds this valley to the S., and extends eastwards to the banks of the Kizil Irmaḳ.

The distance of Ḥasan O'ghlán from Angora is about 15 miles, the heights of Ḥoseïn Ghází * bearing from the village S. 77° W. There are numerous fragments of ruins at Ḥasan O'ghlán, more especially wrought stones, columns, and capitals, but we found no inscriptions.

The soil for several miles is here composed of primary schists. These form the hill slopes and undulating country on which large flocks of Angora and other goats are pastured. This line of country is interrupted by Ḥoseïn Ghází, a lofty dyke of trachytic rocks, which, for miles in width, has altered the adjacent formations, and extends either in continuous masses or detached eruptive dykes and veins. Of the former is the Bághlún or Baulus Ṭághí, the Mount of St. Paul, an upland district with a rounded outline and occasional cliffs, extending to the N. of Angora. Of the latter are the hills of Angora, four in number, on one of which stand the castle and town: the second is called Khedrelez, or St. George,* and has on its summit a modern sepulchral chapel and ruins, supposed by some travellers to have belonged to a temple of Jupiter: there are also two smaller hills to the north.

There has been some discussion between two travellers * whether or not the castle is commanded by an adjacent height: Kinneir affirming the fact, Brown denying it. The hill upon which the castle is situated is separated from that of Khedrelez by a narrow ravine, which is about 100 yards in width, and contains the rivulet of Tabhanah (corrupted from Debbágh Khánah Cháï or Tannery river). The height of the castle above the valley, by our measurement, is 432 feet. The hill of Khedrelez is higher than the castle by a few feet, and the distance of the two crests is about 270 yards.

The rivulet of Tabhanah comes from the S.E., and at its passage

---

* J. N. 643.

† Khedrelez, a colloquial corruption of Khiḍr Iliyás or Khiẓr Iliyás, is the name of a Turkish saint and hero, confounded by the Turks with St. George and the Prophet Elias.—F.S.

‡ Bell's Geography, vol. iv., p. 119.

between the castle-hill and that of Khedrelcz, has been dammed in by a strong work, which was also defended by towers, coming down from the castle (of which there are also others defending every possible rocky ascent), and connected by a wall which is carried over the next hill. Flowing onwards circuitously between Ajit Tághí and Khedrelez, the rivulet curves round to the S. of W., is partly lost in irrigating the vegetable-gardens of Angora, and then flows N.W. till it joins the Chibúḳ Ṣú, immediately below the bridge of Aḳ Köprí (White Bridge), a modern structure, misnamed with respect to its colour, but reposing upon blocks of an older and better construction. The Chibúḳ Ṣú, so named from a village where pipes (Chibúḳ) are made, flows along a narrow valley of trachytic rocks, where are many farms (Chiftliḳ) and gardens, till it enters the valley of Angora, a mile above the monastery of St. Paul, and ½ a mile farther passes under Aḳ Köprí, N. 50° W. of the castle summit. The united stream is only 10 yards in width by about 2 feet in depth, but contains a good deal of fish.

It is needless to enter into details concerning the modern town of Angora, formerly Ancyra, and a city of considerable note in all ages of history. It has been visited and described by many travellers of ability, who have left little that is new to be explored.

Pococke and Tournefort have described the remains of Roman architecture, among which stands foremost a temple in honour of Augustus, of which the Latin inscriptions were copied by Pococke, and those in Greek by Mr. W. I. Hamilton, and copies are also said to exist in the Vatican at Rome. Several massive but irregular ruins of temples, guard-houses, or other public buildings, besides numerous inscriptions in the castle, and some rather rudely sculptured lions, belong probably to the Roman era, if they do not also illustrate partly the state of arts among the Galatians; but of that period few, if any, well authenticated remains appear to have been found.

Remains of Byzantine architecture are by far the most frequent: a column of little pretensions to beauty, and which imagination has dedicated to Lainius, the conqueror of Maximin; numerous sculptures in the walls of the castle; some inscriptions, and various Greek tombs and monuments, illustrate this period. Mr. Hamilton laboured hard among these interesting relics, and it is to be hoped that his researches will assist in elucidating the history of Angora.

Amid ruins of a more modern date are the castle as it now exists, a church of doubtful antiquity, and a subterranean viaduct of some extent; and in a small castle which occupies the highest part of the castle rock, are some old coats of mail of rude fashion and execution.

The modern town of Angora is divided into eighty-four

quarters, each having its great mosque or jámi'. There are from seventeen to eighteen kháns, and only three Ḥamáms. There was formerly a handsome bezestán or market for fine goods, but it is now in ruins. The reports of various travellers concerning the population of Angora vary from 20,000 persons to 100,000, which is not surprising considering the difficulty of obtaining correct information. The kádí, chief justice of the place, whose authority upon such a point ought to be considerable, estimated the population at 54,000, of whom 5,000 only are Christians. But it may be doubted whether the Turks themselves keep any correct account.

From the Christians we obtained an estimate which came to nearly the same thing, 8,000 houses of Mohammedans, 3,200 Armenian Catholics, 1,200 Greeks, 800 Roman Catholic Armenians (called schismatic by their brethren), and 300 Jews. This estimate appeared to all of us to be nearly one half above that of the probable population. According to an Armenian doctor long established in the town, the population consists of 10,000 Mohammedans, 5,000 Christians, and 200 Jews. It appears from the new commercial arrangements entered into between Turkey and Great Britain, and the internal reforms that have been projected in consequence of this by the sublime Porte, that a regular statistical report of the empire is intended to be made as preliminary to other internal changes.

The length and softness of Angora goats' hair is evidently to be attributed to an extreme climate. Cold winters (complained of by Manlius; and in the latter part of December, 1838, we found the snow upwards of a foot in depth, and the minimum temperature, $-16°$ cent., or $+3°$ Fahr.) have everywhere the effect of lengthening the hair or fleece of animals, or of supplying them, as in the uplands of Thibet, with an under down; while the hot summers give to the hair its silky lustre and softness. It is remarkable that not only the cats, but also the shepherds' dogs of the Angora breed have long and fine hair. That well-known breed of cats has lately much diminished, their fleeces having been used to adulterate furs. The circumscribed limits generally assigned to the country of this breed of goats are, as far as we have yet seen, correct: they are not met with to the E. of Kizil Irmák. The quantity of wool annually exported amounts, it is said, to 500,000 ókahs, but of this only 200,000 ókahs, or about 500,000 lbs., are of the more valuable fleece.

The other articles of commerce are yellow berries, the fruit of the *Rhamnus catharticus,* which is much cultivated, more particularly in stony places, as on the hills of St. Paul, and even on that of Khedrelez. The amount of produce is stated at 25,000 lbs. The roots of madder (*Rubia tinctoria*), for red dyes, mastic, tragacanth, and other gums, also form articles of commerce, as well as wax and honey. But the chief trade is in wool, merino twist, and goats' hides. The demand for British goods and manufactures is universally admitted to be very considerable.

The older European commerce of Angora was always very great. The tombstones in the burial ground of St. Paul's attest how many of our countrymen must have been engaged in it. It is difficult to account for its decline, unless from the want of success experienced, or from unwillingness to open communication in a place where consular protection has been abrogated now for a period of fifteen years. Almost abandoned by both English and French, who have now only a few native agents in the place, the Armenians have had the courage to establish a house of their own in London. When the Oriental question is settled, and tranquillity ensured to the country, there is little doubt but that commerce will spring up anew; and with the more activity as the resources of Anatolia, and of Western Asia in general, will be made better known.

The longitude of Angora was determined by lunar distances; and by two sets of observations with chronometers, to be 32° 50′ E. of Greenwich, and the latitude by mean of two mer. alts. of the sun 39° 56′ 30″ N. The mean of the barometer was 26·922, giving an approximate elevation of 2,750 feet. The variation of the compass was 9° W. The magnetic intensity of the earth's surface was also observed.

[Mr. Ainsworth's MS. is accompanied by a map of his route, in eight sheets, on a scale of an inch to four miles; by a plan of Heraclea; with several plans of castles and other buildings by Mr. Russell; an itinerary by Mr. Pulsford from Za'farán Bólí direct to Angora; numerous inscriptions, and a list of the names of places in the Arabic character, with their explanation by Mr. Rasám; and also the whole of the astronomical observations on which the several positions depend, as laid down in the map.]

X.—*Notes on a Journey from Jerusalem by Hebron, the Dead Sea, El Ghór, and Wádí 'Arabah to 'Aḳabah, and back by Petra; in April,* 1838. By the Count De Bertou.

Jerusalem.—After having visited Jericho and the northern end of the Dead Sea; also the monastery of St. Sabas, the Mount of the Ascension, and other remarkable spots in and around the Holy City, we left it on the morning of the 28th March, 1838, and, following the usual road in a general S.S.W. direction, we reached Hebron in the evening; the distance being about 22 miles.

April 1.—During the last three days we were occupied in examining the ruins of the ancient town; in sketching the mosque in which the tombs of Sarah, of Abraham, and of Jacob, are said to be found; and in making a bargain with the Arabs of the Jáhilí tribe,* to conduct us by the southern extremity of the Dead Sea, through the valley of El Ghór and the Wádí 'Arabah to El 'Aḳabah, at the head of the Ælanitic Gulf, and back to Hebron, by Petra; this, after much talking and wrangling, they consented to do for 3600 piastres (36*l.*); and at ½ past 7 this morning we started, under the escort of the Sheïkh 'Abdallah.

The road leads in a S. by E. direction, along the side of a hill about 300 feet high, between olive-groves and vineyards on either hand, but after a short distance cultivation disappears, and the country presents the aspect of complete sterility.

At 6 m. we find the vestiges of an ancient road near the Ḏhahraṭ el Zíf, at the foot of which are some wells. At 9 m. are the ruins of a castle called by the Arabs Karmel; the superstructure is evidently of the middle ages, but the large and well-cut stones of the foundations would seem to belong to an earlier period. The ruins scattered around mark the site of a town, doubtless that of Carmelia. From this spot the road turns off in a S.S.W. direction, and we shortly came to a rapid descent of 20 minutes, which brought us to the wells of Ḳaryateïn (the two villages), near which we saw the encampment of our Arabs, with their black tents stretching over the plain: to the eastward are the Jebel el Zo'arah† and Jebel Esdúm, being the mountains which form the western boundary of the lake Asphaltites. Esdúm is the Bedowin name for salt, and synonymous‡ with the Arab Milḥ. All these mountains, we were told, are covered with salt; and our Sheïkh, 'Abdallah, spoke of a large cave in them; but it would have lengthened our journey too much to visit it.

* Called in the plural number the Jáhilíyín, i. e. Jáhilí Arabs.

† This probably should be spelt Zoweïrah, and has no relation to the Zoar (Ṣo'ar) of Scripture.

‡ It is probable M. de Bertou misunderstood his Bedowin informers: Esdúm, or rather Sádúm itself, is evidently borrowed from Christians or Jews.

At Bír el Karyateïn are the remains of buildings, probably an Arab village; 2 miles farther we reached the camp of our Bedowin friends, and were hospitably received by the great Sheïkh Músa Abú Dahúk, chief of all the Bedowins of the mountains of 'Abraham the beloved;' his people, divided into three encampments, live happily under his laws. In summer, they told me, each camp seeks out fresh pasturage for its flocks and herds, which forcibly recalled to my mind that Esau and Jacob separated from each other for the same purpose; at every step in this country one finds a striking resemblance between the account given in the Bible of the patriarchs of old and the manners and customs of its present inhabitants—nothing has changed.

April 2.—After a fresh attempt on the part of the Sheïkh Músa to get more money from us, we started at noon, and, pursuing a general S S.E. direction over an undulating plain, with the grass dried up for want of water, we crossed the three Wádís of El Lobar, El Hadeïbí and Es Sayál, leaving on our right a small hill called Tel-el-'ard. At 8 miles we reached the ruins of Mask Es-sedíd, where there are fragments of columns rudely sculptured; crossed the valley of Omm el Budún, and at the end of 10 miles had the first glimpse of the Dead Sea, at the outlet of a deep valley on our left. Descending rapidly over ground covered with salt, and occasionally flints, we passed the cave of Magharat el 'Azírat, and shortly before 6 o'clock encamped in the midst of a country bearing the stamp of the most complete desolation. Nearly opposite to us, on the eastern side of the Dead Sea, a long, low, white peninsula projects some distance to the N.N.W.*

April 3.—We continued our descent in a S.E. direction, over calcareous ground, following the dry bed of a torrent called Wádí Zo'arah. At 4 miles passed on the right, in the precipitous face of a hill, the square entrance of an excavated tomb, named Magharat El Dábúrah (Deborah's Cave); our Arabs said it was the work of the Franks. At 8 miles we reached the ruined castle of Zo'arah; at the foot of which passes a torrent bearing the same name; tamarisks and acacias cover the ground, and there is a fine view of the mountains of Arabia in the distance. At 10 miles we reached the foot of the descent: here the waters of the Wádí Zo'arah spread out over a plain, called by the Arabs El Nafíleh, from the quantity of shrubs of that name with which it is covered.

From this point we changed our direction to S. by E., and tra-

* M. de Bertou's MS. is accompanied by a sketch of the Dead Sea, and the mountains on either side, taken from this spot; as also by a drawing of the channel of Wádí 'Arabah, taken 7 miles from its outlet into the valley of El Ghór; and a sketch of the Wádí Gharendel.—Ed.

velled parallel to, and within about 500 yards of, the Dead Sea on our left.

In the limestone hills on our right is a grotto named Magháraṯ Esdúm (Sádúm), whence gushes a salt stream; the Arabs say that the cave may be followed for some miles. The hills are covered with blocks of salt, 7 feet in length by 3 in depth.

At 15 miles we reached the southern end of the Dead Sea, entered Wádí el Ghór,* which is from 2 to 3 miles wide, and travelled over a plain covered with salt, at the foot of salt-hills; these hills diminish in height to the southward, and form the foreground to higher ranges behind them; they are in every part furrowed by salt torrents, which flow in winter, and inundate the plain over which we were travelling, in a direction a little to the W. of S.

At 20 miles crossed Wádí el Fuḳreṯ, coming from the hills to the W., and flowing, as well as numerous other streams, to the Dead Sea. We gradually left the salt-hills, and approached the hills to the E., crossing a marsh, in which the water was not quite so salt as before. At 22 m. we reached the chain of low hills, which since the morning had appeared to us the limit of El Ghór, and to close it up by uniting the mountains of salt to those of Arabia. These hills, which are from 60 to 70 feet in height, are of a whitish and very friable sandstone; they form the buttresses or outwork (contreforts) of the desert, which stretches to the S., and is known by the name of Wádí 'Arabah: they are channelled by numerous small streams, which fall into El Ghór, and eventually into the Dead Sea. A little further on we reached 'Ain el 'Arús (the fountain of the betrothed), a warm spring, of the temperature of 95° Fahr., while that of the air was 88°: the taste of the water was a little sulphureous. On looking back, the direction of El Ghór from this spot was N. 15° E. by compass. Here were some dwarf palms. Just before reaching the eastern hills, our guides turned suddenly to the right, and cried out "Wádí 'Arabah, Wádí 'Arabah," and we entered this celebrated valley, which at first had the appearance of the bed of a great river; and, if its slope were not visible towards the Dead Sea, one would exclaim on seeing it, "this is really the bed of the Jordan:" it is, however, the bed of a torrent which flows in an opposite direction, viz., from S. to N., and falls into El Ghór. At present here is no water; its breadth, which is from 250 to 300 yards, is filled with tamarisks, which the camels devoured eagerly; it extends in a S.S.W. direction, and is bounded by almost vertical banks of grey freestone, about 150 feet in height. We halted for the night at 3 miles from the entrance, and hid our camp as much as possible, in order to escape the lynx-eyes of

* Properly Ghaúr.

the Arabs of Kerek, who inhabit the eastern hills which command the Wádí.

April 4.—In attempting to describe the scenes which we had yesterday beheld, I feel the utter inadequacy of words to express my feelings. I had wandered through the Alps, the Pyrenees, and many other mountains: I had seen countries blasted by the curse of the Almighty, the plains of Moab and the land of Ammon, but had hitherto seen nothing to compare with the mountains of Zo'arah and of Esdúm. Here is desolation on the grandest scale, and beyond what the imagination of man could conceive: it must be seen—to describe it is impossible. In this striking and solemn waste, where nature is alike destitute of vegetation and inhabitants, man appears but an atom:—all around is enveloped in the silence of death—not a bird, not even an insect is seen! The regular step of our camels returned a dull sound, as if the earth were hollowed beneath their feet; the monotonous chant of the camel-driver accompanied at times the step of this inhabitant of the desert, but was suddenly stopped, as if he feared to awaken nature. Three Bedowins went before us to examine the road, for we have to fear meeting with Arabs who might be enemies of our tribe. The sun concealed itself by thick clouds, and seemed unwilling to shine upon the land cursed by the Almighty. We saw the traces of several wolves. Everything seemed to combine to make the landscape a scene awfully sublime.

We started at 7 this morning in a S.S.W. direction; the hills on the right are much more furrowed than those on the left. At 8 miles we got sight of the mountains of Nabí Hárún (the prophet Aaron) or Mount Hor, on the horizon, to the south-eastward, offering a picturesque outline. As we advanced, the Wádí became wider, and assumed the aspect of a desert; the hills on each side decreased in height, and the plain seemed to ascend.

At 9 miles the banks of the valley to the left disappeared, and on the right we perceived distant mountains in the S.W. At 10 miles we passed the opening of Wádí Afdel, which comes from the S.E., and shortly afterwards that of Wádí Koseïb (Rush Valley) from the S.W. At 15 miles we halted for a couple of hours at 'Aïn-el-Ḥafíreh, a small spring of tolerable water for the desert; it is situated at the spot where the road from Hebron to Petra crosses the Wádí 'Arabah. When Moses says, "Then we turned, and took our journey into the wilderness, by *the way of the Red Sea*,"* he certainly speaks of the Wádí 'Arabah;—no epithet can be more applicable—besides, it appears, on examining the configuration of the ground, that Moses had no other road to take in order to go from Mount Hor into the desert of Sinaï; he would be afraid to recross the mountains inhabited by his

* Deut. xi. 1. See also ch. i. 2, 40; Numb. xiv. 25, and xxi. 4.

enemies, and the people, carrying with them their flocks and herds, could not choose a better route; it is a smooth road, of vast width, on which even carriages might travel.

Continued our journey again for two hours, when we halted for the night; the outline of the eastern mountains is highly picturesque.

April 5.—Quitting our camp at 7 this morning, we travelled in a S.S.W. direction, through the valley, which is here about 1500 yards wide. At 3 miles passed on our left a small hill, resembling a truncated cone, called Dubbeṭ-el-Boghlah (Mule's-tail), and shortly after we saw Nabí Músạ, (the Prophet Moses,) at a great distance, bearing S.E. by S. At 5 miles we halted at 'Aïn-el-Ghamar (El Ḳamar?) a spring of the temperature of 15° Cent. (59 Fahr.); both the smell and taste of the water were detestable; but, as we should get no other till the following evening, we were compelled to fill our water-skins. A rock near this spring, of a soft reddish freestone, 70 feet high, is covered with the names, or rather the marks, of the Arabs who pass by this road; we added ours to the number. A female gazelle crossed our path, which the Arabs stopped by imitating the cry of the male. The ground is covered with flint pebbles; all vegetation has disappeared, and the Wádí is gradually lost in the slightly undulating plain, which extends towards the mountains in the E.

April 6.—During the morning we passed on our right the Wádí Ṭalḥ * (Acacia Vale), which extends to the westward, and which the Arabs pointed out to us as being the road to Egypt; in fact it is the route which Burckhardt followed in 1812, when he went from Wádí Músạ to Cairo. From the junction of the Wádí Ṭalḥ, the Arabs give the name of 'Akabah (the Ascent) to the southern prolongation of the Wádí 'Arabah, so that this spot would seem to be the line which separates the waters flowing to the Dead Sea from those discharged into the Red Sea. Indeed it is impossible to mistake the two slopes; one to the north, the other to the south. Hitherto we had seen no insects, but now we met myriads of beetles; we also killed a snake with a flat-pointed head.

From this point we turned off S.S.E., the ground still covered with small black flints, and with large but withered roses of Jericho (Anastatica Herichuntina), named nokdah by the Arabs. Shortly before noon, about 2 miles on our left, we passed an encampment of the Bedowins, Ḥasan ibn Jád, min 'Arab el Ḥoweïṭáṭ, at a place called El Ḳá'; and at the same time we had, at a few paces on our right, the tomb of Sheïkh Rejem Abú Dahíj, near which is also a hollow with water in it named Khaṭíb el Ḍhaï'ah. We passed Wádí Shádḥem on our left, and 2 miles beyond it halted at the

* Ṭalḥ, the Acacia gummifera.

junction of the Wádí Gharendel (Karendel?) with El 'Arabah, [?] while our Bedowins went to fetch water at the springs of Gharendel, about 4 miles distant.

At 5 P.M. continued our journey against the semúm (poisoned) wind, which brought with it a quantity of fine sand with which the plain is covered, and at 7 o'clock encamped for the night.

Continued our journey on the following day by the dry bed of a stream, bordered by tamarisks. At sunrise we had a magnificent view of the mountains of Hormah in the E. At 15 miles halted to get a supply of water at the spring of Ghadiyán, which is strongly impregnated with sulphur. At 3 m. beyond passed a spot named Rejem el hadíd (the iron cast); the valley is spread out into a great plain covered with small gravel of porphyry and granite.

April 8.—At ½ past 9 this morning we got the first sight of the Red Sea. The mountains on either side, which bound the Ælanitic Gulf, have a very picturesque outline; the soil we passed over is furrowed by the dry beds of torrents and covered with tamarisks. We soon after perceived the palm-trees of the Oasis of 'Akabah, and at 11 o'clock pitched our tents in the court of the castle. The luxury of having fresh water in abundance, after having been obliged for eight days to drink water impregnated with brimstone, and exhaling an odour of rotten eggs, and for the last two days even to have occasionally wanted that, is not easily to be imagined by those who have not experienced it; and when we saw a sakká (water-carrier) come to water the ground both within and without our tents, we could not help exclaiming at the apparent waste of so precious a fluid.

April 12.—During a stay of four days at 'Akabah we employed ourselves in sketching the scenery, &c., and in making a journey to Kasr-el-Bedawí (desert castle), about 3 miles from 'Akabah, on the E. side of the gulf. We had intended to have continued our journey to Sinaï, but, finding great difficulty in procuring a guide, &c., we determined to return by Wádí Músa. At noon we quitted the castle of 'Akabah, and, retracing our steps to 'Aïn Ghadhyán, encamped there for the night. On the following day turned off in a N.N.E. direction, and came to Wádí Deleghah, whose waters flow towards the Red Sea; the ground here is covered with mallows. Passed the ruins of the castle of Ká' on our left, and halted for the night at the line separating the Wádís 'Akabah and 'Arabah, called by the Arabs El Sath (the roof), or the culminating point.

April 14.—After an hour's journey we reached the channel of the waters of Wádí Ma'áferah, which flow into El 'Arabah; the hills, which here extend E. and W., reach to the eastern mountains, and thus form the separation between the two basins of the

Dead Sea and the Ælanitic Gulf. The track now led among the offsets of the eastern mountains, and we shortly entered Wádí Abú Kasheïbeh and continued along it to the eastward for 2 miles, when we resumed our direction to the N.N.E.; in this valley are almost trees of oleander (Nerium oleander), and abundance of partridges. The ascent became so steep that we dismounted from our camels, and continued our journey on foot, passing round the base of the peak known by the name of Mount Hor, and saw distinctly the ḳobbeh, or cupola, which covers the tomb of Aaron. Our road now led directly E., through the Wádí Hárún (Aaron's valley), which receives the waters of the Wádí Músa, and then turning N.E. for about 2 miles, we pitched our tents amidst the ruins of Petra.

During a stay of five days among these splendid remains, we occupied ourselves diligently in examining the sites, sketching the temples, and copying inscriptions, and we fully agree with all former travellers in the magnificence of this city of the dead, and can bear ample testimony to the accuracy of M. de Laborde's plates.

April 19.—Quitting Petra at 2 h. 30 m. P.M., we retraced our steps and halted in sight of Mount Hor, where we left our caravan, and, portfolio under the arm, put ourselves under the guidance of Abdallah, and set off to visit the tomb off Aaron.

Mount Hor is a mass of reddish sandstone which rises 1500 feet above the level on which it is placed; there is nothing worthy of the name of road by which to reach the summit; it was by climbing from rock to rock, and after an hour and a quarter of hard work, that we reached the Turbah of Hárún; we entered by the western door the ḳobbeh or dome which covers this ancient and venerable tomb, and found ourselves in a square chamber 9 yards long by 6 wide, having the roof supported by double arches which join in the middle and are supported by pillars: the first object which struck us on entering was a sarcophagus covered with drapery of painted cloth under which are some green rags: we thought at first that it was Aaron's tomb, but the Bedowins of the country, who had joined our party, said that it was only the tomb of his horse. An inscription in the ornamented Arabic or Kufic characters would doubtless tell in whose honour this tomb was erected, but as night was coming on I had not time to copy all of it:* the Bedowins lighted a lamp, which is left there for visitors, and they conducted us to a small staircase at the N.W. angle of the chamber. We descended fourteen steps and

* Unfortunately, the name of the person who erected this monument is in the three last lines which I had not time to copy; but the form of the character sufficiently indicates that this monument is very ancient. Dr. Lowe thinks of the time of 'Omar.

found ourselves in a narrow dark corridor, about 1 yard wide by 2 long, at the end of which we saw the tomb of the brother of the great law-giver. To describe the various sensations that pressed upon me at this moment, would be impossible; I will therefore confine myself to a brief description of what we saw: the mausoleum has the form of a quarter of a sphere placed on a pedestal; fragments of ornaments in wood, a small stone column, and two copper railings, which formerly surrounded the tomb, and are now hung up to the roof, are the only objects to be seen in this cave. The Arabs, who are always in expectation of finding treasure, have searched the stonework; and broken stones, and even some torn out at the base of the tomb, bear witness to the cupidity of this savage people.

We observed with astonishment that the Bedowins paid more respect to the tomb of the horse than to that of the prophet himself. We returned to the upper chamber, where several capitals and bases of small Corinthian columns are lying about. One shaft of a large column still stands upright behind the sarcophagus of the *faras* (horse), and others similar are visible in the external wall. Who could be the people to erect here a monument adorned with Corinthian columns? The Jews under the protection of the Romans or the Nabatheans?

Two large caldrons are left in the ḳobbeh for the use of the faithful who come hither to sacrifice to the prophet; one is destined for the offerings of sheep and goats, the other for that of camels; several *ex voto* offerings are suspended to the roof or to the columns of the upper chamber, such as shreds of cloth, ostrich eggs, glass beads, &c. After having seen everything well, M. Montfort, my travelling companion, and I, set to work to make drawings of both the interior and exterior of the tomb; nor did we leave the summit of Mount Hor till half an hour after sunset, and it was not without difficulty that we regained our tent. We had the satisfaction of possessing our two drawings of the tomb: but we are to start to-morrow morning before daylight, as our Bedowins have a dreadful fear of the Ḥaweiṭát—they are no longer in keïf.*

April 20.—Quitting our camp at Nakbel Robá', at the western entrance of Wádí Hárún, we set out in a N.N.W. direction, passing by the Wádí-el-Robá' into the Wádí el'Abyadh (white valley), and thus in a slanting direction across Wádí 'Arabah towards Hebron and Gaza. El 'Arabah is here very wide, and the ground covered with flints. About sunset we halted to fill our water-skins at the springs of El Weïbeh, or El Lúbiyeh,†

* "Good-humour."

† As lúbiyah (French bean) is a Persian or Hindí word, it is not likely that it would form part of a Bedowin name, such being usually genuine Arabic.—F. S.

which form a little oasis of palm-trees and reeds, and half an hour beyond it we encamped for the night, as our Jáhilín feared to stop at the well on account of the tribes in the neighbourhood with whom they are on bad terms.

April 21.—Continuing in a nearly N. direction we crossed not less than ten nameless Wádís, at intervals of as many miles, their dry beds all leading to the equally dry channel of El 'Arabah, about 3 miles on our right; these small valleys abound in thorny acacias. A little beyond, a low range of hills named Jebel el Ḳofaïḳifeh intervenes between El 'Arabah and our road, which, following their western foot, inclined to the N.N.W. for a few miles as far as Wádí Ḳoṭeïf, after passing which we left about 3 miles on our left an isolated small hill named by our guides Ḳádeseh or El Madárah; the name in this locality is remarkable —may it not be the Kadesh of the Scriptures? The roads to Gaza and Hebron separate here; following the latter, we soon after reached the Wádí Fukreh, which leads out of Jebel Yamen (Right-hand Mountain); this range, which had bounded our northern horizon since the morning, is the beginning of the mountains of Judea, which we now enter by a deep defile, the mural hills on either side rising from 150 to 200 feet; on reaching the end of the valley or gorge, we commence a steep ascent up the mountains, which are about 1000 feet high, winding by a very rough track through a wild and rocky defile, till at the summit we come out again upon an apparent desert. I had hoped and expected a better country: a little beyond this we found traces of vegetation, and halted for the night at the spring called 'Aïn Yamen.

April 22.—Our road this day crossed wide sandy plains, with bushes scattered here and there. After passing the Wádí Mezlíḳah, we traverse some low eminences and pass a river named Ḳurnúb on the left; here is a fountain and two fragments of buildings, which the Bedowins say was the work of the Christians. Farther on we passed the Wádí Abú Ṭarfá (Father Tamarisk), a small hill called Ḳubbet el Baul, the ruins of a town, and followed the course of the Wádí Shehábí in a westerly direction as far as a small ruined fort, where we turned due N. This was the route travelled by the pilgrims from Hebron to Ma'án on their way to Mecca at the time that Moḥammed 'Alí forbade them the route by 'Aḳabah.

At sunset we reached our former encampment at Bír Ḳaryateïn, where our Arabs were joyfully received by their own tribe of Jáhilín, but who had now left the valley and were encamped on the neighbouring hills. The Sheïkh Músa came forward to give us the Arabic salutation of welcome, Má rabá el Ḥamdu-l'illah! and it was not without a feeling of much satisfaction and pleasure that we saw our tent again pitched in the midst of a friendly people after an absence of four weeks, and felt grateful for our safe return from a journey of much fatigue and anxiety, not unattended with danger, and in which we had the gratification of knowing that we were the first Europeans who in modern times had traversed the whole extent of the Wádí, from the Dead Sea to 'Aḳabah, and have proved that, in the present state of things, the river Jordan never could have flowed into the Ælanitic Gulf. On the following day we returned to Hebron.

[It is due to M. de Bertou to state that the above account has been very much abridged from his detailed narrative, in which he gives a very minute itinerary, with compass-bearings nearly every quarter of an hour. These have been made use of in laying down his route in the accompanying map. M. de Bertou also gives some barometric, and several thermometric observations, in order to determine the elevation by the temperature of boiling water; but, as he himself states that his instruments were out of order, it is thought better to omit the results. M. de Bertou's original MS. is preserved in the library for reference.]

XI.—*Extract from a Notice on the Site of Ancient Tyre.* By the Count J. DE BERTOU.

Beïrút, 21st November, 1838.

IN addition to the account of my journey from Jerusalem to 'Akabah, I have now the pleasure to present to the Geographical Society of London the result of my researches concerning the site of ancient Tyre. My inquiries were based upon a series of questions proposed to me by M. Poulain de Bossay, of the Geographical Society of Paris,* and I hope some light may be thrown by them on the archæological and historical discussion raised in the learned world on the subject of the ancient metropolis of Phœnicia.

Perfectly aware of the difficulty which exists in making ancient topographical researches without the assistance of an exact representation of the ground, I hoped to do a useful thing in surveying and measuring, with great exactness, the contours of the peninsula upon which the village of Ṣúr is now built, accompanying this plan, made on a large scale, with a second, indicating both the respective positions and distances of the different localities which may have given rise to discussion.

* Bulletin de la Société de Geographie. Janvier, 1838, p. 47.

The questions that I intend to examine are these:—

I. On which point of the continent stood the first city of Tsúr or Ṣúr, founded by the Sidonians 240 years before the construction of Solomon's Temple?

II. Does the little peninsula upon which modern Ṣúr is built include all the island upon which Tyre stood before Alexander's conquest?

III. Could the harbours which exist now, even in restoring them to their ancient limits, be sufficiently large for such a maritime power as Tyre?

Following the order of these questions, I will examine, first, where is Ṣúr to be placed; and if I can only arrive at probabilities, the insufficiency of my researches will be fully explained by the words of the prophet Ezekiel: "Though thou be sought for, yet shalt thou never be found again, saith the Lord God." xxvi. 21.

[I. On a comparison of various passages of ancient writers, M. de Bertou is of opinion that the most ancient city of Tyre was on the continent around, and upon an eminence now known by the name of Marshúk, where are still some large pieces of rose-coloured granite columns, at a quarter of an hour's distance from the island of Ṣúr, in which island was formerly the temple of Jupiter Olympius. This temple, according to Dio, was joined to the city by Hiram, son of Abibalus, by raising a causeway. The causeway was again destroyed by the Tyrians in the reign of Nebuchadnezzar, when *they put the sea between them and their enemies* to escape from his arms; and from that time till the conquest of Alexander the inhabitants of Tyre confined themselves to the island.]

II. Does the little peninsula upon which modern Ṣúr is built include all the island upon which Tyre stood before Alexander's conquest? After having measured the peninsula, I came back to Beïrút, and there having collected all the historical information that I could procure respecting Tyre, I again asked myself the question, how could the narrow space that I have measured possibly afford room for a city so powerful as Tyre? In turning to Rollin's description of the siege of this city by Alexander, I was struck by the passage in which the historian informs us that the besiegers "avaient remparés le pied de la muraille de grosses pierres pour en empêcher l'abord" (Rollin, vol. vi. p. 101); and that when the assailants had taken away these stones, the base of the wall being cleared, the ships could easily approach. Now, at the present day, the peninsula is so surrounded with shoals and rocks that the smallest boat cannot approach it: if the same obstacles had existed in the time of Alexander, this "remparement" of the walls would have been superfluous, and the galleys of the besiegers could never have come to the foot of the ramparts. This passage opened a door to the probability of an interesting discovery, and I began to think that Maundrell had concluded justly in supposing that the greatest part of the island which Alexander besieged is now under water. I saw a further probability in favour of this supposition in the quantity of pillars which are found upon a rock on a level with the water (*vide* plan No. 9), and for the existence of which on this place I cannot account but upon the hypothesis, that what is at present a shoal was formerly part of the town, and that these columns belonged to a monument, the lighter materials of which, as well as the earth which covered the rock, were swept away by the sea, or rather "scraped," according to the word of the prophet Ezekiel; "I will also scrape her dust from her, and make her like the top of a rock." (ch. xxvi. 4.) This conjecture once formed, I mounted my horse and returned to Ṣúr, in order to arrive at a certainty, and thereby confirm or destroy it. Again on the spot, I took a boat, and soon found, both by repeated soundings and by information furnished by my guides (divers), that a bank of rock existed, which I have indicated on the plan by a grey tint. I do not think that it is venturing too far into the field of conjecture to suppose that this bank of rock, now under water, anciently formed part of the islands which were inhabited after Nebuchadnezzar had destroyed the town situated on the continent. So that the word of the prophet is again literally accomplished.

III. The third question which I have proposed to myself is this: Can the harbours which now exist, even on restoring them to their ancient limits, have been sufficiently large for such a maritime power as Tyre?

As I have already observed, it is impossible, on visiting the peninsula upon which the miserable village of Ṣúr is constructed, not to be struck with the contrast presented by the narrowness of its limits when compared with the grandeur and power of the city which was the parent of Carthage and Cadiz, and whose lofty walls arrested for so long a time the conqueror of Asia. But if the traveller cannot understand how so powerful a city could be crowded into so narrow a compass, his astonishment is still greater when, after having in vain sought for the ports necessary to shelter the innumerable vessels which covered the sea, he only finds miserable basins, which could never have contained more than 250 or 300 small galleys. In the face of such facts one is compelled to doubt the exactness of historians, or to seek by conjecture to reconcile their narratives with the truth. If the discovery of the ledge of rock, which I have before spoken of, the height of the houses of Tyre mentioned by Strabo, and the very natural supposition that this city possessed considerable settlements on the continent, account for the small extent of the actual peninsula, the other problem still remains to be solved—where are the ports which sheltered the innumerable vessels of this rich emporium?

Perplexed by this difficulty, I ardently sought for a solution, which I did not think afforded either by the extent of the limits of the northern port, as indicated on the plan by red dots, or by the discovery of the southern basin, which appears to me to have been used rather for the construction of galleys than to serve as a port, unless the wall of 25 feet thickness, which incloses it towards the sea, be the remains of that inclosure of 150 feet in height which surrounded Tyre when Alexander came to besiege it; and that the space comprised between this wall and the present shore formerly constituted a part of the city. The shallowness of the basin, and the great quantity of pillars and other materials which are seen under water, render this last supposition in some degree probable. Where, then, is the southern port, designated by Strabo under the name of the Egyptian port? I think I have found it. The sponge-divers, who lent me their boat to visit the shoals which encompass the peninsula, informed me of the existence of a submarine bank which extends to a distance of 2 miles in a S.S.W. direction towards Cape Blanco: this bank being covered by water to the depth of 1 to 3 fathoms, it was impossible for me to ascertain whether it is natural or artificial; that it exists is certain, and that it continues at least 2 miles in a perfectly straight line. I could perceive the bank very distinctly, especially by throwing oil on the surface of the water, and I estimate its breadth, which appeared to be always the same, at from 12 to 14 yards. If this embankment (digue) be artificial and was constructed to form a port, there is no longer any exaggeration possible as to the power of the Tyrians, and the riches of their commerce are attested by this great fact. The space comprised between the shore and this bank would form one of the largest ports known. In order to arrive at a certainty it is necessary to have a diving-bell at my disposal. I have requested the President of the Geographical Society at Paris to petition the Minister of Marine to cause the necessary instruments to be put on board the first ship of war which may come to this coast: then it would be easy to verify the fact. If this bank be artificial, or being natural, if it bear traces indicating that it has served as a foundation for a mole, the discovery appears to possess so high an historical interest that I hope, when known in Europe, some of the powers who may have vessels in the Mediterranean will furnish me with the means of verifying its exactness.

Shoals extend for ½ a mile to the north of the peninsula, and form a roadstead, in which the ships that now come to Súr anchor: may not a wall have existed also upon these rocks, and have formed a northern port nearly similar to that of the south? These two long moles projecting to the right and left of the island would correspond perfectly with the following simile, which I find in an Italian translation of Telemachus, and which probably may have been furnished to the Archbishop of Cambray by an historical document:—"Due gran mola simili a due robuste braccia, avanzandosi nel mare, formano un porto, a cui non può recare oltraggio l'impeto de' venti." If this phrase be not a mere figure, it is perfectly descriptive of the two moles, which I have supposed to have existed, and cannot be applied to the small ports, which, far from advancing into the sea like two vast arms, are, on the contrary, indented and almost invisible.

The most minute researches did not enable me to discover any inscription among the ruins of Tyre; I learnt from old men of the country that several marble tablets covered with Frank characters had been removed to 'Akká by Jezzár Páshá when he built his great mosque; and the Arabs, who have a vague idea of the celebrity of the place which they inhabit, added, that these inscriptions were of the time of Alexander. Another *stèlè* was carried off, they informed me, a short time ago, by an Austrian ship-of-war. But information given by Arabs is entitled to very little confidence. I have often experienced this, and very recently I uselessly made a journey of 36 miles in search of an inscription at Rás-el-Musherref, where I found nothing but an uneven surface of rock, discoloured by the action of the sea-water. I mention this circumstance, because it may save a fruitless journey to other travellers who might be disposed to allow themselves to be carried away as I was, by the hope of a discovery.

Except some pillars of grey granite, the greatest part of which are under water, few ruins are seen at Tyre. Alexander, in building the causeway, accomplished the words of Ezekiel, chap. xxvi. v. 12, "And they shall lay thy stones, and thy timber, and thy dust, in the midst of the water." So that all contribute to efface the vestiges of this city, which is to be "sought, not found." It is hardly a century since Maundrell said, "You see nothing here but a mere Babel of broken walls, pillars, vaults," &c. At this day these walls and vaults have disappeared under the sand which is constantly thrown up by the sea, and which would entirely bury the town but for the precautions the inhabitants take every year to secure the ground they occupy against its encroachments.

Altogether the prophecies against Tyre have been accomplished, even to the minutest details, and the best description of its actual state is that given by the prophets themselves.

The numbers of the *Journal des Savants*, of Nov. and Dec., 1835, contain two articles by M. Letronne, in which he critically examines the latest researches on the situation of Carthage by Capt. de Falbe and M. Dureau de la Malle; I find therein striking points of resemblance between the colony and its metropolis: and it may not be without interest to show that the founders of Carthage were guided by recollections of their country, both in the choice which they made of the site of the city and in the works which they executed. Tyre was first built upon the peninsula formed by the continent before the junction of the island with it by the causeway of Alexander, and this favourable situation procured for this city secure and large ports (Rollin, vol. vi. p. 90). Carthage was also built upon a peninsula, and the following words, borrowed from Appian, are equally applicable to Tyre and Carthage:—'Carthage was situated at the extremity of a gulf, and greatly resembled a peninsula' (App. Bell. Pun. ch. 95). Tyre had two ports; if they were formed by the two great moles of which I have spoken, they must have communicated by the strait which Alexander filled up. If these two ports were only the two basins represented on the plan, I am of opinion they must have passed from the northern to the southern port by the canal which separated the two islands, these two ports would have the same narrow entrance which we know was closed by an iron chain. Appian, speaking of the two ports of Carthage, says: "The ports were so situated, that it was necessary to pass from one into the other, and that there was but one entrance for both on the side of the sea; this was 70 feet wide and closed by iron chains." By a remarkable coincidence, both Alexander and Scipio constructed large causeways, the former to seize upon the mother, the other to reduce the daughter; and the inhabitants of the two cities, alike despising these works, at first supposed them to be impracticable, and only commenced any opposition after they had witnessed their success. This parallel might be pushed much farther, but I confine myself to pointing out such circumstances as appear to me most striking.

Although I fear I have already exceeded the limits of a letter, I cannot close my observations upon Tyre without mentioning the hydraulic works which the ancients executed at Rás-el-'Aïn, and noticing the prodigious quantity of water furnished by the different sources or subterraneous channels, the produce of which, confined in reservoirs, acquires a level sufficiently elevated to flow against the natural slope of the ground by means of aqueducts, which still exist as far as Marshúḳ.

There still remain at Rás-el-'Aïn seven reservoirs which furnish water: I will state in a separate note the dimensions of each. The water of the principal basin has been, and still is, employed as a moving power, as is shown by the construction of the basin itself; but one may perceive that it has also been carried by a watercourse to the basin which still supplies the aqueduct which went to Tyre, but at present stops at Marshúḳ after passing through the rock of that name. The dimensions of the channels correspond to the greatness of the city which it was to supply. An interesting question presents itself on the subject of this aqueduct: as I have just said, it stops at Marshúḳ, *i. e.*, at 2700 yards from Ṣúr; but it may be traced by the ruins, which are not yet covered by sand, to the towers marked on the plan by No. 19. These towers cover reservoirs of fresh water, which amply supply the wants of the 1200 inhabitants of the town, and also suffice to water the gardens around.

Volney supposes that the water is carried there by a subterranean channel wrought in the foundation of the aqueduct; I thought so myself for a moment, but a circumstance, which I had not observed at my first visit to Marshúḳ, appeared to destroy this opinion. In modern times a sugar-mill has been constructed on that site, and precisely at the spot where the aqueduct is interrupted a cave has been dug to receive the mechanism of the mill; this cave would have met the subterranean channel if it had been made in the foundations of the aqueduct, and if so, the water would no longer have reached Ṣúr. I have in my possession two bottles of water, one taken at Rás-el-'Aïn, and the other at the reservoir at Ṣúr, which I will cause to be analyzed; as to the temperature, there was 1° of difference in the two places. At Rás-el-'Aïn, 2 o'clock, P.M., the temperature of the air being 82° Fahr., that of the water 68°. In the reservoir at Ṣúr, temperature of the air the same, of the water 67°.

The temperature of the air being more elevated than that of the water of the reservoirs at Rás-el-'Aïn, it does not appear to me probable that this water is the produce of wells obtained by boring as several travellers have supposed, since a number of facts prove that the temperature of such wells is always higher than that of the surface of the earth.

Both Mr. Moore, the British consul at Beïrút, who accompanied me to Tyre, and myself, bought from the Arabs several intaglios of pretty good workmanship; my travelling companion possesses a Hercules beautifully executed, but the head of which is unfortunately a little damaged, and also a scarabæus of white agate in perfect preservation, bearing an inscription in Phœnician characters terminated by a date. I also purchased besides some coins, a Minerva Medica of fine workmanship and in perfect preservation, and a head, which from the features and the Phrygian bonnet, I fancy to be that of Æsop, and lastly, a Cupid, whose particular attributes I cannot make out.

A fellá̤h of Ṣúr some time ago found a marble torso of a young man; although the workmanship is not bad, it did not strike me as sufficiently good to merit a place in an European collection.

I have much pleasure in acquainting you with a discovery of a monument which affords a fresh confirmation of the accuracy of Herodotus respecting the expedition of Sesostris. Reading a short time ago this learned historian, I came to the following passage, after several paragraphs upon the same subject:—" La plupart des colonnes (*stèlè?*) que Sésostris fit élever dans les pays qu'il subjugua ne subsistent plus aujourd' hui. J'en ai pourtant vu dans la Palestine de Syrie, et j'y ai remarqué les parties naturelles de la femme, et les inscriptions dont j'ai parlé plus haut."

This obscene representation, which indicates the cowardice shown by the people in the defence of their territory, brought to my recollection that I had heard of the existence of similar representations in this neighbourhood, but which were attributed to certain rites of the Ansáríes. I thought I remembered that it was on the rocks of the Necropolis of 'Adelún they had been seen, and the following day at 3 o'clock in the morning, I rode thither with the hope of finding both these emblems and some monuments of Egyptian workmanship, which might attest their origin: at 4 o'clock in the afternoon I arrived at the monument described by Herodotus.

I visited with great care the whole of the Necropolis without finding any other sculptures than that alluded to. I may remark that the frame of this monument differs entirely from those of Nahr-el-Kelb, and does not bear, as they do, the emblem of a winged globe. Although the inscription on the above monument is very indistinct, I can affirm that it is written hieroglyphics. The rock of 'Adelún is hard and of a grey colour; in a word, similar to that of the defile at Nahr-el-Kelb; and if this monument is more defaced than those which are better preserved in the latter place, it is to be attributed to its being less sheltered, and more exposed to the westerly winds which prevail on this coast. The Necropolis of 'Adelún is situated 3 hours to the N. of Tyre, and the river Ḳásimíyeh (Leontes) flows at about an equal distance between these two places. Between the rock where the subterranean excavations exist and the sea, there is a plain of about 850 yards wide, covered with ruins which attest the existence of a city. Several names being rudely engraved on the rock in Greek characters, as well as Greek and Latin crosses on some of the tombs, I am led to suppose that it was inhabited by Christians. The Egyptian stèlè is situated 50 paces to the N. of a cavern which the traveller cannot fail to perceive in following the road from Sidon to Tyre. The Necropolis of 'Adelún contains more than 2000 excavations; all those which I entered were made to receive three bodies; the furthest compartment, being probably reserved for the head of the family, is invariably larger than those at the sides.

I sought in vain on the rocks and in the tombs for the allegorical images seen by Herodotus; but I have just learnt that those of which I had been informed, exist at a small distance (towards the S.) beyond the place where I discovered the Egyptian stèlè, and that they cover the sides of a small temple cut in the rock and in which is still seen an altar, and an inscription in Greek characters, probably the dedication of the temple to Venus. I shall not fail to visit with care this monument. Would it not be curious to find there some vestiges of Egyptian sculpture which would give to the allegory of which I have before spoken an origin which was probably unknown to those who wrote the Greek inscription?

To what city did the Necropolis of 'Adelún belong? It could not be Sarepta, whose name is preserved in that of Ṣarfand, a little Arab village, the position of which is well known. The city of Leontopolis must have been on the banks of the Leontes, where ruins are still seen; the Necropolis then must have depended on Ornithopolis, which Strabo places between Sidon and the Leontes.

References to the Plan of Ṣúr.

1. Bogáz, or entrance of the port, now obstructed by columns.
2. Forts.
3. Wall which enclosed the port.
4. Second wall or breakwater.
5. Space, thirty yards wide, in which are many columns under water.
6. Fort, with a pharos (built by the Crusaders ?)
7. Rocks on part of which the wall is built.
8. Column of beautiful rose-coloured granite.
9. Rocks and many grey granite columns. Line of a wall of circumvallation, amongst the ruins of which there is a considerable quantity of broken pottery: it extends from No. 7 round the coast to 16.
10. After excavating at this spot to the depth of nearly four yards, through ruins, I reached a pavement of slabs so large that I could not succeed in moving them.
11. Small hills, rising about four yards above the general level of the ground, which is about two yards above the sea.
12. Appears to be the highest point of the peninsula.
13. Hypogeum (square chamber, five yards by four).
14. Sarcophagus.
15. Granite column, still standing.
16. The stones which formed the base of the wall have marks which show that they were connected by iron cramps.
17. Two masses of masonry covered with sand. The space between them probably formed a landing-place.
18. Numerous columns on the beach and in the sea, possibly the site of the Temple of Hercules.
19. Square towers, above tanks of fresh water used by the inhabitants.
20. Port or building-dock, with 6 feet depth of water.
21. Walls nine yards wide, rising a little above the surface of the water, and usually taken for rocks. The external part of the walls is of fine broad stones, the interior of rubble and broken pottery.
22. Gate of the town.
23. A great number of columns under water.
24. Islets on the level of the sea.
25. An embankment or bank under water, extending 2 miles in a S.S.W. direction.
26. Columns of grey granite, marking the site of the Cathedral built by St. Eusebius.
27. Rocky bank which extends west one mile, with fourteen fathoms water at its outer extremity.
28. Angle of the ancient wall of circumvallation, probably the limit of the island?
29. In digging between these two lines, salt water is found at 6 feet depth.
30. Sandy bottom, with from 6 to 8 fathoms water, throughout the space comprised between the bank and the beach.

XII.—*Extracts from a Journal of Travels in Palestine &c., in 1838; undertaken for the Illustration of Biblical Geography.* By the Rev. E. Robinson and the Rev. E. Smith. Drawn up by the Rev. Edward Robinson, D.D., Professor of Theology in New York.

I. From 'Akabah to Jerusalem, through the Western Desert.

It had been our intention to go directly from 'Aḳabah to Wádí Músa along the great valley El 'Arabah; but circumstances induced us to change our plan; and we determined to keep our good Towárá guides, and take the road across the Great Western Desert in the direction of Gaza and Hebron,—a route as yet untrodden by modern travellers. Besides our five Ṭowárá Arabs under the direction of Tuwëileb, who had travelled with Rüppell, Laborde, and Lord Lindsay, we took two Arabs of the 'Amrán, a tribe living around 'Aḳabah and to the S.E. of that place, as the Ṭowárá were not acquainted with the route we proposed to follow. We left 'Aḳabah late in the afternoon of April 5th, 1838, and, recrossing the plain of Wádí 'Arabah, began to ascend the western mountains by the great Hajj route. We soon encamped for the night; and from this point we had seven long days' journey with camels to Hebron. The ascent soon becomes steep and difficult. The way is almost literally strewed with the bones of camels, and skirted with the graves of pilgrims; all testifying to the difficulty of the pass. On reaching the summit, we soon came out upon the great plateau of the Desert, probably from 1200 to 1500 feet above the sea, and found ourselves higher than the mountain-peaks which we had seen from below, and through which we had just ascended. Not far from the top of the pass, we left the Ḥajj route; and, turning off in a direction about N.N.W., we launched forth again into "the great and terrible wilderness."

For the first two days the general character of this desert was similar to that between Cairo and Suez,—a vast unbounded plain, a hard gravelly soil, irregular ridges of limestone hills in various directions, the mirage, and especially the Wádís or watercourses. Our Arabs gave to this part of the Desert the name Et-Tíh, the Desert of Wandering. The Wádís are here frequent: at first they all ran N.W. into the main watercourse of this part of the Desert, Wádí Jeráfeh; which, having its head far to the south, runs in a N.E. course to join the valley El 'Arabah nearly opposite to Mount Hor. We crossed Wádí Jeráfeh about the middle of the second day, and were struck with the traces of the large volume of water which apparently flows through it in the winter season. On the morning of the third day we reached the watershed of the Desert; after which all the Wádís run in a westerly direction into the great watercourse which drains the more western part of the Desert, and flows down to the sea near El 'Araïsh.

Almost from the time we entered upon this vast plain, we had before us, as a landmark, a high conical mountain, apparently isolated, along the western base of which we were to pass. It bears the name 'Aráif en-Nákah; and a lower ridge extends from it eastward. For nearly three days this mountain of the Desert was before us. As we approached it on the third day, the country became undulating and uneven, and the hills more frequent. I estimated the height of Jebel 'Aráif above the plain at about 500 feet; it is composed wholly of limestone, covered with pebbles of flint, and has no traces of volcanic action. It forms the south-western corner or bulwark of the mountainous region which extends hence to the northward; and from it a ridge stretches east, terminating in a bluff called Makrah, near El 'Arabah and opposite Mount Hor, as we saw, on a subsequent journey, from the pass of Nemellah.

The general elevation of the great plateau continues nearly the same, except where traversed by the Wádís; and the gradual ascent to the water-shed is not perceptible, and can only be discovered by the course of the streams in the valleys.

To the S.S.W. of Jebel 'Aráif is a mountain called Ikhrim, lying between our route and Wádí 'Araïsh, and farther to the north we saw the mountains Yelek and El-Helál. After passing Jebel 'Aráif, our course turned more towards the N.N.E., and the character of the Desert was changed. On our right was now a mountainous district, composed of irregular limestone ridges, running in various directions, and occupying the whole region quite to Wádí 'Arabah; as we had afterwards an opportunity of observing. This mountainous district is penetrated by none of the roads which lead from the vicinity of the Red Sea to Gaza or Jerusalem; but these roads all fall into the one we were travelling before reaching Jebal 'Aráif, or not far from that mountain. All these circumstances go to show that our route could be no other than the ancient Roman road from Aïlah to Hebron and Jerusalem; which also, like the present road, could not well have been anything more than a caravan route for beasts of burden.

The road passes along the western side of this mountainous district, crossing many broad Wádís which flow down from it westward, with elevated ridges of table-land between them. We made frequent and minute inquiry after the names of places or stations which are known to have existed anciently upon this Roman road. Of the more southern ones, Rosa and Gypsaria, we could find no trace. Early on the fourth day we crossed a broad Wádí called El-Lisán, marking perhaps the site of ancient Lyssa; but we could discover no trace of ruins. In the forenoon of the fifth day we diverged a little to the left, to visit ruins which had been described to us under the names Anjeh and 'Abdeh, and which are doubtless the remains of ancient Eboda. They consist of the walls of a large Greek church, and an extensive fortress, both situated upon a long hill or ridge overlooking a wide plain. Connected with the fortress are cisterns and deep wells walled up with uncommonly good masonry. On the south side of the hill and below are the ruins of houses, surrounded by traces of extensive ancient cultivation.

We were now crossing a more sandy portion of the Desert; and in the afternoon of the same day we had our first specimen of the simoom or southern wind of the Desert. It came over us with violence like the glow of an oven, filling the air with fine particles of dust and sand, so as to obscure the sun and render it difficult to see objects only a few rods distant. We encamped in Wádí Ruḥeïbeh, where we had never heard of ruins; but, on ascending the hill on our left, we discovered the remains of a city not much less than 2 miles in circuit. The houses had been mostly built of hewn stone; there were several public buildings and many cisterns; but the whole is now thrown together in unutterable confusion, as if the city had been suddenly overthrown by some tremendous earthquake. What ancient city this can have been, I have not yet been able to learn. The Arabic name suggests the Rehoboth of Scripture, the name of one of Isaac's wells (Gen. xxvi. 22), but the other circumstances do not correspond.

We now approached a more fertile region. Towards noon of the sixth day we reached Khulásah, the site of ancient Elusa. It was a city of at least 2 miles in circuit. The foundations of buildings are everywhere to be traced; and several large unshapen piles of stones seem to mark the sites of public edifices. Fragments of columns are occasionally seen, but no cisterns. A public well, still in use, seems to have supplied the city.

After crossing another elevated plateau, the character of the surface was again changed. We came upon an open undulating country: all around were swelling hills, covered in ordinary seasons with grass and rich pasturage; but now arid and parched with drought. We now came to Wádí Seba'; and on the N. side of its watercourse we had the gratification of discovering (April 12th) the site of ancient Beersheba, the celebrated border-city of Palestine, still bearing in Arabic the name of Bír Seba'. Near the watercourse are two circular wells of fine water, more than 40 feet deep. They are surrounded with drinking-troughs of stone for the use of camels and flocks; such as were doubtless used of old for the flocks and herds which then fed on the adjacent hills. Ascending the higher ground N. of the wells, we found these low hills strewed with the ruins of former habitations, the foundations of which are distinctly to be traced. These ruins extend over a space half a mile long by a quarter of a mile broad. Here then is the spot where Abraham and Isaac and Jacob often lived! Here Samuel made his sons judges; and from hence Elijah wandered out into the southern Desert, and sat down under the Rethem, or shrub of broom, just as our Arabs sat down under it every day and every night! Over these swelling hills the flocks of the patriarchs roved by thousands;—we now found only a few camels, asses, and goats.

From Bír Seba' to Hebron we travelled 12½ hours, here equivalent to about 30 miles. The general course was N.E. by E. After 1½ hour we came out upon a wide open plain, covered with grass, but now parched with drought. Fields of wheat and barley were seen all around; and before us were hills, the beginning of the mountains of Judah. At Dhoheríyeh, the first Syrian village, the hills around were covered with mingled flocks of sheep and goats, and herds of neat cattle, horses, asses, and camels, in the true patriarchal style of ancient days. At this place our good Towárá left us: we took other camels and proceeded to Hebron. Here the pool over which David hung up the assassins of Ishbosheth still remains, and fixes the site of the ancient city. The cave of Macphelah cannot well have been within the city; and therefore the present mosque cannot cover its site. We could not but notice the fertility of the neighbouring valleys, full of corn-fields and vineyards yielding the largest and finest clusters of all Palestine; and likewise the rich pasturage of the hills, over which were scattered numerous flocks and herds. Yet to a careless observer the country can only appear steril and forbidding; for the limestone rocks everywhere come out upon the surface, and are strewn over it to such a degree, that a more stony or rocky region is very rarely to be seen.

We took the direct road to Jerusalem. It is laid with stones in many places, and is doutless the ancient road, which patriarchs and kings of old have often trod; but it is only a path for beasts; no wheels have ever passed there. The distance to Jerusalem is about 21 miles, on a course between N.N.E. and N.E. by N. We hurried onward, and reached the Holy City at sunset, April 14th, just before the closing of the gates on the evening before Easter Sunday.

II. Jerusalem. Our journey *to* Palestine was now completed; and our researches and travels *in* Palestine were to begin. In respect to these we adopted for our future guidance the two following principles, viz., 1. To direct our researches chiefly to those parts of the country which former travellers had never visited; and, 2. To obtain information, as far as possible, not from the legends of monks and other foreigners, but directly from the native Arabs of the land. We remained at first more than three weeks in Jerusalem; and afterwards made that city the central point from which to set off on excursions to different parts of the country. In the mean time we diligently explored every part of the city, and even here saw or heard of several things which to us at least were new.

On entering Jerusalem I was prepared, from the descriptions of many travellers, to find the houses miserable, the streets filthy, and the population squalid; but in all these respects I was agreeably disappointed. The houses are better built, and the streets cleaner, than those of Alexandria, Smyrna, or Constantinople. The hills and valleys which marked the different quarters of the ancient city are still distinctly visible. The valley of the Tyropœum may be traced from its head near the Yaffá gate to its foot at the pool of Siloam. The hills of Zion, Akra, Bezetha, and Moriah, are yet distinct and marked. The latter, on which stood the ancient Temple, is now occupied by the mosque of 'Omar and the extensive court or area around it.

One of the earliest objects of our attention was naturally this area, in reference to its antiquity and connexion with the ancient Temple. It is an elevated plateau or terrace, nearly in the form of a parallelogram, supported by and within massive walls built up from the valleys or lower ground on all sides. The southern wall is about 60 feet high. The upper part of these external walls is obviously of modern origin; but it is not less easy to perceive that the lower portions, for the most part, are of an earlier date. These are composed generally of very large stones, many of them 20 feet or more in length by 5 or 6 feet thick, hewn in a peculiar manner. At the first view of these walls, I felt persuaded that these lower portions had belonged to the ancient Temple, and were to be referred back at least to the time of Herod, if not to the days of Nehemiah or Solomon. This conviction was afterwards strengthened by our discovering, near the S.W. corner in the western wall, the remains, or rather the foot, of an immense arch, springing out from the wall in the direction towards Mount Zion, across the valley of the Tyropœum. The traces of this arch are too distinct and definite to be mistaken; and it can only have belonged to the bridge which, according to Josephus, led from this part of the Temple-area to the Xystus on Zion; thus proving incontestably the antiquity of that portion of the wall from which it springs.

We then examined the remarkable tower in the citadel near the Yaffá gate, which, even to the unpractised eye, bears strong marks of antiquity. Some former travellers have already regarded this as the Hippicus of Herod; and we found every reason to assent to this conclusion. So far as we could discover, the lower part of the tower is wholly solid, as described by Josephus; at least there is no known or visible entrance to it, either from above or below.

The present walls of the modern city were built about 300 years ago, as appears from numerous Arabic inscriptions. Remains of the former wall, which probably existed in the time of the crusades, are still visible on the outside, N.W. of the Yaffá gate; also on the N. side of the city, and in the interior of the N.W. corner. Of the *ancient* wall around Mount Zion, traces may yet be seen for some distance in the scarped rocks below the S.W. brow of Zion. On the high ground N. of the N.W. corner of the city we discovered evident traces of what must have been the *third* or exterior wall described by Josephus in this quarter, erected after the time of Christ. Here must have stood the tower Psephinos; and from this point we were able to trace the foundation of the same ancient wall for a considerable distance further in a N.E. direction.

Of the *second* wall of Josephus, which at the time of the Crucifixion was the exterior wall of the city on this side, we could find no remaining traces, unless it be two square ancient towers which we discovered connected with the wall inside of the Damascus gate, one on each side of the gate. These towers are built of large stones precisely like those mentioned above as belonging to the ancient Temple walls. They have been much injured in building the modern wall of the city, but are evidently ancient, and apparently older than Hippicus; they were, most probably, the guard-houses of an ancient gate upon this spot; and this could well only have belonged to the said second wall. If this hypothesis be correct, it will go far to decide the question as to the site of the church of the Holy Sepulchre, which must then have fallen within this wall, and so within the ancient city. Indeed, the church stands upon the very ridge of the hill Akra, which, according to Josephus, and to every probability, must have formed part of the lower city, and been enclosed within the second wall.

Another object of our attention was the supply of water in and around the city. At the present day Jerusalem is supplied almost wholly by rain-water, preserved in cisterns cut in the rock on which the houses stand. Almost every house has one or more cisterns; that in which we resided had no less than four very large ones. The ancient city was probably supplied in the same manner. Indeed, with a little attention, there can never be any want of water within the walls. The aqueduct which comes from Solomon's pools beyond Bethlehem brings water only to the mosque of 'Omar. Outside of the city, besides the ancient reservoirs, there are wells in various places, some with water and some without. The brook Kidron, in the valley of Jehosaphat, flows only when the rain-water descends into it from the adjacent hills. Fountains of running water exist only in this valley; and of these there are three, viz.:—1. the fountain of the Virgin, or of Siloam, just south of the site of the Temple; 2. the pool of Siloam, just within the entrance of the Tyropœum; 3. the well of Nehemiah, or of Job, opposite the entrance of the valley of Hinnom. This last is a deep well of living water, which in the rainy season overflows: it is, beyond doubt, the En Rogel of Scripture. The pool of Siloam is wholly artificial, and receives its waters from the fountain of the Virgin, through a subterraneous channel cut through the solid rock. We crawled through this channel and measured it. The fountain of the Virgin is also evidently an artificial excavation in the rock; but whence the water is derived is a mystery. It has a sweetish, slightly brackish taste; and flows irregularly, or only at irregular intervals. We were witnesses of this irregular flow; and were told by the women who came for water that sometimes, during summer, it ceases to flow for several weeks; when, on a sudden, the water comes gushing out again in abundance.

Ancient writers have spoken of a fountain of living water as existing under the Temple; though their assertions have, in general, obtained little credit. Soon after our arrival in Jerusalem, we were told of a similar fountain under the mosque of 'Omar, the waters of which were used to supply a bath in the vicinity of the mosque. We went to the bath, and found two men drawing water from a deep well. They told us that the water flows into the well from a passage cut in the rock, and leading under the mosque, where is a chamber and a living fountain. In summer, when the water is so low as not to flow out into the well, they go down and bring it out by hand. The taste of the water is precisely like that of the fountain of the Virgin in the valley below. We made all our preparations to descend into the well and examine the fountain, but were hindered at the time, and were unable afterwards to resume the investigation. Is, perhaps, the water of this fountain brought down by a subterraneous channel from some higher point? Is there a connexion between this fountain under the mosque and that in the valley below; and is the irregular flow of the latter in some way dependent on this circumstance? These questions may, not improbably, at some future time, be answered in the affirmative.

When we arrived at Jerusalem war was raging in the north between the Druses and the forces of the Páshá; and, as if we were to have a specimen of all the evils of the Oriental world, in a few days after our arrival the plague broke out; at first doubtfully, then decidedly, though mildly. Other travellers left the city immediately; and some who were on their way thither turned back. We continued our investigations without interruption; and a kind Providence preserved us from the danger.

III. From Jerusalem to Gaza, Hebron, and Wádí Músá.

On returning to Jerusalem, from an excursion of eight days to Engeddi and the Dead Sea, we found the plague slowly but constantly increasing; and it was rumoured that the city was soon to be shut up. We therefore remained but a single day, in order to make preparations for our longer journey to Wádí Músá. We set off, May 17th, on horses and mules; and, on May 19th, the city was shut up, and none suffered to go out, without first performing a quarantine of seven days. Our excursion occupied in all 23 days.

We made at first a slight detour, in order to pass by Beït Jala, a Christian village, half an hour N.W. of Bethlehem; and then continued S.W. across the mountains to the direct ancient road from Jerusalem to Eleutheropolis and Gaza, through a region as yet unvisited by modern travellers. At a distance on our right was the deep valley of Turpentine (so called by monks and travellers), or, as the Arabs name it in this part, Wádí Surár, which runs in a S.W. direction, until it opens out into the great plain between the mountains and the Mediterranean. On our left was another similar valley, Wádí Sunnet. The region is full of ruined sites and ruined villages, some deserted and some partially inhabited; among which are still found several ancient names. On our right, beyond Wádí Surár, we could see the hill and ruined village Sohá, which it has pleased the monks to assume as the ancient Modin, the burial-place of the Maccabees, against the express testimony of Eusebius and Jerome. We came at night to Beït Netíf, a large village on a high part of the ridge between the two valleys above mentioned.

The next day was devoted to a visit to Beït Jibrín, the ancient Betogabris of Greek and Roman writers, of which and its fortress we had heard much from the Arabs; and to a search for the site of ancient Eleutheropolis. From the elevated spot where we lodged, the sheïkh of the village pointed out to us several places still bearing, in Arabic, names corresponding to their ancient Hebrew appellations, and celebrated in Scripture as the scenes of Samson's exploits and history: such were Zorah, Timnath, Sokho, and others. Four places were also pointed out, respecting which Eusebius and Jerome have specified their distances from Eleutheropolis, viz., Zorah and Bethshemesh, towards Nicopolis; and Jarmah and Sokho, on the way to Jerusalem. Following out the specified distances along the ancient road, we came directly upon Beït Jibrín, which lies among hills between the mountains and the plain. Here are the remains of a large Roman fortress of immense strength, which was built up again in the time of the crusades: around it are the traces of an extensive city.

We had received the impression that we must look for Eleutheropolis further west upon the plain; and accordingly turned our course that way to Ṣafíyeh, a conspicuous village, lying on an isolated hill. Here, however, we could find no trace of any ancient site. We then proceeded to Gaza; whence, after two days, we returned by a different route, searching diligently for the sites of ancient Lachish, Gath, and other cities, but finding none except Eglon, on a mound strewed with stones, still called 'Ajlán. Again arrived at Beït Jibrín, we visited several very singular artificial caverns in the vicinity. Eusebius and Jerome mention also Jedna and Nazib as being distant from Eleutheropolis, one 6 and the other 7 miles, on the way to Hebron. These names still exist; and, taking the Hebron route, we found Jedna to be just 6 miles from Beït Jibrín. Nazib lies yet a little further on another parallel road. This circumstance seems to decide the identity of Beït Jibrín with Eleutheropolis. The former was the ancient name; the latter was imposed by the Romans, and has been since forgotten, as in so many other instances. It is also remarkable that those ancient writers who speak of Eleutheropolis do not mention Betogabris; while those who speak of the latter are silent as to the former. Rejoicing in this result, we pursued our way to Hebron; and, after a steep and toilsome ascent on a ridge between two deep valleys, we rested for a time at Taffúh, the Beth Tappuah of Judah; and arrived at Hebron in about 6 hours from Beït Jibrín. Here, dismissing our muleteers, we engaged camels for Wádí Músá from the sheïkh of the Jehálín, a Bedawí tribe inhabiting the territory S.E. of Hebron.

We had long before formed the plan to proceed to Wádí Músá by way of the south end of the Dead Sea, and so southwards along Wádí 'Arabah, in the hope of being able to solve the pending question, whether the Jordan could ever have flowed through this valley to the Gulf of 'Aḳabah. Here, too, we had hoped again to have been the first; but were anticipated by the Count de Bertou, who preceded us by three or four weeks, and whom we had seen at Jerusalem after his return. After being detained two days at Hebron, we set off, May 24th; and, passing in sight of ancient Ziph on the left, and Yutta (ancient Jutta) on the right, and near the ruins of Carmel and Maon, we continued across an undulating desert in a S.E. direction, and came, towards the close of the second day's journey, to the brow of the steep descent leading down to the Dead Sea. This descent is in all not less than 1500 feet; but here, and far to the south, it is divided into two parts or offsets of nearly equal height: between these lies a terrace or plain nearly three hours broad, the surface of which is covered with low ridges and conical hills of soft chalky limestone, verging into marl. At the foot of the second descent is a small deserted Turkish fort, in the narrow Wádí Zuweïreh (not Zoar), which leads out to the sea in about half an hour. We reached the shore not far from the northern end of Usdúm, a low, long mountain ridge, running here from N.N.W. to S.S.E., and giving the same direction to the shore of the sea. This ridge, Usdúm, is, in general, not far from 150 feet high, and continues to run in this direction for two hours to the southern extremity of the sea, where it trends to the S.S.W. for an hour more, and then terminates. The striking peculiarity of this mountain is, that the whole body of it is *a mass of solid rock-salt;* covered over, indeed, with layers of soft limestone and marl, or the like, through which the salt often breaks out, and appears on the sides in precipices, 40 to 50 feet high, and several hundred feet long. Often also it is broken off in large and small pieces, which are strewed like stones along the shore, or fallen down as débris.

The south end of the sea is very shallow, and the shore continues quite flat for some distance further south; so that there are traces of its being overflowed by the sea for two or three miles south of the water-line, as we saw it. The western side of this southern valley, or Ghor, is wholly naked of vegetation; but on the eastern side, where streams come down from the eastern mountains, there is a luxuriant vegetation and some tillage. We continued on the western side, along the base of Usdúm, crossing several purling rills of transparent water, flowing from the mountain towards the sea, but salt as the saltest brine. Before us, as we advanced southwards, appeared a line of cliffs, 50 to 150 feet high, stretching across the whole broad valley, and apparently barring all further progress. These cliffs are mentioned by Irby and Mangles, who supposed them to be sand-hills. We approached their western end in 2½ hours from the south end of the sea. They proved to be of marl, and run off from this point in a general course S.S.E. across the valley. All along their base are fountains of brackish water oozing out, and forming a tract of marshy land towards the north. Our route now lay along the base of these cliffs; and, after resting for a time at a fine gushing fountain, we came, in two hours, to the mouth of Wádí Jíb, a deep valley coming down from the south through the cliffs, and showing the latter to be only an offset between the lower plain which we had just crossed, and the higher level of the same great valley further south. The name El Ghór is applied to the valley between the Dead Sea and this offset; further south the whole of the broad valley is called El 'Arabah, quite to 'Akabah. These apparent cliffs are not improbably the 'Akrabbim of Scripture. The Wádí Jíb begins far to the south of Mount Hor, beyond

Wádí Gharandel, and flows down in a winding course through the midst of El 'Arabah, draining off all its waters northward to the Dead Sea. Where we entered Wádí Jíb, at its northern side, it is half a mile broad, with precipitous banks of chalky earth or marl, 100 to 150 feet high; and exhibiting traces of an immense volume of water in the rainy season, flowing northwards. It may be recollected that the waters of Wádí Jeráfeh, in the western desert, which drains the S.E. part of that desert, far to the southward of 'Aḳabah, also flow northwards into El 'Arabah, and so, of course, through Wádí Jíb. Hence, instead of the Jordan flowing southwards to the Gulf of 'Aḳabah, we find the waters of the desert further south than Aḳabah flowing northwards into the Dead Sea. The nature of the country shows, without measurement, that the surface of the Dead Sea must be lower than that of the Red Sea or the Mediterranean.

We continued our course up the Wádí Jíb southwards for several hours, its banks becoming gradually lower, and at length permitting us to emerge from it. We were now not far from the eastern mountains, nearly opposite the broad Wádí Ghuweïr, while before us was Mount Hor, rising like a cone irregularly truncated. We turned into these mountains at some distance north of Mount Hor, in order to approach Wádí Músá from the east, through its celebrated ancient entrance. A long and steep ascent—the pass of Nemellah—brought us out upon the plateaus of the porphyry formation; above which are still the hills of sandstone among which Petra was situated. The entrance to this ancient city, through the long chasm or cleft in the sandstone rock, is truly magnificent; and not less splendid and surprisingly beautiful is the view of the Khazínah, or temple hewn in the opposite rock, as the traveller emerges from the western extremity of the passage. Then follow long ranges of tombs hewn in the rocky sides of the valley, with ornamental façades, in a style of striking though florid architecture. What we sought in Wádí Músá was more the general impression of the whole; since the details have been correctly given by the pencil of Laborde. We examined particularly whether any of these excavations were perhaps intended as dwellings for the living; but could see no marks of such design—nothing but habitations of the dead, or temples for the gods. There was, indeed, no need of their being thus used; for the numerous foundations of dwellings show that a large city of houses built of stone once stood in the valley.

We had nearly completed our observations, and were preparing soon to set off on our return by way of Mount Hor, when the old sheïkh of Wádí Músá, Abú Zeïtún, who caused so much difficulty to Mr. Bankes and his companions in 1817, came down upon us with thirty armed men, demanding a tribute of a thousand piastres for the privilege of visiting his territory. We declined payment of course; but, after long and repeated altercation, it came to this result, that, unless we paid this full sum, he would not suffer us to visit Mount Hor. We attempted, nevertheless, to set off in this direction, our own sheïkh leading the forward camel; but the hostile party closed around, and swords were drawn and brandished; which, however, among these Arabs, means nothing more than to make a flourish. As it was in vain for us to use force against so large a party, we decided to set off on our return by the way we came. This took the old man by surprise, and thwarted his plans. Messengers soon followed us, saying we might return for the half; and, at last, for nothing. We replied, that he had driven us from Wádí Músá, and we should not return, but should report his conduct at Cairo. The old man then came himself, to get our good-will, as he said, which was worth more to him than money. We thought it better to keep on our way; and suffered no further interruption. It was probably the fear of the Páshá of Egypt alone that withheld these miscreants from plundering us outright; and we afterwards received compliments from the Arabs in and around Hebron for the boldness and address with which we had extricated ourselves from the old sheïkh's power.

Descending the pass of Nemellah, we struck across El-'Arabah in a W.N.W. direction, travelling the greater part of the night. In the morning we reached Wádí Jíb, here quite on the western side of El-'Arabah, and stopped for a time at the fountain El Weibi. Other fountains occur at intervals along the valley at the foot of the western hills, both north and south of El Weïbí. From here a path strikes up the western mountain in the direction of Hebron, which is used by the southern Arabs. Our guides took a more northern road, leading up a very steep pass called Sufáh, over a broad surface of shelving rock extending nearly from the bottom to the top, an elevation of 1000 or 1200 feet. This is probably the hill Zephath, afterwards Hormah, where the Israelites attempted to enter Palestine, but were driven back, and were also attacked by the king of Arad; Num. xiv. 40, seq., xxi. 1, seq., Judges i. 17. Some miles N.N.W. of this pass is a conical hill still bearing the name of Tel Arad, probably the site of the ancient town. All these circumstances lead me to place the site of Kadesh in the great valley below, near the fountain El Weïbí or one of the neighbouring springs. Here it would be near the border of Edom, opposite to a broad passage leading up through the eastern mountains, and in full sight of Mount Hor. That the Israelites must have approached Palestine through the Wádí 'Arabah, is a necessary conclusion from the mountainous character of the district on the west of this valley, through which no road has ever passed. But no trace of the name Kadesh is to be found, neither in the valley below nor on the table land above.

Our further way to Hebron led us by the sites of 'Ararah, the Aroer of Judah; and Melh, where is a fine well and the traces of a town, not improbably the ancient Moladah or Malatha. Nearer to Hebron we passed Semú'ah, perhaps the Hebrew Sema; and Yuttah, the ancient Sutta, the probable birthplace of John the Baptist, and still a town of some importance. At Hebron we remained a day and a half, being obliged to send for horses to Jerusalem.

We left Hebron again on the 6th of June, taking now a S.W. course by the large village Dúrah, the Adora of Josephus; and descending the mountain to El Burj, a ruined castle of which we had heard much, but where we found nothing of antiquity. Hence we bent our course northward among the hills; and passing again through Jedna, rested for a time at Terkumieh, the Tricomias of former ages, leaving Beït Jibrín on our left. We lodged a second time at Beit Neïtíf; and the next morning descending N.N.W., we came to the site of the ancient Bethshemesh in the opening of Wádí Surár into the plain. Here are evident traces of a large city. From this point we turned our course N.W. into the plain, in search of the ancient and long-lost Ekron. After travelling in this direction for four hours, we came to the large village 'Aḳir, an Arabic name corresponding to the Hebrew Ekron. The situation corresponds also to the accounts of Eusebius and Jerome. Nothing of antiquity remains; perhaps because the ancient houses, like the modern hovels, were built, not of stone, but of earth.

From Ekron to Ramleh is two hours: here we lodged; and the next day proceeded to Jerusalem by the camel-road, which also is the ancient Jewish and Roman way, over Ludd (Lydda), Gimzo, Lower and Upper Bethhoron (now Beït-U'r), and Jib or Gibeon. The pass between the two villages of Bethhoron is a steep and rugged ascent of some 1500 feet, up the point of a ridge between deep valleys. It is the ancient road which the Roman armies ascended, and has in several places steps cut in the rock. The present shorter and less practicable route between Ramleh and Jerusalem appears not to have been in use in the time of the Romans. Looking down from Upper Bethhoron, a broad valley is seen in the S.W. issuing from the mountains and hills into the plain; while on the ridge that skirts its S.W. side, is seen a village called Yálon, the Arabic form for the Hebrew Ajalon. Here then is probably the spot where Joshua in pursuit of the five kings, having arrived at Upper Bethhoron, looked back toward Gibeon, and down upon the valley before him, and uttered the command: "Sun, stand thou still on Gibeon; and thou, Moon, in the valley of Ajalon!"

We found Jerusalem still shut up on account of the plague

and therefore pitched our tent in the olive-grove north of the city, before the Damascus gate. * * * *

In other excursions from Jerusalem, and on our subsequent journey northward to Beïrût, we visited the villages and sites of Anothoth, Gibeah, Micmash, and Bethel, all N.E. and N. of the Holy City, and still bearing in Arabic the names Anâtah, Jeba', Mükhmás, and Beit-ín. The extensive ruins of the latter place, Bethel, lie 45 minutes N.E. of Bíreh, just on the right of the Nablús road. Farther north we turned aside to Jifnah, the Gophna of Josephus; and also to Seïlúm, the site of ancient Shiloh, which Josephus also writes Silun. (Σιλοῦν.) Along the western shore of the Lake of Tiberias, we made minute and persevering inquiries after the ancient names Capernaum, Bethsaida, and Chorazin; but no trace of them remains among the Arab population. If former travellers have heard these names, it must have been from the monks of Nazareth or their dependents.

On the way from Şafet to Tyre, nearly two hours N.W. of Şafet, we passed near the crater of an extinct volcano; which was probably the central-point of the great earthquake of Jan. 1st, 1837, by which Safet and the adjacent villages were destroyed.

Extract from a Letter of Professor Berghaus.

Potsdam, April 30, 1839.

Sir,—In accordance with former communications from Professor E. Robinson, of New York, he will transmit to you this day one of the maps, which are the fruits of the travels of himself and his companion, the Rev. E. Smith, in Arabia Petræa, and the Holy Land. I cannot permit the opportunity to pass by without accompanying this sketch with a few words.

In the course of my life I have had in my hands many documents in reference to geographical, and especially cartographical objects, and from them have acquired the conviction that, among all oriental travellers since the time of Niebuhr, the prize is due to the late lamented Burckhardt, so far as it respects minute attention, even to things apparently indifferent, and also accuracy in the measurement of bearings and angles, and in the specification of time for the determination of distances. This conviction I have expressed publicly and unreservedly, perhaps, in other words, in the Memoirs accompanying my maps of Asia.

This view, however, I must now essentially modify, after having carefully examined the Journals of Messrs. Robinson and Smith during their travels in the peninsula of Mount Sinai, through the great desert Et-Tih, and in Palestine. The observations of these two travellers are so full and comprehensive, their notes upon the form and the features of the country so exact and definite, that the geographer is in a situation, on the basis of these specifications, to construct a special map of the territory, which may perhaps leave little more to be desired.

In support of these views, I may refer to the accompanying sketch-map of the route of the travellers from 'Akabah through the desert Et-Tih to Hebron, which I have constructed, and which we, Professor Robinson and myself, would commend to the favourable notice of the Royal Geographical Society.

Professor Robinson has placed his journals at my disposal, and, on my recommendation, has prepared from them an abstract in a tabular form, particularly adapted to serve as a foundation for the construction of the route-maps.

You will be able to appreciate the impatience with which I entered upon the construction of these maps, if you will have the goodness to call to mind that I published a few years ago a map of Syria, which was so fortunate as to meet the approbation of your Society. [*Journal*, vol. vii. p. 183.] My attention was more especially drawn to the tour from 'Akabah to Hebron, because the travellers have here passed through a real *terra incognita*, which is now for the first time represented on a map. The original sketch of this route is three times larger than the copy herewith forwarded to you. I would also mention that several points, the position of which was determined by Messrs. Robinson and Smith, are not inserted, because they fall without the margin of the copy. These points are Jebel Ikhrim, Jebel Yelek, and Jebel el-Ḥelál, all lying westward of the route, and adapted to determine the course of the Wádí el-'Aráish, and its distance from the line of travel.

In my map of Syria I have assumed Hebron to be in long. 35° 12′ 25″ E. from Greenwich, according to the Azimuth of Jerusalem, supplied by Seetzen's very rough map. (*Syrian Memoir*, p. 34.) But I find this Azimuth to be erroneous, since the itinerary of Messrs. Robinson and Smith gives the long. of Hebron at 34° 57′ 13″ E., a difference of more than a quarter of a degree. Yet I would remark, that even this position can only be regarded as a *first approximation* to the true longitude of Hebron, inasmuch as the subsequent routes of the travellers afford the means of corroborating the determination from both Jerusalem and Ramleh, and, above all, from Gaza, which lies nearly on the same parallel with Hebron, or El-Khulil.

The most inaccurate part of my map of Syria is the topographical delineation of Judea, because no traveller within my reach had then examined this portion of the Holy Land with the same attention which Burckhardt had bestowed on the country E. of the Jordan, and around Mount Lebanon. It is therefore eminently creditable on the part of Messrs. Robinson and Smith, in respect to Biblical geography, that they have visited and investigated the territory of Judea in all directions—this land of the earliest of all history for the nations of Christian civilization. In this way they have become real discoverers in the field of topography and history.

Their examination extends over the whole country between the shores of the Mediterranean and the Dead Sea, from the parallel of about Nablús to the south end of the Asphaltic Lake, and from these points, on the one hand southwards as far as to Wádí Músa, and on the other northwards as far as to Safed, Saide, and Beïrút.

On the Mount of Olives, Professor Robinson determined the length of a base of 1425·72 yards, by means of which, and the application of their very numerous bearings, I have been able to construct and calculate a net of triangles, extending N. to Taiyíbeh, E. to the Dead Sea, S. (as yet) as far as to the Frank Mountain, and which we may probably be able to carry on as far as to the region of Ramleh and Gaza, and perhaps also to Hebron, Carmel, and 'Aïn Jiddi. I find the distance between the Mount of Olives and the N.W. corner of the Dead Sea to be 29093·5 yards, or 14·34 geographical miles. At 'Aïn Jiddi they measured a second base, in order to determine the breadth of the Dead Sea. This I find to be 15953·1 yards, or 7·86 geographical miles.

I have already constructed a portion of the itineraries in Judea, on a scale three times, and in some parts six times, larger than that of the accompanying sketch map. This was necessary, in order to exhibit in full all the details.

I am of opinion that it would be a great loss for geography, were the materials collected by Messrs. Robinson and Smith not to be used for the construction of a map on a large scale. Their journey, undertaken solely for the interests of Biblical geography, would be deprived of its finest fruits, and the many hardships they must have endured will have been in a measure superfluous, should the results of their measurements and observations be published only in the form of a journal; for this can never produce the lively impression that is felt in looking at a good and accurate map, and this great desideratum I hope shortly to take in hand.

In conclusion, I would remark, as to the accompanying route-map from 'Akabah to Hebron, that the direction of the Wádís between Bírsheba and Hebron may perhaps require some slight correction, when all the itineraries of the travellers shall have been fully constructed.

I have the honour to be, &c.

HEINRICH BERGHAUS,

For. Hon. Mem. R. G. S. of London.

To Captain Washington, R.N., Secretary.

XIII.—*Translation of a Portion of the Account of a Pilgrimage to Jerusalem, Terra Santa, Mount Sinai, and Egypt, begun at Oppenheim on the 23rd of April,* 1483. *Written in Latin by* BERNARD DE BREITENBACH, *Dean of Mentz, one of the Pilgrims; edited in Spanish by* MARTIN DAMPIES, *and printed at Saragossa in* 1498. Communicated by BARTHOLOMEW FRERE, Esq.

[The original work was published in Latin at Mentz, 1486, entitled "Opusculum sanctarum Peregrinationum in Montem Syon, ad venerandum Christi Sepulcrum in Jerusalem, atque in Montem Synai ad divum virginem et martyrem Katharinam;" and again at Spires in 1490, with the title "Peregrinatio Hierosolymitana ad Sepulcrum Domini et Katariniana ad Montem Sinai, per varias partes Orientis cum iconibus." See Biog. Univ. "Breydenbach." There is likewise a copy in Dutch in the British Museum printed 1486, and bound up with a Latin copy; and there is also a copy in German printed 1488; and, lastly, a Latin copy on vellum in the King's Library at the British Museum, printed 1486.* A folio MS. of this work, written on paper, near the close of the 15th century, is mentioned by Mr. Hunter in his Catalogue of the MSS. in the library of the Hon. Soc. of Lincoln's-Inn, where he states that a French translation was printed at Lyons in 1489.]

BEFORE describing this pilgrimage, it is well to give the form of the contract which we made with the rulers of Jerusalem for our safe conduct to Mount Sinai and thence to the city of Cairo. This contract must be made, before they depart, by all pilgrims who wish to return by the vessel which brought them out; that the captain may wait for them at Alexandria: and this must be done in his presence and in that of the guardian of Mount Sion, because, otherwise, those who remained behind would have difficulty in making a contract, as the Moors would deal arbitrarily and unreasonably with them. Thus we agreed with the Sultan's lieutenant in Jerusalem, whose name was Naydon, and with the chief Calino, who is styled master and knight of the hospital of pilgrims, who was named Sabathytaneo. The contract was made on the 17th day of July.

The first condition was, that they were to give us a safe conduct till we arrived at Mount Sinai and in the monastery of the saint, and from thence to Cairo: in which journey the Calino himself should accompany us, at his own expense, as far as Gazera;

* This copy contains the following prefatory notice by J. West:—"This most rare book of the travels of the religious to the Holy Land, printed on vellum, contains the oldest views engraved that I have ever seen: they seem to have been taken upon the spot. The book was printed at Mentz, 1486, which is seven years before the printing of the Nuremberg Chronicle, which has always been supposed to have been the oldest printed book with charts and maps. I bought this book at the sale of the Harleian printed books. In my collection of ancient engravings there is a historical print with the date 1466."

and should provide for all costs, charges, and duties; otherwise the pilgrims would be thrown into great confusion by the wicked race of the infidels.

It was stipulated in the contract that each of the pilgrims and of their guides, whom they call Mucrelos, should be provided with an ass, and should have provisions for their subsistence till their pilgrimage was completed and they reached Cairo. It was stipulated also that he should take charge of conveying all the provisions and baggage as far as Gazera, excepting the wine which we might choose to carry with us: he was bound to find camels for the conveyance of all the above articles.

It was moreover agreed that from Gazera to Mount Sinai and Cairo he should send with us his deputy, whom they call the under Calino, at our expense. He was bound to provide us a tent well equipped, in order that, if in the midst of that desert we required rest, we might have shelter from the burning heat of the sun; and he was to provide certain skins for carrying water, otherwise we should have many journeys without finding any that we could drink. By the same conditions each of the pilgrims was bound to pay twenty-three ducats, half of it immediately in Jerusalem, the other half when we should have arrived at Gazera, and he should have provided the camels and provisions before-mentioned.

All the above articles of agreement were ratified with the seals of the lieutenant and of the chief Calino; but they are wont to be ill observed, according to the custom which they have of old: and for this provision we paid down two ducats upon the spot.

On the festival of the Magdalen, which is the 22nd of July, our fellow-pilgrims departed to the port of Joppa, because the ships were waiting for them. Only the eighteen under-mentioned remained in Jerusalem in order to proceed to Mount Sinai. We waited for thirty-three days, apprehending the great heat of the sun, and in this interval we frequently visited the accustomed places and stations.

Here are named the knights and pilgrims who desired to visit the virgin martyr St. Catherine: may they be more worthily recorded in the Book of Life!—

El Sr. Don John Count of Solms, Lord of Myntzenberg; the last in point of age, but the first in rank and spirit.

El Sr. Bernard de Breitenbach, Dean of Mayence and chamberlain, the principal author of the present work.

El Sr. Mossen Philip de Bilrem. With these went Erhard Rervich of Utrecht, a very skilful painter, who by his talent painted the scenes here represented well and to the life.

Mossen Maximin of Rapenstein, Lord of Boinech.

Don Fernando de Mernarve.

Mossen Gaspar de Bulach.

Mossen George Marx.

Mossen Nicolas, mayor en hurt (?), with whom (besides their servants) there went two friar minors, Paul and Thomas, who understood many languages.

Mossen Henric, of Schauenberg.

Mossen Caspar de Sienli.

Mossen Sigismund de Marserbach.

Mossen Pedro Velsech.

Mossen John Lazino, archdeacon and canon of the church of Transilva in Hungary.

Father Felix Fabri, of the order of predicadores, lecturer of Ulm, and a very accomplished preacher, who had before been in the Holy Land, a learned man in many ways.

All the above persons, being assembled together, having collected our provisions, on the 24th of August, the festival of St. Bartholomew, set out from the Holy City of Jerusalem, and arrived after sunset at Bethlehem, where we remained two days, visiting the holy places already described. All the troubles and distresses which we underwent in this short journey, from the treachery of the infidels, I have thought fit to suppress.

On the 27th of August, early in the morning, we left Bethlehem; and in the evening arrived at the ancient Ebron, where we saw the Damascene Plain,* the Double Cave,† and other places already described. We then entered the monastery of St. George, where are some Greek monks; and they have there a chain which belonged to the said Saint, which has thirty-nine rings or links, and we put it round our necks, seeing that the Moors did so; because, if any one who is troubled with an evil spirit or with madness puts it on, he is cured immediately, and so they hold it in great veneration: here they show still the footmarks of the horse of this St. George, of the earth of which we carried away a portion.

We set out before daybreak on the morrow, which was the 28th of August; and the sun had set when we arrived at a very large and solitary building, where we remained through the night. Here the mountains come to an end; and a very extensive and delightful plain commences. In the course of this journey we passed near a castle, which they call the St. Samuel; thence inclining to the left is a city, as it is said, of the holy Abraham, where the Moors have established a great hospital, where they distribute food to as many as come there, be they of what nation or sect they may. 1200 loaves are baked there daily

* This plain is mentioned in an earlier part of the work as the spot where Adam was created.

† The burying-place of Adam and Eve and the Patriarchs.

for the poor, and the yearly expenses amount to 24,000 ducats, all of which goes in charity. On the following day, which was the 29th of the said month, we departed for the city of Gazera, and arrived soon after noon, but our entry was delayed till night. This was done by the advice of our Calino, that the ruler of the city might not take away our wine, as he had at other times done to pilgrims. On this day we came to many cisterns for water, which the Moors draw from thence with much labour, and give it to the pilgrims for the love of God. Near to this city of Gazera there was a very large and ancient fig-tree, which bears its figs seven times in a year—they call it the fig-tree of Pharaoh. As soon as night arrived we entered Gazera, where with much trouble we found a covered place to lodge ourselves in, for the hospital of pilgrims was too small for so many persons: at length we were placed in a house, with bad walls, and full of mud, where we continued twelve days under great hardships and much sickness arising from the unwholesome air. Neither were there wanting many frauds and vexations which were put upon us by our Calino and the infidels, which it would be tedious to enumerate.

The above city is twice as extensive as Jerusalem, but in respect of its edifices is far from being equal to it. It is two Italian miles from the sea, and was formerly called Gaza, the capital of the Philistines, spoken of in Judges xvi. &c.

* * * * * * *

On the 9th of September we went into the plain without the city, where we found ready twenty-four camels; and, this being fulfilled, our head Calino took his leave of us, leaving the other, his deputy, according to the agreement we have already mentioned.

The following day we took our way through plains, which were bounded on our right by the sea, and on the left by the Holy Land: straight before us to the S. lay the Great Desert. We halted for the night at Lebhem, a small village one mile from Gazera; this was on account of our sick, who were ill able to travel: here we found no fuel but some poor roots of broom. There was a well of great depth, but it was empty, from which it is said that the blessed Virgin drank in the flight into Egypt.

We departed from Lebhem on the 11th of the said month, over a great plain of which we could see no end, excepting to the westward, where the main sea is its boundary. At length we reached a country called in Arabic Carvatha, and in Latin Cades: here we remained for the night, having pitched our tents under the hollow of a small hill—as ill off for fuel as the first night. Our water was brought from so far that it was fitter to poison than refresh us. It rained a little, as it is wont to do in the evening in those parts; and as far as we could see in the course of our route, the more desert and uninhabitable the country became, so that we arrived that day in a wilderness where never man nor the son of man had dwelt, for there were neither woods, nor forests, nor trees, but all was desolation and the gloom of death, full of ravines and aridity from the burning heat of the sun. We often saw in that desert an appearance of smoke, as if proceeding from a fire; but afterwards it was ascertained by experience to be the movement and revolution of fine sand and dust borne in the air, for there are small hill-ridges entirely formed by accumulated dust, so that, where there is one day an open track, on the next there is a mountain of sand, and there can thus be no fixed or certain route across this country.

Travelling thus at a foot's-pace, we reached on the following day a placed called Gayon, and there we pitched our tents upon a very white soil in a very wide ravine: no water was to be found, nor any wood, nothing but the roots of herbs. The provisions brought from Gazera failed us, except the biscuit.

The next day, the 13th of September, we struck into the bed of a torrent or very broad ravine, shut in on each side by the mountains, and called by the Arabs Vadalar,* and here we found much colocynth.

On the following day, which was the 14th of the above month, we entered upon a terrible wilderness, more so than on the first of the preceding days, where we saw no living creature, bird or beast, but ostriches, which are the only inhabitants of that desert; and proceeding along, we came amongst mountains of very great height, utterly barren, covered with nothing but stones and crags, and they are called by the Arabs Gebelhelel,† from their barrenness. At night we pitched our tents in a place called Magare.

On the morrow following, in a very cold desert, unusual in the eastern regions: we slept at night in a place called in Arabic Mynschene, where the soil was white, and composed of rubble, and the stones like flakestones of sand, which is called creta.

After passing the night we took our way through a region, the eastern extent of which is unknown, so much so that a man even travelling on horseback could not, in nine days, arrive at any human habitations. Some think that it is the uninhabitable zone, called the torrid, which extends up to the terrestrial Paradise. On this day, which was the 16th September, along the confines of the desert, we arrived at Alberoh.

We proceeded on the next day, the 17th of the month aforesaid, to Mesmar, at the foot of a very high mountain, which they called Caleb, which looks as if formed artificially.

Setting out from thence on the next day, the 18th of the month,

* Wádí Ailah.—Tr.

† El-Helál.—Tr.

we passed through a country where the hills and valleys and sands, and even the stones, had the appearance of salt. Thus we arrived at a ravine which is called Aguado (a watering-place); but it belied its name, for there was no water.

Starting the following day, we came to a hill from which we saw the Mountains of the Lord, Horeb and Sinai, on our left, and on our right the Red Sea; but they were yet four days' journey distant from us. From thence, having first refreshed ourselves, we descended on foot, the way being dangerous, with narrow precipices full of stones. At the bottom we remained for a night and day, sheltered in the caverns, to allow our sick to repose. Here was neither water nor wood, nor any green thing.

In the morning of the day following, which was the 20th September, pursuing our journey we entered upon a very steep and rugged mountain, where there was no other description of tree but a thorny one, full of flowers of a sweet odour, which refreshed us. Some say that the crown of our Lord's passion was woven of those thorns: we therefore took some of them with us. This mountain was not sterile, but beset with sharp-pointed rocks and stones, which shone as if they had been anointed with oil, and their colour was a mixture of black and red. There we saw a beast larger than a camel, which our Calino said was a unicorn; and a shepherd came up to us whistling and following his flock, which was a marvel to us, who for so long a time had seen no trace of a human being nor a tame animal. When night came on we set up our tent in a place called Scholïe.

Throughout all the toils and perils of these journeys we had always this consolation, through the intercession of the blessed virgin Saint Catherine, that after midnight a star brighter than the rest showed itself to the south, which they call the star of St. Catherine, which, being over Mount Sinai, showed us the direct way. Whenever we found that we deviated from it, we were sure that we were in a wrong course.

On the 21st of this month we entered a valley to the S., and thence we saw the holy mountain of Sinai, which stood out above the other mountains near it: at the sight of which we much rejoiced, though it was still far from us; but we could not recognise any particular features. Our sight could discern nothing more than a mass, as it were, enveloped in obscurity; and, as the sun was now setting in the W., we pitched our tent in a place called Abalharoh. It is a great plain, surrounded on all sides by lofty mountains, where they say that Moses kept the flocks of the Gentile, Jethro, his father-in-law, who was a priest of Midian; and it is believed that the temple where he kept his idol was not far from thence. In this place we saw a deeply-excavated rock, in which they say Moses often slept, and abode there, from whence he could see over the whole plain, and keep good watch over his sheep.

The next day, the 22nd of September, very early in the morning, we began to enter the desert by a narrow stony valley, and then came upon a wide and long plain, which extended to the foot of Mount Sinai. The sands which covered it were red, as were the mountains which border it. Here it is believed that the children of Israel dwelt whilst Moses conversed with the Lord on Mount Sinai. We saw a place upon a rock raised like a chair or pulpit, from which it is said that this holy man promulgated the law in presence of the people. When we began to ascend the valley, some Arabs joined our company, and we received no injury from them. Thus journeying along this great plain, near the acclivities of very lofty mountains, we turned into a deep, rugged, and narrow valley, full of stones, which we traversed very slowly, on account of its danger. On our right the monastery of our lady the martyr, St. Catherine, presented itself, situated upon the acclivity of Mount Sinai; the sight of which was a cause of much rejoicing, so many days having passed without our having set eyes upon any dwelling, and now we beheld the end which fulfilled the object of our desire.

[On the thirteenth day after their departure from Gazera, on the first of which they only travelled one mile to Lebhem.]

XIV.—*Note on Some Names of Places on the Shores of the Red Sea.* By A. Thomson D'Abbadie, Esq.

Musawwa', 10*th March*, 1838.

The following attempt to correct the orthography of names of places in the Red Sea, is, as yet, partial and incomplete. But my approaching journey to Abyssinia will put it out of my power to follow up any longer a subject which becomes more arduous as it advances towards perfection. It had been my wish to examine how far the nomenclature of the ancients would agree with my present corrected chart: but I am sensible that many travellers have not done their duty in withholding their information with a vain desire to give the last polish to a work which might often be better finished by others. Besides, I have but one manuscript, and the difficulties of my intended tour may deprive me of that.

I had the good fortune to meet an eminent Oriental scholar at Jiddah, M. Fresnel, who had taken up his quarters there in order to pursue his researches in the elucidation of Arab tradition prior to the age of Mohammed. He kindly laid aside his manuscripts, collected the pilots of the Red Sea, and taught me to distinguish the niceties of Arabic orthography. As it was very often impossible to make the pilots understand the words pronounced, as we supposed them to have been written after the English fashion, it became necessary to perform an imaginary tour from place to place. When among islands we often needed the help of an intelligent pilot from Ḥadramaut, who, being accustomed to consult maps, could better explain, in sea-terms, the relative positions and distances. Every name when received from the pilot was successively pronounced by each one of us, the Arabs present being consulted as to the distinctions between س [s] and ص [ṣ], ح [ḥ] and ه [h], etc., which are not always attended to by the lower class of seamen, to which the pilots generally belong. Even with all these precautions I could not avoid such faults as depend on a difference of pronunciation. Thus, I have written *seil* in some places, and *seyyil* in others.

We have been greatly puzzled by the island of Nahelej, which, in my copy of the chart, is laid down as being totally distinct from Nórah; our Dahlak pilot assured us that there was no large island to the N.E. of Nórah, and that Nahelej, or a name like it, is a spot *in* the isle of Nórah. He likewise knew nothing of a distinct island called Mahoon, although it was evident that he never hesitated in giving the names in their proper order. This uncertainty prevented us from recognizing several small islands near Nórah.

It may not, perhaps, be uninteresting to communicate the names of the points of the compass, which I also owe to M. Fresnel. They are used by every pilot in the Red Sea.

North . .	Jáh.	E. by S. . .	Maṭla' el jauzá.
N. by E. .	Maṭla' el firḳid.	E. S. E. . .	—— el tír.
N. N. E. .	—— el na'sh.	S. E. by E. .	—— el iklíl.
N. E. by N.	—— el náḳah.	S. E. . .	—— el 'aḳrab.
N. E. . .	—— el ayyúḳ.	S. E. by S. .	—— el ḥimáreïn.
N. E. by E. .	—— el wáḳi'.	S. S. E. . .	—— el soheïl.
E. N. E. .	—— el sumák.	S. by E. .	—— el sallibár.
E. by N. .	—— el thurayyà.	South . .	—— Ḳoṭb.
East . .	—— Mṭala'.		

Maghríb signifies *West*, and is applied in place of Maṭla, in order to designate the opposite points of the compass: thus, Maghríb el 'Aḳrab, means S.W., &c. These names belong to constellations, but I have not been able to identify more than one or two.*

As the present mode of sailing in the Red Sea has probably been followed during a long course of ages, I shall give you the

* They are explained in the Jihán Numá, p. 61.—F.S.

names of the spots in which we anchored during our voyage from Jiddah to Musawwa', with a northerly wind. Starting at daybreak, Feb. 8th, we halted near sunset at Mersa Raghwán (Rugguan, Engl. Ch.): Feb. 9th, at Mersa Ibráhím, which is the port N.E. of Líth; Feb. 10th, at Fará (Farrar Islands); on the 11th, after 3 hours' stay at Konfodah, we anchored in Halí; the 12th, after having coasted as far as El Barak (el Burk), we crossed the Sea, passing near Fásiliyyat (Warsaleat Island), and on the 19th our pilot recognised the land near Kandállái (Gandalite). As there is no good anchorage on this coast, we went on to Musawwa', where we arrived on the 16th.

I have also written down all the names from Yembo' to Mokhá.

Names of Places in a Coasting Voyage from Sawákin to Mokhá.

*According to a Dahlak Pilot.**	*English Chart.*	*Remarks.*
Fejj Rá'í-l-kasab	(Omitted)	*i. e.* valley of the Sugar-cane Shepherd.
Mersá Lekák Hindí	Mersa Legakinde Mersa Hadoo	Both names are used ?
——— Kiláb 'Alí		
——— Hádhiyú (Hádhihu?)		
——— Hádhinú (Hádhínhu?)		
——— Al Shabak	Al Shubuc	
——— Sheïkh Sa'd	Mersa Sheikh Saad	
——— 'Aïkah		
——— Lahm		
——— Kofút		
——— Sha'b Sunbul		
Jezírat el Amír		
——— el Hádí		
——— Melákiyát		Three small islands N. of Ras Mugda.
Rás Mikdam	Ras Mugda	
Mersá Turunj hálet	Trikatatah	
——— Kat'ah Jábinú (já beïn-hu?)	Guttat Tromba	Here follow seven headlands: they have no names.
Rás 'Asís		
Jezerírat Amarat	Amarat	
Rás Shekab		W. of Amarat.
Jezírat Koban		
Rás el Sáj		S.E. of Rás Shekab.
Jezírat el 'akík, or Bahdúr	Aggeeg, or Badour	
——— hajar		
Rás el Debír	Ras Deber	
——— el 'abíd		
——— Abú Lábis (-l Abis?)	Aboo Yahbis	
——— or Mersá-lkassár	Ras Casar	Also named Ras Kassár.
Mersá Samad 'Allám		
——— Mandalú (Mandal-hu?)	Mudaloo	
Rás Teraubah	Serabar (probably)	

* It must be borne in mind that in the names taken from the English chart of the Red Sea the vowels are to be sounded as usual in English; in those from the Dahlak pilot, as in Italian, or as in the English words *father, there, fatigue, cold, rude.*—Ed.

According to the Dahlak Pilot.	*English Chart.*	*Remarks.*
Mersá Rúrat	Rarrat	
—— Kabru-sh-sheïkh	Gubroo Sheikh	
—— El berr reyyim		This is the general name of this part of the coast; Karn is synonymous with Rás.
Karn, or Rás } 'Adaf	Garna Duff	
—— Kandelláyí	Gundalite	
Jezírah Difneïn	Diffnane	
Mersá Mobárak	Mersa Moobaruck	
—— Ibráhím	Mersa Ebrahim	
—— Inte'silah	Indesilee	
—— Abú Ruba'	Aboo Rubah	
—— Mahallah		
—— Kuba'	Coobach	
—— Ughayyarú		An extensive valley N. of Rás Harb, and which contains fresh water.
Rás Harb	Ras Hurub	
—— Kurkusum		
Mersá Dákhiliyyah (Inner Harbour)		
Rás 'Abd-el-kádir		The headland that projects most N. of Musawwa'.
Jezírah Sheïk Sa'íd		S. of Musawwa'.
—— el tawálet (Long Island)		S.W. of Musawwa'.
Harkíkó, or Arkíkó	Argeego	Called Dakhanú (Sorghum Dochna) by the Habáb.
Musawwa',		Called Bati' by the Habáb.
RásHarár, N. of Musawwa'	Massowah	Called by the Habáb, Bati' Point of land N. of Musawwa.
—— Modar, N. of Musawwa'		W. point of Musawwa' Island.
Jezírah Deseï	Dissee	
—— Haudhah (Cistern Island)		Triangular island, N.E. of Deseï Island.
Rás Haudhah (Cistern Head)		First headland E. of Deseï.
Jezírah 'asákir (Army Isles)		Two islands N. of Rás Hawádet.
Rás Kúrelah		Second headland E. of Deseï.
—— el Dellemah		Third headland, ibid.
Jezírah Umm-en-námús (Mother-of-Law Island)	Larmoose	
Kurún Duluh, Rás Horeïrah }		S. of Umm-en-námús Island.
Jezírah Delkus (Dhá-lkuss?)		
—— Del'íd (Dhá-l'íd, Festival Island)		
—— al 'ajúz (Old Woman)	Adjuce	
Jebel Hawákil	Howakil	
—— Derkamán	Delgamna	
—— Bak'ah	Jibbel Bucker	
—— Seïl Bahár	Sarbo	
—— Abú 'Okbah, or 'Akabah		
—— Omm-es-sahríj, or Mersá Endeddeh }		This name appears to have been divided and applied to two distinct islands.

According to the Dahlak Pilot.	*English Chart.*	*Remarks.*
Rás Kurmud kebír ——— Ṣaghír		This name is applied to two capes distinguished from each other by the adjectives Kebír and Ṣaghír.
—— Maurakh Ḥoyó (Ḥaïhu ?)		
Jebel Benát (el Wá) —— Ḥandú	Banat' lar	
—— Dúrefrós	Duroro ?	
—— Ḥanfilah	Amphila	
Rásu-r-rá 'atán	Ras Ourata	*Rás* Ará'tá ought to be more northerly ?
Rás Kaṣṣár	Ras Cussar	
(Rás er-rebáṭ ?)	Ras Seerboot	
Ghubbaṭ Weleleh		This name is given to the whole space from Rás Kaṣṣar to Kurdúmiyát Islands.
Jezíraṭ Kurdúmiyát	Coordomeat	
'Eïd	Edd	
Jebel Kudd 'Alí	Coordarlee	
—— Abá 'il	Jibbel Abbelat	
Jibál Rakhamah	White Quoin Hill (?)	
Berr as-ṣúlah, Rás Beïlúl		
Dahlak Bender Mokhá		

Berr Aṣúlah is the name of the coast which lies between Jibál, Raḥmet, and Beïlúl, the Ras Billool of the English chart. As the Arab vessels sail from Rás Beïlul to Bender Mokhá, my Dahlak pilot could not give me any more names on the African coast.

Dahlak, and the Net of Islands which surround it.

According to the Dahlak Pilot.	*English Chart.*	*Remarks.*
Dahlak	Town	
Memlaḥ (Saltern)	Memlaḥ	
Adhal, or Edhel		Town, or village W. of Memlaḥ.
Durbeshah, or Durbeshát	Derboshat	
Erwá	Erwa	
Dúbellú	Doobelloo	
Rás Kusum	R. Goosum	
	Goobanee	
Kunbíbeh	Cumbeeber	
Dásaghau (Dhá-s-saghw ?)		Between the sea and Kúnbibeh, which ought to be more to the E.
I'bárah (Ibárah ?)		W. of Säl'eït.
Säl'eït	Salat	
Jamheïleh		N.W. of Dúbellú.
Jezíraṭu-n-nokhrá		Nokhrá island at the entrance of the principal bay of Dahlak.

The preceding belong immediately to Dahlak island.

Jozíraṭ Kádó	Kaddo	
Sherm Sáyilah Bádirah	Sale Badera	
—— Hárrah	Harrat	
—— El Abú	Laboo	
—— Nórah (Naúrah, so pronounced)	Nora	

According to the Dahlak Pilot.	*English Chart.*	*Remarks.*
Rás Kubárá (Ḥubárá—Bustard?)		N. point of Nórah island.
—— Dúlbahút	Dulbahout	
—— Berr 'addah	Buradoo	
—— Doḥol	Dohul	
—— Dahreh	Dahrel	

CLUSTER OF ISLANDS AROUND DAHLAK.

According to the Dahlak Pilot.	*English Chart.*	*Remarks.*
Omm 'Alí ('Alí's mother)	Ommarlee	Two islands.
Deheris Melek		First island N.W. Omm 'Alí.
Antúḥ		Second ditto.
Abú Sheráyi'aṭ		Third ditto, or most northerly.
Enta' fúsh	Tookfush	
Ṭurkub		Two islands N. of Enta'fúsh.
Ṭanan	Tunnum	
Wusṭah	Wooster	
Esrútau	Suratoo	
'Awálí Shaúrah	Howallee Shorah	
——— Ḥuṭum (Ḥutub?)	——— huttoob	
Enta' sănú	Intensnoo	
Esbáb	Usbob	
Hurmíl	Harmeel	
Ḥukáleh	Hukally	
Seïl 'Anber	Sale Amber	
Ante' untur		E. of Ḥukáleh.
Ghabbiḥu		S. of the preceding.
Dulhalam (Dhú-l'Helem)	Dulhulum	
Enta'ídel	Entadell?	
Ṭabániyó		Not identified, N.E. near Dulhalam.
Adásí	Hadassee	
Dahreh		W. of Adásí.
Adhgher (Azkár?)	Askar	
Dahreṭu-n-núreh		S. of Azgher.
Seïlu-n-núreh		S. of the preceding.
Seil Bal'ah		W. of Seïl Núreh.
Durr es-surúm	Durrafroos	
Duru'tam	Bettah	
Ento'ghodaf	Entogaeluf	
Dallemet		
Möseïl		Near the preceding island—position not identified.
Adbáret	Hadbar	
'Ukúsh		Between the N.W. point of Dahlak Islands and Duḥu.
Dulbu'úd, or Dulkush		Between the same N.W. point and the 'Ukúsh.
Dubinnes		Near the preceding ones.
Derujruj	Derridjeree	
Sarad	Sarod	
Ḍarghelleh	Darghelee	
Durkaham	Durghaum	
Kundábílú }		
Endabir }		N. of Ente terra'.
Enteráhiyá }		
Enteterra'	Euteurah	

According to the Dahlak Pilot.	*English Chart.*	*Remarks.*
Mandút		Long sandy island, which terminates the shoal N. of Deseï.
Delfídel (Dhú-l fídal)		
Muta'dhebn	Mursateban	
Síl Seïdthan		Near Muta'dhabn.
Durkamán		Two islands W. of the preceding.
Diladhí 'ah	Dilladeah	
Mahún	Mahoon	
Dehábir		Two islands, one smaller than the other.
Deróm	Derome	
Toweïrah (Little Bird)		N. of the following
Dulkus (Dhú-l kús ?)	Dulkoos	
Akrab	Agrub	
Seïl 'Arabí	Sale Arabee	
Dunnafárik (Dhú-n nafárik)		
Gharíb	Gurreet	
Dhú-l akl	Dulgold ?	
Dhú-l yighaf	Dulgoof?	
Hawátib Kebír	Howatib	
——— Saghír		
Dubí 'ah	Duldeah ?	
Dinísheb		S. of Duldeah of the chart.
Yermálkó	Jermalho	
Sinna'í	Senach	
Ferj saghálah	Dahret Segarla	
Rajiyum	Rajuma	
Rákah	Rackah	
Musta' milah	Mustarmila	
Weld Muhíreb		S. of Rajúm.
Dhauber	Soober	
Salmá	Salma	
U'kán	Oucan	
Mojeïdí	Moghady	
Durkurúsh (Dhú-l Kurúsh)	Dulgrosa	Dhú-l Kurúsh? i.e. he who has Kurúsh (piastres).
Mashílghá	Mashilgar	
Belhessú (Bá-l Hasú?)	Bolhessoo	
Hatítú	Howate ?	
Tehór el jebel	Tor	
Tuhúr el yed (Dhohúr el yed)		W. by S. of Tehúr el jebel.
Derakahu-l bahr	Derakah	
——— berr		W. of the preceding.
Umm-en-náyim		Somewhere near the preceding.
Hawít	Howate	
Medhbúhah		Close to Hawít.
Musári'	Moosmaree	
Seil Umm 'Alí	Sale Amber	

The first mountain S.W. of Harkíkó is named Jebel Kadam: this is, I suppose, the Geedan of the English chart. Several pilots, questioned by me, knew nothing of *Goob Duenoo*. Ansley Bay appears to have no well-known name. A Dankalí pilot, born in the Isle of Deseï, called this bay Kobb el Káf, or the Velvet Gulf; but, as it is little frequented by those who trade at Musawwa', this name is not understood. Those who answer my inquiries here, call it Baḥr Búrí. The little headland named *Quoin* in the chart, is called here Maḳa 'nilíyah, from a populous neighbouring village. Somewhere in the south of the bay is Gembúthleh. There are a great many anchorages in the bay: that near Zullah is called Mersá Dólá.

The insulated rock east of Deseï is called Seïl kebír; Seïl Ṣaghír is between the preceding one and Deseï. In this latter isle are 40 spots bearing names: I have collected only the following:—

The principal anchorage bears the name of the island; coasting thence towards the south are Mersá Soránḳólah, M. Arakómah Seïl Arakómah (insulated rock), Arakómah Kebír (a little headland), M. Kadedheïnah, M. Lahóshalítah, Rás Górsétúleh (south point of the island), Seïl Górsétúleh, M. Hanḳil Soghaïr, M. Hanḳil Kebír, Rás Hanḳil: Seïl Rokúbeleh, is a small cluster of rocks south of Deseï. In the phraseology of the Red Sea, Seïl, sometimes pronounced Ṣeyyil, is an insulated rock emerging from the surface of the water. Mersá is an anchorage, *i. e.* a harbour or roadstead. Ṭahlah is a shoal barely covered: if very extensive, it is named Roḳúḳ. Baḥáyir is a shoal in deep water.

[Besides the above list, Mr. D'Abbadie has had the kindness to allow many other names of places on the Society's copy of the Charts of the Red Sea to be corrected from his own copy of that Chart; and Mr. Renouard has corrected the rest.

It is due to Captains Moresby and Elwon of the Indian Navy, who executed this laborious survey of the Red Sea, to state that several names of places mentioned in their "Memoir on the Survey" are unaccountably omitted in the Charts: these names are now being engraved on the plates.

It is much to be regretted that before publishing such valuable Charts, the orthography of the Arabic words was not corrected and reduced to some standard. To remedy this inconvenience as far as possible, the names of places in Arabic, as written by Mr. Raṣam, have been engraved, and are now printed on the Charts; and it is intended at the close of the Sailing Directions for the Red Sea, now publishing by order of the Court of Directors of the East India Company, to give a list of all the names of places in the Arabic character, also in the Roman character reduced to one standard of orthography, and the corresponding name on the Chart.—Ed.]

XV.—*An Account of Arctic Discovery on the Northern Shore of America in the Summer of* 1838. By Messrs. Peter Warren Dease and Thomas Simpson. Communicated by I. H. Pelly, Esq., Governor of the Hudson's Bay Company.

Fort Confidence, Great Bear Lake, 15th Sept., 1838.

Hon. Sirs,—It now becomes our duty to report the incomplete success of the expedition to the eastward this summer, in consequence of the extraordinary duration of the ice. Much, however, has been done to prepare the way for another attempt next year; and our hopes, instead of being depressed, are elevated by the knowledge so painfully acquired this season.

On the 6th of June our boats were conveyed on the ice to the mouth of Dease's river (then just open), the ascent of which was commenced the following day. With some assistance from Indians, we reached the portage leading to the "dismal" lakes (discovered by Mr. Simpson last winter) and carried the boats across it without accident. The ice on these lakes was still perfectly solid; and we were provided with iron-shod sledges for the passage. On these we fixed the boats, and, the wind being fair, hoisted sail, which greatly aided the crews on the hauling ropes. In this manner these frozen reservoirs, which are fully 30 miles long, were passed in two days, and we reached our provision station at "Kendall" river on the 19th. There we had the satisfaction to find the two men (left there by Mr. Simpson in April) well, and their Hare Indian hunters successful in the chace. Two of these active fellows consented at once to accompany us along the coast; and proved not only good voyagers, but, during our frequent detentions among the ice, killed so many rein-deer as enabled us to save nearly half our summer stock of provisions. Next day (20th June) we proceeded to the Coppermine river, which we found still fast. It gave way on the 22nd, and we descended all its terrible rapids at full flood, while the ice was still driving. Below the Bloody Fall the river did not clear out till the 26th, and on the 1st of July we pitched our tents at the ocean. Two or three Esquimaux families were seen there, but they took the alarm, and fled over the ice towards some distant islands. Here and on various parts of the coast a fine collection of plants was made by Mr. Dease.

We remained imprisoned in the mouth of the Coppermine, awaiting the opening of the ice, till the 17th of July. Our subsequent progress along the coast was one incessant, we may say desperate, struggle with the same cold obdurate foe; in which the boats sustained serious damage, several planks being more than half cut through. At various points we saw *câches* of the Esquimaux placed upon lofty rocks, out of reach of beasts of prey; but we did not fall in with any of the owners, who seemed to have all gone inland to kill rein-deer, after their winter seal-hunt among the islands. Fragments of Dr. Richardson's mahogany boats were found widely scattered; and many articles left by his party at the Bloody Fall were carefully preserved in the native keepings. On the 29th of July we at length succeeded in doubling Cape Barrow. The northern part of Bathurst's inlet was still covered with a solid sheet of ice; and, instead of being able to cross over direct to Point Turnagain, we were compelled to make a circuit of 140 miles by Arctic Sound and Barry's Islands. On the easternmost of that group Mr. Simpson discovered, at the base of a crumbling cliff, several pieces of pure copper-ore, and the adjacent islands had also the appearance of being strongly impregnated with that metal. A series of specimens of all the principal rocks along the coast was preserved. In order to attain Cape Flinders, we had to perform a portage across an island, and several over the ice. On the 9th of August we doubled that cape; and in a little bay, three miles to the southward of Franklin's farthest encampment in 1821, our boats were finally arrested by the ice, which encompassed them for twenty-two days! so different was the season of 1838 from that of 1821, when Franklin found a perfectly open sea there on the 16th of August. In June, the early part of July, and the middle of August, we had frequent storms, accompanied by snow and frost; but during the greater part of July and the beginning of August storms prevailed, which, together with the severity of the preceding winter, we consider as the cause of the tardy disruption of the ice this season.

On the 20th of August we were obliged to relinquish all hopes of advancing farther with the boats. That our efforts might not however prove wholly fruitless, Mr. Simpson offered to conduct an exploring party on foot for ten days. It was at the same time arranged between us, that, in the event of any favourable movement taking place in the ice, Mr. Dease should advance with one boat. Signals were agreed upon to prevent our missing each other on the way; and, should we unfortunately do so, the last day of August was fixed for the rendezvous of both parties at Boat Harbour. That unlucky spot is situated in lat. 68° 16′ N., long. 109° 2′ W.; variation of the compass 46° E. Mr. Simpson's narrative of his journey and discoveries to the eastward is annexed.

On the 31st of August we cut our way out of our icy harbour, the grave of one year's hopes; and, having the benefit of fair winds, crossed Bathurst's inlet among Wilmot Islands, and safely re-entered the Coppermine river on the 3rd of September. The following day we proceeded to the Bloody Fall, and there secreted our superfluous provisions. The ascent of the Coppermine (hitherto deemed impracticable), to near the junction of Kendall river, was accomplished on the fifth day. We deposited the boats in a woody bluff, where they can be conveniently repaired next spring: then, taking our bundles on our backs, we traversed the barren grounds, and returned to winter quarters yesterday.

Here we had the satisfaction to find everything in good order, the buildings rendered more comfortable, and some provisions collected. Our return so much earlier than we ourselves expected on leaving Point Turnagain has enabled us to commence the fall fisheries in good time; and, though our stock of ammunition and other necessaries for the Indians is reduced very low, want no longer stares us in the face, as it did for several months after our arrival here last year. We are most happy to add that the natives have experienced neither famine nor sickness this season; the only death within our knowledge being that of a blind old man.

Sept. 20.—We have the honour to acknowledge the receipt, this afternoon, of Governor Simpson's despatch of 28th February. As things have fallen out this season, it is fortunate that no party was sent down the Great Fish river to meet us. And, from the experience we now possess of the coast to the eastward, we are of opinion that a retreat by the Coppermine may be effected when the entire ascent of the Great Fish river may be no longer practicable. We feel deeply indebted for the confidence reposed in us, and the ample authority granted by the Governor's circular and previous letters to draw upon the resources of all parts of the country. This power we have hitherto used in extreme moderation; and we are glad to say that we are not reduced to the necessity of exercising it any further. One of our men leaves us in consequence of a bad complaint, and has been replaced by a servant from M'Kenzie's river. To chief trader M'Pherson, the gentleman in charge of that district, we are indebted for valuable assistance in many ways: likewise to chief factor M'Leod, of Athabasca: between them, our order of last winter for an additional supply of pemmican, dogs, sledge-wood, leather, ammunition, guns, axes, and tobacco, has been completed; while the prompt and kind attention of chief trader Ross, at Norway House, has fulfilled the private orders of our people, for a part of which we now send to Great Slave Lake. By the same conveyance we discharge our Chipewyan hunters, as we are unable to provide them any longer in clothing.

Since writing the foregoing we have been obliged to condemn one of our two sea-boats. In its stead we shall transport an inland *bateau*, of a rather superior construction (built by Ritch at Fort Chipewyan two years ago), to the Coppermine next June, by the route followed this year, making up the additional hands required to navigate such a craft with Hare Indian hunters. And, to convey the expedition from this dreary abode at the close of our enterprise, we shall again require the aid of a boat from M'Kenzie's river. We have the honour, &c.

PETER W. DEASE, C.F.
THOMAS SIMPSON.

To the Governor, Deputy Governor, and Committee of the Honourable Hudson's Bay Company, London.

Mr. SIMPSON's *Narrative of a Journey on foot to the Eastward.*

On the 20th of August, the date appointed for the return of former expeditions from these desolate shores, I left our boats still hopelessly beset with ice, to perform a ten days' journey of discovery on foot to the eastward. My companions were five servants and two Indians. We carried a wooden-framed canvas canoe, and nearly the same other baggage as on the journey to Point Barrow last year, with the addition of a tent for the nightly shelter of the whole party on a coast almost destitute of fuel. Each man's load at starting weighed about half a cwt., and our daily progress averaged 20 geographical miles. About the middle of the first day's journey we passed the farthest point to which Sir John Franklin and his officers walked in 1821. Beyond that, the coast preserved its N.N.E. trending to our encampment of the same night, situated on the pitch of a low cape, which I have named Cape Franklin. From W. to N.E. a new land or crowded chain of islands of great extent, in many places high and covered with snow, stretched along at the distance apparently of 30 miles, and led to the apprehension that we were entering a deep sound or inlet. The mainland now turned off to E.N.E., which continued to be nearly its general bearing for the three following days: it is flat in its outline, our path leading alternately over soft sand, sharp stones, and swampy ground. At the distance of from 1 to 2 miles, the coast is skirted by a range of low stony hills, partially clothed with dull verdure, which send down to the sea numberless brooks and small streams. None of the latter at that season reached above our waists, though the deep and rugged channels of many of them showed that in the spring they must be powerful torrents. Two leagues inland a hill (which I have named Mount George after Governor Simpson) rises to the height of 600 feet, and forms a conspicuous object for a day's journey on either side. The ice all along lay immoveably aground upon the shallow beach, and extended in every direction as far as the eye could reach. The great northern land still stretched out before us, and kept alive doubts of our being engaged in exploring an immense bay, which even the increase in the tides, the quantity of sea-weed and shells, and the discovery of the remains of a large whale and of a polar bear, could not altogether dispel. These doubts seemed almost converted into certainty, as we drew near on the fourth evening an elevated cape, and saw land apparently all round. With feelings of bitter disappointment I ascended the height, from whose summit a splendid and most unlooked-for view suddenly burst upon me. The ocean, as if transformed by enchantment, rolled its free waves beneath, and beyond the reach of vision to the eastward. Islands of various shape and size overspread its surface; and the northern land terminated in a bold and lofty cape bearing N.E., at least 40 miles distant; while the coast of the continent trended away to the S.E. I stood in fact on a remarkable headland, at the eastern entrance of an ice-obstructed strait. The extensive land to the northward I have called VICTORIA LAND, in honour of our youthful sovereign, and the eastern extremity Cape Pelly, after the Governor of the Honourable Company. To the promontory where we encamped I have attached the name of Cape Alexander, after an only brother, who would give his right hand to be the sharer of my journeys. The rise and fall of the tide there was about 3 feet, being the greatest observed by us in the Arctic seas. The coast here changes its character, the water becomes deep, the approach easy, and I have little doubt that the islands contain secure harbours for shipping. Next morning, at the distance of 8 or 9 miles, we crossed another high cape, formed of trap-rocks, in lat. 68° 52′ N.; the variation of the compass being 63° E. The travelling had become more and more toilsome, our road now passing over some miles of round loose stones; and then through wet mossy tracts sown with large boulders and tangled with dwarf willows. At our usual camping hour we opened a large bay, studded with islands, which ran in for 5 miles to the S.S.W., and then turned off in a bold sweep of rounded granitic hills (like those near Melville Sound and Cape Barrow), dipping to the sight in the E.S.E. at the distance of 30 miles. To walk round even the portion of the bay in view would have consumed three days; the time allotted for outgoing was already expired; and two or three of my men were severely lame, from the fatigue of their burdens, the inequalities of the ground, and the constant immersion in icy-cold water. I besides cherished hopes that, by making the best of our way back, we might, agreeably to my arrangement with Mr. Dease, meet him bringing on one of the boats, in which case, with an open sea before us, we could have still considerably extended our discoveries before the commencement of winter. I may here remark that we were singularly fortunate in the five days of our outward journey, the weather being so moderate and clear that I daily obtained astronomical observations: whereas, before our departure from the boats, and during our return to them, we had continual storms, with frost, snow, rain, and fogs. Close to our farthest encampment appeared the site of three Esquimaux tents of the preceding year, with a little stone chimney apart. We passed the remains of a larger camp, and several human skeletons, near Cape Franklin; but nowhere throughout the journey did we find recent traces of that few and scattered people.

The morning of the 25th of August was devoted to the determination of our position, and the erection of a pillar of stones on the most elevated part of the point. After which I took possession of the country, with the usual ceremonial, in the name of the Honourable Company, and for the Queen of Great Britain. In the pillar I deposited a brief sketch of our proceedings, for the information of whoever might find it. Its situation is in lat. 68° 44′ N., long. (reduced by C. F. Smith's watch from excellent lunars at the boats) 106° 3′ W.; the magnetic variation being 60° E. The compass grew sluggish and uncertain in its movements as we advanced eastward, and frequently had to be shaken before it would traverse at all. Two miles to the southward of our encampment a rapid river of some magnitude discharges itself into the bay, the shores of which seemed more broken and indented than those along which we had travelled. Independently of Victoria Land, and an archipelago of islands, I have had the satisfaction of tracing fully 100 miles of coast, and of seeing 30 miles farther, making in all (after deducting Franklin's half-day's journey already mentioned) about 120 miles of continental journey. This is in itself important; yet I value it chiefly for having disclosed an open sea to the eastward, and for suggesting a new route, along the southern shores of Victoria Land, by which that open sea may be attained, while the mainland (as was the case this season) is yet environed by an impenetrable barrier of ice. Whether the open sea to the eastward may lead to Ross's Pillar, or to the estuary of Back's Great Fish river, it is hard to conjecture, though the trending of the most distant land in view should rather seem to favour the latter conclusion.

The same evening, on our return, we met the ice at Trap Cape driving rapidly to the eastward. As we proceeded, the shores continued inaccessible, but open water was now everywhere visible in the offing. Several bands of buck reindeer were migrating to the southward along the hills; two which we shot were in far superior condition to those in Bathurst's Inlet and near the Coppermine. A few musk-oxen were also seen; and numerous flocks of white geese (*Anser hyperboreus*), generally officered by large grey ones (*A. Canadensis*), were assembling in the marshes, and taking their aërial flight to more genial climates. At dusk, on the 29th of August (our tenth day), we regained the boats, and found them still enclosed in the ice, which the N. and westerly gales seemed to have accumulated from far and near towards Point Turnagain.

(Signed) THOMAS SIMPSON.

3.

A NEW ENGLISH DICTIONARY OF

UNIVERSAL BIOGRAPHY.

Constructed on an improved and entirely original Plan, and conducted by some of the most distinguished Literary Men of Great Britain and the Continent.

TO BE PUBLISHED

By Messrs. LONGMAN and Co., and Mr. MURRAY, in conjunction.

In active Preparation.

4.

AN INTRODUCTION TO THE LITERARY HISTORY

OF

THE 15TH, 16TH, AND 17TH CENTURIES.

BY HENRY HALLAM, ESQ.

Vols. II., III., and IV., 8vo., completing the Work. Now published.

5.

THE HISTORY OF CHRISTIANITY,

FROM THE BIRTH OF CHRIST TO THE EXTINCTION OF PAGANISM IN THE ROMAN EMPIRE.

BY THE REV. H. H. MILMAN,

Prebendary of Westminster, and Minister of St. Margaret's.

3 vols. 8vo.

6.

HISTORY OF THE POPES OF ROME

THEIR CHURCH AND STATE,

DURING THE XVITH AND XVIITH CENTURIES.

Translated from the German of LEOPOLD RANKE.

3 vols. 8vo.

7.

ANCIENT SPANISH BALLADS.

Translated, with Notes,

BY J. G. LOCKHART, ESQ.

A New Edition, with original and splendid Illustrations. 4to.

8.

THE CORRESPONDENCE OF THE LATE WILLIAM WILBERFORCE.

EDITED BY HIS SONS.

2 vols. Post 8vo.

Printed uniformly with THE LIFE, to which they may be considered as Supplementary.

*** Owing to the extent of the Correspondence of Mr. Wilberforce, it was found impossible to include more than a very small part of it in the Life; a selection, therefore, of the most valuable and important Letters, &c. never before printed, are here given to complete the picture of Mr. Wilberforce's Life and Character.

9.

MAXIMS AND HINTS FOR AN ANGLER, AND THE MISERIES OF FISHING: WITH MAXIMS AND HINTS FOR A CHESS-PLAYER.

BY RICHARD PENN, ESQ.

Second Edition, enlarged. Illustrated with Woodcuts. Fcap. 8vo.

10.

A SUMMER'S DAY AT HAMPTON COURT BEING A GUIDE TO THE PALACE AND GARDENS;

WITH

An historical and illustrative Catalogue of the Pictures according to the New Arrangement, including those in the Apartments recently opened to the Public.

BY EDWARD JESSE, ESQ.,

Surveyor of Her Majesty's Parks and Palaces,—Author of 'Gleanings in Natural History.'

WITH NUMEROUS WOODCUTS. Fcap 8vo.

11.

CONSIDERATIONS ON NATIONAL EDUCATION.

BY SARAH AUSTIN.

Fcap. 8vo. 3*s*. 6*d*.

12.

PENTALOGIA ARISTOPHANICA.

By THOMAS MITCHELL, Esq.

8vo.

This Volume will contain Mr. Mitchell's Text of the Acharneans, Wasps, Knights, Clouds, and Frogs, with short Notes, *more particularly adapted to the Use of Schools.*

In compliance with the suggestions of several of the Conductors of our Great Public Schools, the Notes will be placed at the end of the Text.

13.

REFORM IN 1839, AND REFORM IN 1831.

BY THE RIGHT HONOURABLE

SIR ROBERT WILMOT HORTON, BART., G.C.H.

&c. &c. &c.

8vo. *Ready.*

14.

SCENES OF DOMESTIC LIFE AMONG THE ROMANS.

Translated from the German of

PROFESSOR BECKER, OF LEIPSIG.

With Illustrations. 2 vols. Post 8vo.

15.

THE POETICAL WORKS OF THE REV. H. H. MILMAN;

INCLUDING

THE FALL OF JERUSALEM—BELSHAZZAR—THE MARTYR OF ANTIOCH—ANN BOLEYN—SAMOR, &c. &c.

With Preface and Notes by the Author, a Portrait, and other Illustrations.

3 vols. Fcap. 8vo.

UNIFORM WITH THE WORKS OF SCOTT, CRABBE, SOUTHEY, &c.

16.

A HAND-BOOK for TRAVELLERS in GREAT BRITAIN.

Post 8vo. *In Preparation.*

17.

NARRATIVE

OF A

JOURNEY THROUGH SOME OF THE LESS VISITED PARTS OF THE AUSTRIAN EMPIRE:

INCLUDING

Bohemia, Styria, Istria, and Upper Austria: followed by detailed Notices of the Austrian Government and People; with particular reference to Feudality, Jurisprudence, Religion, Education, Morality, Finance, Civil and Military Administration, and Domestic and Foreign Police.

By PETER EVAN TURNBULL, ESQ., F.R.S., F.S.A.

2 vols. 8vo.

18.

TWELVE SERMONS

Delivered in the

NEW TEMPLE OF THE ISRAELITES AT HAMBURG,

By Dr. GOTTHOLD SALOMON;

And Translated from the German by ANNA MARIA GOLDSMID.

8vo.

Among the motives which have led to the translation of these Sermons, is the hope that from their perusal many of my Christian countrymen may derive a better knowledge of the actuating faith of the Jews.....To these Sermons I confidently ask the attention of the kindly and conscientious Christian. I may add, that many of them may be found available for persons of every religious denomination and sect. If, in religious discussion, men of all creeds would seek, not points of difference, but points of agreement, how much of the strife and bitterness that deform God's earth would disappear!"—*Translator's Preface.*

19.

TRAVELS IN THE HIMALAYAN PROVINCES OF HINDOSTAN AND THE PANJAB,

In Ladakh and Kashmir, in Peshawar, Kabal, Kunduz, and Bokhara.

By Mr. WILLIAM MOORCROFT and Mr. GEORGE TREBECK.

Prepared for the Press from Original Journals and Correspondence,

By HORACE HAYMAN WILSON, M.A., F.R.S.,

Professor of Sanscrit in the University of Oxford.

With Illustrations and a new Map, by John Arrowsmith.

2 vols. 8vo. *Nearly ready.*

20.

THE FROGS OF ARISTOPHANES.

Edited, with English Notes, and adapted to the Use of Universities and Schools,

By THOMAS MITCHELL, A.M.

Forming No. V. of "Mitchell's Plays of Aristophanes." 8vo.

21.

GOETHE'S THEORY OF COLOURS.

Translated from the German,

By CHARLES LOCK EASTLAKE, R.A.

8vo.

"It is not the Translator's intention, nor does he pretend to be competent, to defend this theory i all its bearings. He is aware that men of science in our own country and in Germany have objected to certain conclusions in it. These, however, cannot invalidate the truth of the experiments; and, *as he has never met with an artist acquainted with the work who did not admit its great utility,* he conceived that he might do some service to his profession by translating it."

22.

THE ART OF DEER-STALKING:

ILLUSTRATED BY A NARRATIVE OF A FEW DAYS' SPORT IN THE FOREST OF ATHOLL.

By WILLIAM SCROPE, Esq., F.L.S.

With 12 beautiful Illustrations, by Edwin and Charles Landseer, R.A.

A New Edition, Revised. Royal 8vo., 2*l.* 2*s.*

23.

THE CORRESPONDENCE OF WILLIAM PITT, FIRST EARL OF CHATHAM.

Edited by WM. STANHOPE TAYLOR, Esq., and Captain JOHN HENRY PRINGLE,

The Executors of HIS SON, JOHN, EARL OF CHATHAM

8vo.

The Third and Fourth Volumes, completing the Work.

24.

ILLUSTRATIONS of the NORMANS IN SICILY;

BEING

A SERIES OF THIRTY DRAWINGS OF THE SARACENIC AND NORMAN REMAINS IN THAT COUNTRY.

Folio. *Nearly Completed.*

25.

A NEW CLASSICAL DICTIONARY

Of Ancient Mythology, History, Biography, and Geography

FOR THE USE OF COLLEGES AND SCHOOLS.

Carefully compiled from *Original Authorities,* and exhibiting the results of Modern Researches down to the present time.

By the Rev. JOHN WORDSWORTH, A.M.,

Fellow of Trinity College, Cambridge.

One closely-printed volume, 8vo.

Albemarle Street, July, 1839.

BOOKS RECENTLY PUBLISHED

BY

Mr. MURRAY.

1.

GIBBON'S HISTORY OF THE ROMAN EMPIRE.

An entirely New Edition, Edited, with Notes,

By the Rev. H. H. MILMAN.

In Twelve Volumes, illustrated with original Historical Maps.
8vo. 9*s*. each.

"There can be no question that this edition of Gibbon is the only one extant to which Parents and Guardians and Academical Authorities ought to give any measure of countenance."—*Quarterly Review*, No. CXXIV.

"Gibbon's History, especially as edited by Mr. Milman, is a work for all time and for all classes. It never before was a work which could be safely put into the hands of the young, or of those whose opportunities and means for detecting its perversions were few. Now, however, the errors of this luminous and imposing history have been skilfully and convincingly noted. The poison, if not extracted, has been made palpable."—*Monthly Review*, May, 1839.

2.

THE LIFE OF EDWARD GIBBON,

WITH SELECTIONS FROM HIS CORRESPONDENCE, &c.;

And Notes by the Rev. H. H. MILMAN.

Uniform with Milman's Gibbon. One Volume, 8vo. 9*s*.

"The present volume is a valuable and necessary companion to the *Decline and Fall*. The majority of readers may be excused though they do not make themselves familiar with Gibbon's miscellaneous works: no one who desires to be informed in the most engaging and dignified manner of the most important eras in the world's annals can allow himself to remain unacquainted with the history which Mr. Milman has so admirably corrected, or with the life and correspondence of its very remarkable author."—*Monthly Review*.

3.

AN INTRODUCTION TO THE LITERARY HISTORY OF THE 15th, 16th, and 17th CENTURIES.

By HENRY HALLAM, Esq.

Vol. I. 8vo., 15*s*.

4.

JOURNAL OF THE ENGLISH AGRICULTURAL SOCIETY.

CONTENTS:

1. On the PRESENT STATE of the SCIENCE of AGRICULTURE in ENGLAND.—2. EARL SPENCER on the BREEDING of CATTLE and SHEEP.—3. SIR JAMES GRAHAM on the DEANSTON FREQUENT-DRAIN SYSTEM.—4. On SEVERAL OPERATIONS of DEANSTONIZING, from the HON. R. CLIVE, M.P.—5. CHARLES SHAW LEFEVRE, M.P., on SUB-SOIL PLOUGHING on a DRY GROUND.—6. J. PAINTER, Esq., on GAS-WATER as a MANURE.—7. On the VALUE of SEVERAL VARIETIES of WHEAT, by JOHN MORTON, Esq.—8. REV. W. RHAM, on the ANALYSIS of SOILS.—9. On the PROGRESS of AGRICULTURE in SCOTLAND since the FORMATION of the HIGHLAND SOCIETY, by MR. JOHN DUDGEON.—10. MISCELLANEOUS NOTICES, &c.

No. 1. 8vo. 2*s.* 6*d.*

5.

DISPATCHES AND CORRESPONDENCE OF HIS GRACE THE DUKE OF WELLINGTON.

Edited by Lieut.-Colonel GURWOOD.

Complete in Twelve Volumes, 8vo. 20*s.* each Volume.

6.

HISTORY OF ENGLAND FROM THE PEACE OF UTRECHT to the PEACE OF AIX-LA-CHAPELLE.

By LORD MAHON.

Complete in Three Volumes, 8vo. Vol. I., 16*s.*; II., 14*s.*; III., 18*s.*

7.

ON THE PRESENT UNSETTLED CONDITION OF THE LAW AND ITS ADMINISTRATION.

By JOHN MILLER, Esq., of Lincoln's Inn.

8vo. 4*s.* 6*d.*

8.

THE PROGRESS AND PRESENT POSITION OF RUSSIA IN THE EAST.

A New Edition, with a Map shewing the Encroachments of Russia. 8vo., 6*s.*

9.

THE AFRICAN SLAVE TRADE.

By THOS. FOWELL BUXTON, Esq.

Second Edition. 8vo. 5*s.*, or, with Arrowsmith's large Map of Africa, 8*s.*

10.

Dedicated to the University of Oxford.

THE STATE IN ITS RELATIONS WITH THE CHURCH.

By W. E. GLADSTONE, Esq.,

Student of Christ Church, and M.P. for Newark.

Third Edition, 8vo., 9*s.* 6*d.*

11.

THE PILGRIM'S PROGRESS, WITH A LIFE OF JOHN BUNYAN.

By ROBERT SOUTHEY, LL.D.,

Poet Laureate, &c. &c. &c.

New Edition, illustrated with Portrait and Engravings, post 8vo., 10*s.* 6*d.*

12.

A NARRATIVE,

By Sir FRANCIS B. HEAD, Bart.

Third Edition, with a Supplemental Chapter, on the PROPOSED UNION of the CANADAS, the CLERGY RESERVE QUESTION, &c. 8vo., 12*s.* bound.

The Supplemental Chapter is sold separately. 8vo. 2*s.* 6*d.*

13.

AN INDEX TO THE QUARTERLY REVIEW.

Vols. 41 to 59, inclusive.

Forming Nos. CXIX. and CXX.

8vo. 12*s.*

14.

A LIBRARY EDITION

OF

LORD BYRON'S WORKS.

Beautifully Printed, with a new and copious Index and a Portrait.

8 vols. 8vo., £4. 4*s*.

FOR ILLUSTRATORS OF THE WORKS OF LORD BYRON, a very limited number of Copies of this New Edition in 8 vols., 8vo., have been printed in QUARTO.

15.

OBSERVATIONS

ON THE REV. RICHARD JONES'S

'REMARKS ON THE MANNER IN WHICH TITHES SHOULD BE ASSESSED TO THE POOR RATE."

BY WILLIAM BLAKE, ESQ., F.R.S.

8vo. 2*s*.

16.

THE NORMANS IN SICILY;

Or, Notes of a Journal made in Sicily in the Autumn of 1836, chiefly illustrative of the Architecture and History of the Normans in that Island.

BY HENRY GALLY KNIGHT, ESQ., M.P.,

Author of 'Notes of a Tour in Normandy.'

Post 8vo. 8*s*. 6*d*.

17.

DOMESTIC MANNERS OF THE RUSSIANS.

In a Series of Letters describing a Year's Residence in that Country, chiefly in the Interior,

BY THE REV. R. LISTER VENABLES, M.A.

Post 8vo., 9*s*. 6*d*.

"I cannot venture to add another to the numerous books on Russia which have already appeared, without pleading as my apology that I visited that country under circumstances affording opportunities not usually within a stranger's reach, of observing the habits and character of the people. I am, therefore, induced to publish the following letters under an impression that some account of domestic life in the Interior may be, to a certain degree, interesting from its novelty.

"Being nearly connected by marriage with several Russian families, I accompanied my wife into that country in the summer of 1837, for the purpose of visiting her relations, among whom we spent twelve months, either in private houses in the Interior, or in habits of constant intercourse at St. Petersburgh.

"The court, the capital, the army, the public institutions of the country, with its political position and views, have occupied the pens of far abler travellers: my object has been to give some description of private life, national customs, and domestic habits in Russia."

18.

HANNIBAL IN BITHYNIA:

A DRAMATIC POEM.

BY HENRY GALLY KNIGHT, ESQ., M.P.

Second Edition. Fcap. 8vo. 5*s.* 6*d.*

19.

DEDICATED BY PERMISSION TO HER MAJESTY.

THE PARLIAMENTS AND COUNCILS OF ENGLAND.

Chronologically arranged from the Reign of William I. to the Revolution in 1688.

BY CHARLES HENRY PARRY, ESQ., M.D., F.R.S

8vo. 30*s.*

" Your book will supply a very important desideratum in English historical literature. There is no book which contains all the facts; and historical writers, who prefer facts to theories, will find it a most useful collection of dates and events, which they would spend hours in seeking among the voluminous series which you have consulted, for any period to which these labours refer. I consider the volume one of the most valuable works, in every point of view, that have appeared for a long time."—*Sir Harris Nicholas to Dr. Parry.*

Twelve Copies only on large paper, 4to., £3. 3*s.*

20.

A NEW GREEK GRAMMAR,

FOR THE USE OF SCHOOLS.

BY THE REV. CHARLES WORDSWORTH, M.A.,

Late Student of Christ Church, and Second Master of Winchester College.

12mo., 3*s.* 6*d.* bound.

21.

BUENOS AYRES, AND THE PROVINCES OF LA PLATA.

BY SIR WOODBINE PARISH, K.C.H.,

MANY YEARS H. M. CHARGE' D'AFFAIRES IN THAT REPUBLIC.

With an entirely new Map of South America, by John Arrowsmith, and Plates.

8vo., 18*s.*

22.

MEMOIR OF PEREGRINE BERTIE,

ELEVENTH LORD WILLOUGHBY D'ERESBY,

Commander-in-Chief of Queen Elizabeth's Forces in the Low Countries and in France, and Governor of Berwick.

BY A DESCENDANT IN THE FOURTH GENERATION.

8vo. 8*s*. 6*d*.

23.

CORRESPONDENCE OF WILLIAM PITT, FIRST EARL OF CHATHAM.

Edited by WM. STANHOPE TAYLOR, ESQ., and Captain JOHN HENRY PRINGLE, the Executors of his Son, JOHN, EARL OF CHATHAM.

Vols. I. and II. 8vo., 18*s*. each.

To be completed in Four Volumes.

24.

MURCHISON'S GEOLOGY OF THE SILURIAN REGION, OR BORDER COUNTIES OF ENGLAND AND WALES.

With large Geological Map, Views, Coloured Sections, and numerous Plates of Organic Remains. 2 vols. Royal 4to.

25.

SELECTIONS FROM THE POETRY OF ROBERT HERRICK.

EDITED BY THE LATE CHARLES SHORT, ESQ.

Fcap. 8vo. 5*s*. 6*d*.

26.

THE LIFE OF LORD ANSON,

The Circumnavigator of the Globe.

By Sir JOHN BARROW, Bart., F.R.S.

Printed uniformly with "Barrow's Life of Howe."

Portrait. 8vo., 14*s.*

27.

ELEMENTS OF THE PATHOLOGY OF THE HUMAN MIND.

By THOMAS MAYO, M.D., F.R.S.,

Fellow of the College of Physicians; and late Fellow of Oriel College, Oxford.

Fcap. 8vo., 5*s.* 6*d.*

28.

ACCOUNT OF THE MOST IMPORTANT DISEASES PECULIAR TO WOMEN.

By the late Dr. GOOCH. *Second Edition.* 8vo. 12*s.*

29.

ESSAYS ON THE MOST IMPORTANT DISEASES OF WOMEN.

By ROBERT FERGUSON, M.D.

Post 8vo., 9*s.* 6*d.*

30.

LETTERS ON PARAGUAY,

COMPRISING

An Account of a Four Years' Residence in that Republic, under the Government of the Dictator Francia.

By J. P. and W. P. ROBERTSON.

A New Edition, 2 vols. Post 8vo., 21*s.*

31.

FRANCIA'S REIGN OF TERROR;

The Continuation of Letters on Paraguay.

BY J. P. and W. P. ROBERTSON.

Post 8vo., 10*s*. 6*d*.

32.

A LETTER TO LORD BROUGHAM,

ON THE

POLICE REPORTS and POLICE BILLS.

BY JAMES TRAILL, ESQ.,

Of the Middle Temple, Senior Magistrate of the Police Office, Union Hall.

33.

THE AGAMEMNON OF ÆSCHYLUS.

A new Edition of the Text, with Critical, Explanatory, and Philological Notes designed for the use of Students in the Universities.

BY THE REV. THOMAS WILLIAMSON PEILE, M.A.,

Senior Fellow and Tutor in the University of Durham, and formerly Fellow of Trinity College, Cambridge.

8vo. 12*s*.

34.

AN ENQUIRY WHETHER

THE SENTENCE OF DEATH PRONOUNCED AT THE FALL OF MAN Included the whole ANIMAL CREATION, or was Restricted to the HUMAN RACE;

A Sermon preached before the University of Oxford.

BY THE REV. WM. BUCKLAND, D.D., F.R.S.,

8vo., 1*s*. 6*d*.

35.

THE QUARTERLY REVIEW, No. CXXVII.

CONTENTS:

I. THE REIGN of FERDINAND and ISABELLA of SPAIN.
II. POEMS by R. M. MILNES, M.P.
III. ATHENS and ATTICA, by the REV. DR. C. WORDSWORTH.
IV. OLIVER TWIST.
V. THE SILURIAN SYSTEM, by R. J. MURCHISON, F.R.S., F.L.S.
VI. POPISH PERSECUTION in the TYROL—The EXILES of ZILLERTHAL.
VII. RUSSIA, PERSIA, and ENGLAND.
VIII. WILD SPORTS OF SOUTH AFRICA.
IX. THE HOUSEHOLD and THE MINISTRY.

36.

TRANSACTIONS OF THE ROYAL SOCIETY OF LITERATURE.

VOL. III. PART. II.

Illustrated by numerous Engravings. 4to., 1*l.* 1*s.*, or 10*s.* 6*d.* to Members.

CONTENTS :—

On the ANTIQUITY of the EARLY CYCLES. Second Memoir. By the Rev. Dr. NOLAN.
On the BATTLE of MARATHON. By GEORGE FINLAY, Esq.
On the POSITION of APHIDNA. By the Same.
On the POSITION of the OROPIAN AMPHIARAEION. By the Same.
On the BOOK and HISTORY of JOB. By the Rev. EDWARD DAVIES.
On the REMAINS of a TEMPLE in DAMASCUS. By W. R. HAMILTON, Esq.
On the ANCIENT NAME of EGYPT. By the Same.
On the CEILING of the MEMNONIUM at THEBES. By the Rev. G. TOMLINSON.

37.

A LETTER TO THE MARQUIS OF LANSDOWNE

ON THE

ESTABLISHMENT of a BOARD of NATIONAL EDUCATION.

BY ROBERT ISAAC WILBERFORCE, M.A.,

Vicar of East Farleigh, and late Fellow of Oriel College.

Second Edition. Fcap. Price 8*d.*, or 7*s.* 6*d.* per dozen.

38.

TRAVELS IN THE TRANS-CAUCASIAN PROVINCES OF RUSSIA,

AND

ALONG THE SOUTHERN SHORES OF LAKES VAN AND URUMIAH

With a Visit to the Southern Coast of the CASPIAN SEA.

BY CAPTAIN RICHARD WILBRAHAM,

Employed on a Particular Service in Persia.

WITH MAP AND VIEWS FROM THE AUTHOR'S SKETCHES.

8vo., 18*s*.

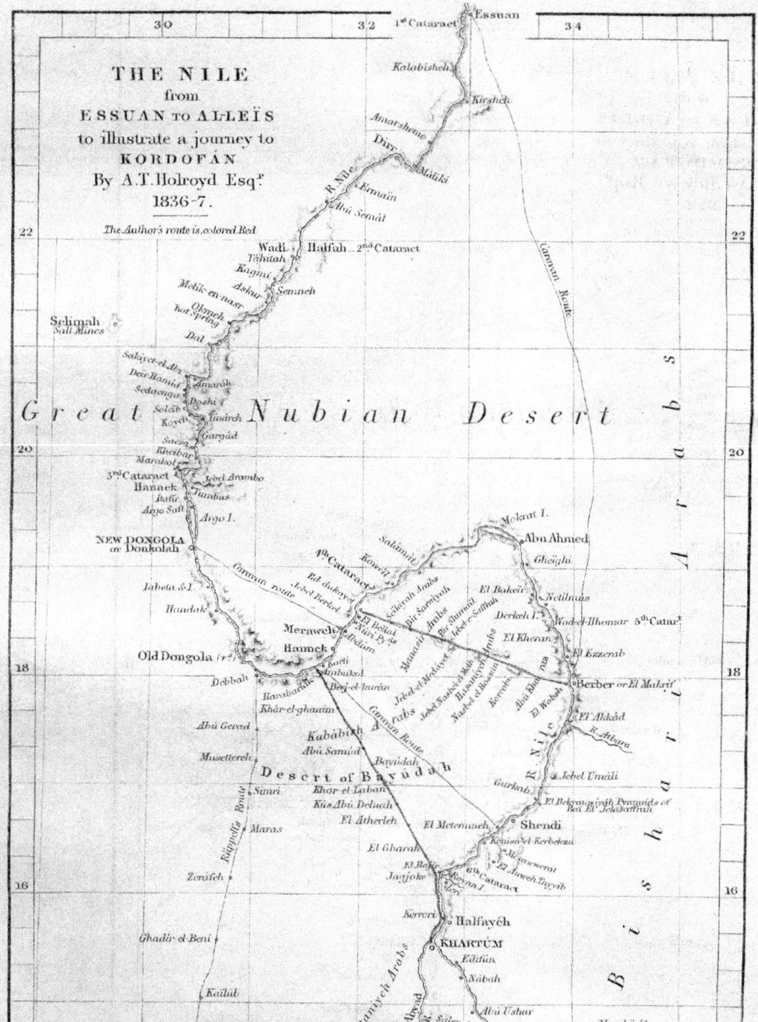

John Arrowsmith

Pubd. for the Journal of the Rl. Geogl. Society by John Murray, London 1839.

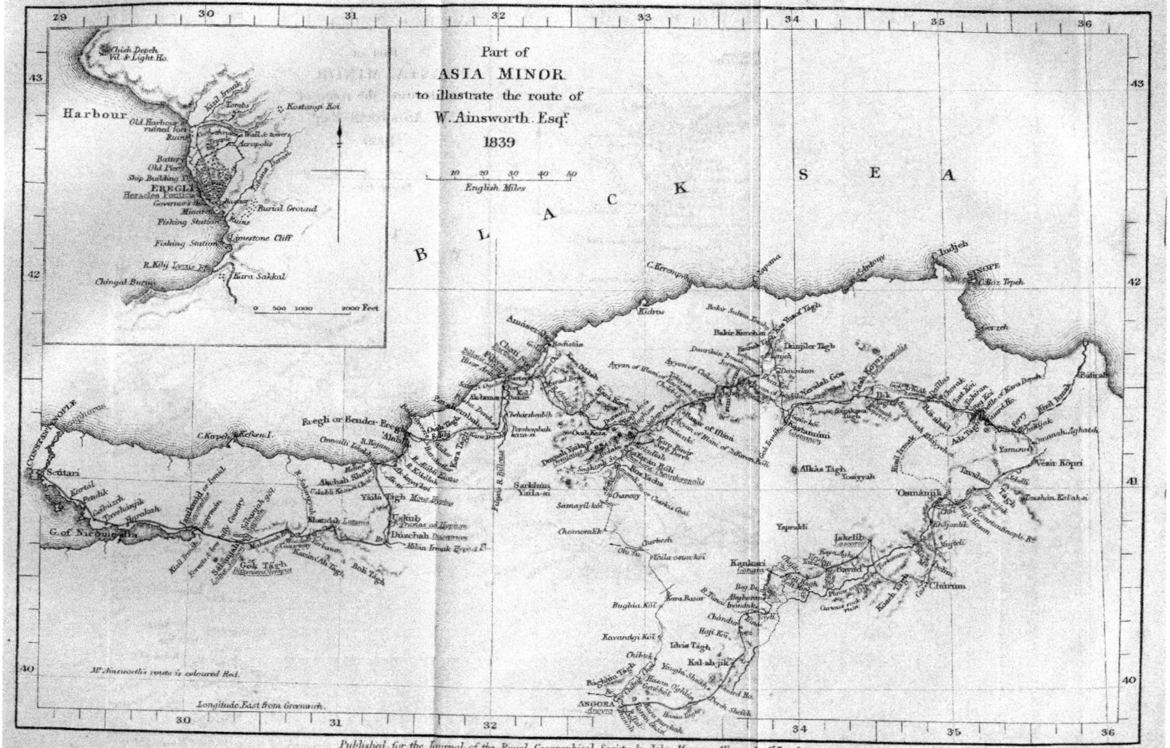

Published, for the Journal of the Royal Geographical Society, by John Murray, Albemarle St. London.

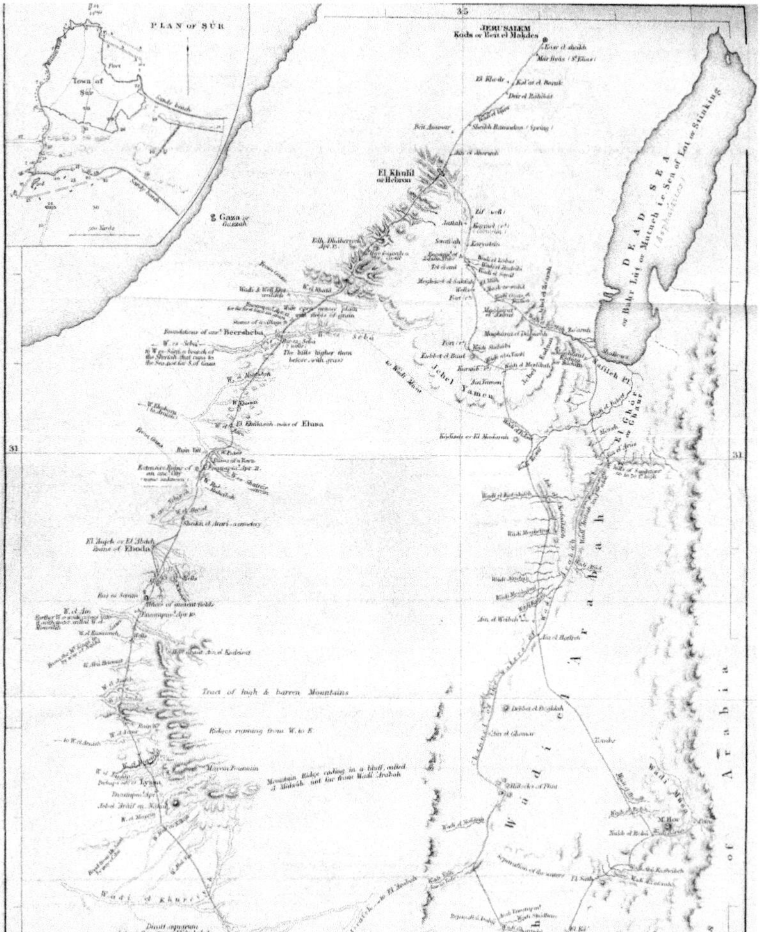

Pub.d for the Journal of the Royal Geographical Society by John Murray, Albemarle St. London 1839.

CONTENTS.

The Third Part of the Ninth Volume will be Published in November next.

London: Printed by William Clowes and Sons, Stamford Street.

Printed in Dunstable, United Kingdom

91219946R00161